Stabilization and Solidification of Hazardous, Radioactive,and Mixed Wastes

Stabilization and Solidification of Hazardous, Radioactive, and Mixed Wastes

Edited by

Roger D. Spence
Caijun Shi

CRC Press
Taylor & Francis Group
Boca Raton London New York

CRC Press is an imprint of the
Taylor & Francis Group, an **informa** business

CRC Press
Taylor & Francis Group
6000 Broken Sound Parkway NW, Suite 300
Boca Raton, FL 33487-2742

First issued in paperback 2019

© 200 by Taylor & Francis Group, LLC
CRC Press is an imprint of Taylor & Francis Group, an Informa business

No claim to original U.S. Government works

ISBN-13: 978-1-56670-444-1 (hbk)
ISBN-13: 978-0-367-39341-0 (pbk)

Library of Congress Cataloging-in-Publication Data

Stabilization / solidification of hazardous, radioactive, and mixed wastes / edited by Roger Spence and Caijun Shi.
    p. cm.
    Includes bibliographical references and index.
    ISBN 1-56670-444-8
1. Hazardous wastes--Stabilization. 2. Hazardous wastes--Solidification. 3. Radioactive waste disposal. 4. Mixed radioactive wastes. I. Title: Stabilization and solidification of hazardous, radioactive, and mixed wastes. II. Spence, R. D. (Roger David). 1948- III. Shi, Caijun. IV. Title.

TD1063.S737 2004
628.4'2—dc22   2004057099

Visit the CRC Press Web site at www.crcpress.com

Library of Congress Card Number 2004057099

**Visit the Taylor & Francis Web site at
http://www.taylorandfrancis.com**

**and the CRC Press Web site at
http://www.crcpress.com**

Preface

"The truth goes through three stages before it is recognized;
in the first stage, it is ridiculed, in the second it is opposed,
and finally it is regarded as self-evident."

That well-worn statement by philosopher Arthur Schopenhauer aptly describes the history of stabilization/solidification (S/S) technology over the last 30-plus years. As an early practitioner, I have witnessed its development from a derided "black art," through resistance from regulators and waste generators, to today's acceptance as a standard environmental tool. In the early days, stabilization publications were largely technical commercials for a particular company or process, or were compilations of the same in book form. The exceptions were in the nuclear industry, where both research and small-scale operations had been conducted since the 1950s, and in end-use applications such as mine backfilling and soil and road-base applications. In the late 1970s and 1980s, good scientific work was being conducted on non-nuclear applications in the relatively new field of hazardous waste treatment, and reported on in technical journals. Practical applications had also begun to mature. The early assortments of equipment and techniques from other technologies evolved into off-the-shelf, task-specific treatment plants and operational procedures specifically aimed at S/S. Recognizing these advancements, my book in 1990 was the first attempt to pull together the science, technology, and application of S/S up to 1989 into a coherent, single-reference source; it is now out of print, and much has happened in the field since then.

In the last 15 years or so, a large body of scientific work on S/S has been published, but primarily as individual journal papers, symposium proceedings, or handbooks sponsored by EPA and other organizations such as the Portland Cement Association. As a peer reviewer for the American Chemical Society's publications, and for others, I have had the opportunity to see some excellent contributions in this field that probably have received little attention by practitioners in the field. *Stabilization and Solidification of Hazardous, Radioactive, and Mixed Wastes* fills a great need in this field — to consolidate the research, technology, and general practice in an up-to-date work, and it does so by utilizing the knowledge and experience of many of the most capable and experienced scientists, engineers, regulators, and teachers in its writing. Most importantly, it concentrates on the primary systems used in S/S today — those based on cement technology — eliminating discussion of the many approaches that have gone by the wayside over the years, or been marginalized, for both technical and practical reasons. Discussions of waste types, principles, and properties of cemented waste forms, and of test methods for evaluating them, are done in depth. However, the technology of cement-based systems also applies broadly to nonportland cement-based processes that undergo similar cementitious reactions and are subject to the same test and evaluation protocols and implementation practices.

Stabilization and Solidification of Hazardous, Radioactive, and Mixed Wastes should become the new standard reference source for anyone working in the waste treatment field, especially for those dealing with the technical aspects of S/S: research, testing, teaching, and regulation.

Jesse Conner
Conner Technologies

About the Editors

Dr. Roger Spence is a Senior Chemical Engineer in the Process Engineering Group of the Nuclear Science and Technology Division of Oak Ridge National Laboratory (ORNL). He received his B.S. in chemical engineering from Virginia Tech in 1971 and his Ph.D. in chemical engineering from North Carolina State University in 1975. He has been at ORNL since 1975, working on advanced separation techniques, advanced nuclear fuel cycles, nuclear fuel reprocessing, nuclear reactor safety, nonproliferation safeguards technology, and advanced decontamination techniques. He has worked in the field of stabilization/solidification since 1985 on development and testing of formulations and matrices, the leaching and leaching theory of solidified waste forms, and techniques for *in situ* stabilization. He has published or presented over 100 papers and reports and is a member of the American Institute of Chemical Engineers, the American Chemical Society, the Air & Waste Management Association, Tau Beta Pi, and Phi Kappa Phi. Dr. Spence edited another book in this field, *Chemistry and Microstructure of Solidified Waste Forms*, published by Lewis Publishers (now CRC) in 1993. He is the current chairman of the ANS-16.1 Working Group that worked on reissuing the standard short-term leach test for solidified radioactive waste, ANSI/ANS-16.1-2003.

Dr. Caijun Shi is a professor of the School of Civil Engineering and Architecture, Central South University, Changsha, Hunan, China, adjunct associate professor of the State University of New York (SUNY) at Buffalo, and president of CJS Technology Inc., Ontario, Canada. He received his B. Eng. and M. Eng. from Southeast University, Nanjing, China, and Ph. D. from the University of Calgary, Calgary, Alberta Canada. His interests and expertise include stabilization/solidification of wastes, waste containment, utilization of industrial by-products and recycled materials, and construction materials. He has published more than 100 technical papers and has been invited for presentations on different subjects in many countries. He has received several awards from different organizations. Dr. Shi has invented a few patented technologies and products. One of his inventions, the self-sealing/self-healing barrier, has been used as a municipal landfill liner in the world's largest landfill site in South Korea. He is a fellow of the International Energy Foundation and an active member of many technical committees in the American Society for Testing and Materials (ASTM), American Concrete Institute (ACI), Canadian Standard Association (CSA), and Professional Engineer Ontario (PEO), Canada. He has developed technologies and provided consulting services to many governmental organizations and private companies.

Contributors

Abir Al-Tabbaa
Department of Engineering
University of Cambridge
Cambridge, U.K.

Frank Cartledge
Vice Provost of Academic Affairs
Louisiana State University
Baton Rouge, LA

Jesse Conner
Conner Technologies
Pittsburgh, PA

Andrew C. Garrabrants
Civil and Environmental Engineering
Vanderbilt University
Nashville, TN

Steve Hoeffner
Clemson Environmental Technologies
  Laboratory
Clemson, SC

Guo H. Huang
Department of Environmental Systems
  Engineering
University of Regina
Regina, Saskatchewan, Canada

Paul D. Kalb
Brookhaven National Labaoratory
Upton, NY

David S. Kosson
Civil and Environmental Engineering
Vanderbilt University
Nashville, TN

Christine Langton
Savannah River Technology Center
Aiken, SC

Earl W. McDaniel
Consultant
Oak Ridge, TN

Robert C. Moore
Sandia National Laboratories
Albuquerque, NM

A.S. Ramesh Perera
Department of Engineering
University of Cambridge
Cambridge, U.K.

Xiaosheng S. Qin
Department of Environmental Systems
  Engineering
University of Regina
Regina, Saskatchewan, Canada

J. Murray Reid
Viridis
Berkshire, U.K.

Amitava Roy
Center for Advanced Microstructure
  and Devices
Louisiana State University
Baton Rouge, LA

Caijun Shi
CJS Technology Inc.
Burlington, Ontario, Canada

Darryl D. Siemer
Idaho National Engineering and
  Environmental Laboratory
Idaho Falls, ID

Roger D. Spence
Oak Ridge National Laboratory
Oak Ridge, TN

Julia A. Stegemann
Department of Civil and Environmental
  Engineering
University College
London, U.K.

Gerald W. Veazey
Los Alamos National Laboratory
Los Alamos, NM

Arun Wagh
Argonne National Laboratory
Argonne, IL

Table of Contents

1 Introduction

Roger Spence and Caijun Shi

CONTENTS

Stabilization/solidification (S/S) is typically a process that involves the mixing of a waste with a binder to reduce the contaminant leachability by both physical and chemical means and to convert the hazardous waste into an environmentally acceptable waste form for land disposal or construction use. S/S has been widely used to dispose of low-level radioactive, hazardous, and mixed wastes, as well as remediation of contaminated sites. According to the USEPA, S/S is the best demonstrated available technology (BDAT) for 57 hazardous wastes.[1] About 30% of the Superfund remediation sites used S/S technologies, according to a USEPA report in 1996.[2] Of all the binders, cementitious materials are the most widely used for S/S. Compared with other technologies, cement-based S/S has the following advantages:[3,4]

- Relatively low cost
- Good long-term stability, both physically and chemically
- Good impact and compressive strengths
- Documented use and compatibility with a variety of wastes over decades
- Material and technology well known
- Widespread availability of the chemical ingredients
- Nontoxicity of the chemical ingredients
- Ease of use in processing (processing normally conducted at ambient temperature and pressure and without unique or very special equipment)
- High waste loadings possible
- Inertness to ultraviolet radiation
- High resistance to biodegradation
- Low water solubility and leachability of some contaminants
- Ability of most aqueous wastes to chemically bind to matrix
- Relatively low water permeability
- Good mechanical and structural characteristics

- Good self-shielding for radioactive wastes
- Long shelf life of cement powder
- No free water if properly formulated
- Rapid, controllable setting, without settling or segregation during cure

1.1 DEFINITION OF STABILIZATION AND SOLIDIFICATION

"Stabilization" refers to techniques that chemically reduce the hazard potential of a waste by converting the contaminants into less soluble, mobile, or toxic forms. The physical nature and handling characteristics of the waste are not necessarily changed by stabilization.

"Solidification" refers to techniques that encapsulate the waste, forming a solid material, and does not necessarily involve a chemical interaction between the contaminants and the solidifying additives. The product of solidification, often known as the waste form, may be a monolithic block, a clay-like material, a granular particulate, or some other physical form commonly considered "solid."

The terms solidification/stabilization and stabilization/solidification are often used interchangeably and are referred to as S/S. Other commonly used terms were fixation and chemical fixation, which have generally been replaced by S/S despite some objections to the use of "monolith" in the USEPA definition of S/S.[3,5-7] Solidification and stabilization can be accomplished by a chemical reaction between the waste and solidifying reagents or by mechanical processes. Contaminant migration is often restricted by decreasing the surface area exposed to leaching or by coating the wastes with low-permeability materials.

1.2 A BRIEF SUMMARY OF THE ORIGIN OF S/S TECHNOLOGY

The origin and development of S/S technology were described by Jesse Conner, first in his benchmark book in 1990, followed by the review in 1998 co-authored with Steve Hoeffner.[3,5] S/S technology was first used for treatment of radioactive wastes in the 1950s. At early stages, liquid radioactive wastes were solidified using portland cement in drums or other containers, then buried at government-controlled disposal sites or at sea. High cement content was required to solidify the water in these liquid wastes. Mineral adsorbents such as vermiculite were used with cement to help absorb the high water content, reduce the amount of cement required, and prevent formation of bleed water. The large volume increases from solidifying what was essentially water led to subsequent high disposal costs. This led to calcination and vitrification becoming the preferred techniques, because of the large mass and volume reductions resulting from vaporizing the water and densifying the solids.

Prior to the Resource Conservation and Recovery Act (RCRA), disposal of liquid waste, other than by underground injection, was regulated under the authority of the Clean Water Act or controlled through state laws. Little documentation can be found on S/S of hazardous wastes prior to 1970. From 1970 until the passage of RCRA in 1976, few companies investigated the S/S of industrial liquid wastes. Chemfix Inc. discovered and patented a method using sodium silicate solution and portland cement for S/S of mine drainage sludge. During the same time period, Conversion System Inc. (CSI) developed a lime–fly ash process for the treatment of sludge from power plant flue gas desulfurization. Dravo Corporation used blast furnace slag for solidification of high-volume sludge from electric power plants. Crossford Pollution Control (later Stablex) in England began commercialization of its cement–fly ash process for solidification of inorganic waste streams at central treatment sites. The USEPA funded a few projects to evaluate S/S processes for hazardous waste treatment. Despite the lack of regulations, a considerable amount of waste was solidified during this time period. Reportedly, Chemfix solidified more than 375 ML (100 million gallons) of sludge from 1970 to 1976. CSI and Dravo installed several flue-gas desulphurization gypsum (FDG) treatment systems and treated much larger quantities of waste. In 1974, about 47,000 tonnes of sludge containing mercury were solidified and then disposed in the ocean (ocean disposal of treated radioactive, hazardous, or mixed waste has been banned by most countries since this time).

1.3 POST-RCRA DEVELOPMENT OF S/S TECHNOLOGY AND ORGANIZATION OF THE BOOK

After the passage of RCRA in 1976, S/S began to receive attention from governmental agencies, waste generators, and engineering firms. The USEPA began to fund research and development on S/S technologies not only at USEPA laboratories, but also with contractors and universities. During this time period, extensive research and development activities on various aspects of S/S were conducted. Many inorganic and organic binders, including those listed in Table 1.1, were developed for

TABLE 1.1
List of Binders Used for S/S

Inorganic Binder Systems	Organic Binder Systems
Portland cement	Bitumen
Portland slag cement	Urea formaldehyde
Portland pozzolan cement	Polybutadiene
Portland cement–silicate system	Polyester
Polymer modified cement	Epoxy
Masonry cement	Polyethylene
Lime–pozzolan cement	
Calcium aluminate cement	
Alkali-activated slag cement	
Alkali-activated pozzolan cement	
Phosphates	
Gypsum	
Sulfur polymer cement	
Alkali silicate minerals	

S/S. Inorganic binders, such as cement, are effective in immobilizing heavy metals through chemical and physical containment mechanisms, but are not as effective in immobilizing most organic contaminants. Many substances in the wastes significantly affect the setting and hardening characteristics of binders, especially cement-based cementing systems.[8,9]

A variety of processes and equipment were also developed. The mixing of wastes and binders can be carried out through either an *ex situ* or *in situ* process. For *ex situ* mixing, pugmills, mortar mixers, or concrete mixers are often used. *In situ* methods are widely used for remediation of contaminated sites and can be classified into the following three categories: backhoe-based methods, drilling/jetting/augering/trenching methods, and shallow area methods. Selection of the mixing method is based on the depth of the contamination and the characteristics of the contaminated media.

The USEPA sponsored a series of seminars on S/S, known as the "Immobilization Technology Seminars." The American Society for Testing and Materials (ASTM) sponsored a series of symposiums beginning in 1981, initially focusing on test methods and later expanding to all aspects of S/S. Gilliam, with co-editors,[10–12] edited three books of the papers presented at three of these symposiums. Spence[13] edited a similar book of papers from an American Chemical Society meeting. An extensive number of publications, reports, and presentations have been made on S/S, plus websites covering S/S, its application, and items of interest for S/S professionals. From roughly 1990 until the present time, S/S processes have been demonstrated and have been reasonably accepted by regulators, industries, and environmentalists. In 1999, the USEPA published a resource guide, which included recently published materials such as field reports and guidance documents that address issues relevant to solidification/stabilization technologies.[14] This and other technology-related documents are available over the Internet at the Hazardous Waste Clean-Up Information (CLU-IN) website at http://clu-in.org.

This book is intended to provide the regulatory and scientific basis for the use of S/S processes, a description of different S/S systems, a description of the testing and evaluation of the materials before and after treatment, and finally a summary of some previous field applications of S/S. Chapter 2 discusses the general guidelines for developing a waste form for a given application, giving two decision flow path schematics. Chapter 3 discusses characterization and classification of waste, an important preliminary step in treating waste or remediating sites. Binders are discussed in Chapters 4 (cement), 5 (polymers), and 6 (phosphate, sulfur polymer cement, gypsum, and hydroceramic). A variety of additives or sorbents are available to minimize interference with the hydration of cement or to enhance the immobilization of contaminants. Common additives or sorbents include activated carbon, zeolites, clays, carbonate, oxidizing agent, reducing agent, sulfides, organoclays, iron, and aluminum compounds. Chapter 7 discusses interactions between contaminants and binders, and Chapter 8 discusses some of the additives used to enhance binder properties or contaminant stabilization. Chapter 9 discusses the microstructure of S/S waste forms. Chapter 10 discusses the leachability from S/S waste forms. Chapter 11 discusses the evaluation of waste forms, their durability, and the test methods used. Chapter 12 discusses QA/QC for S/S. Chapter 13 presents four applications of S/S to real-world problems, one going back three decades that is still relevant today, one a more recent hazardous site, and two at USDOE sites.

This book can be used as a textbook for advanced courses in environmental engineering, or a reference book for students, laboratory workers, S/S technology vendors and buyers, engineers, scientists, regulators, and environmentalists.

References

1. USEPA. Technology Resource Document — Solidification/Stabilization and Its Application to Waste Materials, EPA/530/R-93/012, June 1993.
2. USEPA. Innovative Treatment Technologies: Annual Status Report, 8th Edition, EPA/542/R-96/010, November 1996.
3. Conner, J.R. *Chemical Fixation and Solidification of Hazardous Wastes.* Van Nostrand Reinhold, New York, 1990.
4. IAEA, Improved Cement Solidification of Low and Intermediate Level Radioactive Wastes, Technical Report Series No. 350, International Atomic Energy Agency, Vienna, 1993.
5. Conner, J.R. and Hoeffner, S.L., A critical review of stabilization/solidification technology. *Critical Reviews in Environmental Science and Technology.* 28(4):397-462, 1998.
6. Cullinane, M.J. and Jones, L.W., Stabilization/Solidification of Hazardous Waste, Cincinnati: USEPA, Hazardous Waste Engineering Laboratory (HWERL), EPA/600/D-86/028, 1986.
7. USEPA. Onsite Engineering Report for Waterways Experiment Station for K061, Washington, D.C., 1988.
8. Mattus, C.H. and Gilliam, T.M., A Literature Review of Mixed Waste Components: Sensitivities and Effects Upon Solidification in Cement-Based Matrices, ORNL/TM-12656, Oak Ridge National Laboratory, Oak Ridge, TN, 1994.
9. Trussell, S. and Spence, R.D., A Review of Solidification/Stabilization Interferences, *Waste Management*, Vol. 14, No. 6, pp. 507–519, 1994.
10. Cote, P. and Gilliam, T.M. (Editors), *Environmental Aspects of Stabilization and Solidification of Hazardous and Radioactive Wastes*, Volume 1, STP-1033, American Society for Testing and Materials, Philadelphia, PA, 1989.
11. Gilliam, T.M. and Wiles, C.C. (Editors), *Stabilization and Solidification of Hazardous, Radioactive, and Mixed Wastes*, Volume 2, STP-1123, American Society for Testing and Materials, Philadelphia, PA, 1992.
12. Gilliam, T.M. and Wiles, C.C. (Editors), *Stabilization and Solidification of Hazardous, Radioactive, and Mixed Wastes.* Volume 3, STP-1240, American Society for Testing and Materials, Philadelphia, PA, 1996.
13. Spence, R.D. (Editor). *Chemistry and Microstructure of Solidified Waste Forms*, Lewis Publishers, Boca Raton, FL, 1993.
14. USEPA, Solidification/Stabilization Resource Guide, EPA/542-B-99-002, U.S. Environmental Protection Agency, Washington, D.C., April 1999.

2 General Guidelines for S/S of Wastes

Caijun Shi and Roger Spence

CONTENTS

2.1 REGULATORY BASIS FOR USE OF S/S

The Solid Waste Disposal Act (SWDA) of 1965 was primarily the first federal attempt to manage municipal solid waste (Title 40 CFR Parts 240-257). Those regulations were primarily directed at the management of municipal solid waste (Subtitle D facilities) and established the initial framework for future solid and hazardous waste management. The Resource Conservation and Recovery Act (RCRA) was passed in 1976 and established statutory requirements and the basis for the management of "hazardous" wastes. This Act was actually an amendment to SWDA and imposed requirements for tracking solid and hazardous wastes from generation through final disposal (i.e., cradle-to-grave). The Hazardous and Solid Waste Amendments (HSWA) of 1984 amended RCRA and corrected some of its initial shortcomings. Some important programs developed as a result of the HSWA amendments, including (1) Land Disposal Restrictions (LDR) program; (2) Underground Storage Tank (UST) regulations; and (3) a switch in the testing of certain "toxic" wastes by the Extraction Procedure to the Toxicity Characteristic Leaching Procedure (TCLP).

1-56670-444-8/05/$0.00+$1.50
© 2005 by CRC Press

The Land Disposal Flexibility Act was passed in 1996 and allows characteristic wastes to be treated in Clean Water Act (CWA) systems or disposed of in Class I injection wells.

The primary goal of RCRA is the protection of human health and the environment. Other goals include waste reduction and the conservation of natural resources, the reduction or elimination of hazardous waste, the promotion of recycling, and the encouragement of state programs for RCRA. The main purpose of the LDR program is to prohibit activities that involve placing untreated hazardous wastes in or on the land. The U.S. Environmental Protection Agency (USEPA) has developed a treatment standard for most hazardous wastes (listed and characteristic). The treatment standards set either (1) concentration limits for treated wastes below which wastes may be safely land disposed or (2) specific technology-based standards that must be used on the waste. Stabilization/solidification (S/S) is the best demonstrated available technology (BDAT) for many contaminants and applications.

The EPA hazardous site cleanup program, referred to as Superfund, was authorized and established in 1980 by the enactment of the Comprehensive Environmental Response, Compensation, and Liability Act (CERCLA), Public Law (PL) 96-510. This legislation allows the Federal government (and cooperating state governments) to respond directly to releases and threatened releases of hazardous substances and pollutants or contaminants that may endanger public health or welfare or the environment. Prior to the passage of PL 96-510, Federal authority regulated hazardous contaminant release mainly through RCRA and the CWA and its predecessors. The general guidelines and provisions for implementing CERCLA are given in the National Oil and Hazardous Substances Contingency Plan (NCP) (Federal Register, 40 CFR 300, 1982).

Three classes of actions are available when direct government action is called for:

1. Immediate removals are allowed when a prompt response is needed to prevent harm to public health or welfare or to the environment. These are short-term actions usually limited to 6 months and a total expenditure of $1 million.
2. Planned removals are expedited, but not immediate, responses. These are intended to limit danger or exposure that would take place if longer term projects were implemented and responses were delayed.
3. Remedial actions are longer term activities undertaken to provide more complete remedies. Remedial actions are generally more expensive and can only be undertaken at sites appearing on the National Priorities List of the NCP.

Remedial actions may involve technically complex problems that are expensive to resolve. The selection of technical measures takes place only after a full evaluation of all feasible alternatives based upon economic, engineering, environmental, public health, and institutional considerations. Off-site transportation and disposal of waste is generally an expensive option and is justified only when proven cost-effective, and then only in facilities that comply with current hazardous waste disposal regulations under Subtitle C of RCRA.

Waste stabilization is specifically included in the NCP as a method of remedying releases of hazardous materials and controlling release of waste to surface water. Solidification and encapsulation are mentioned as techniques available for on-site treatment of contaminated soils and sediments. Under the general requirement to evaluate all alternatives for remedial action, it will be necessary to evaluate the cost-effectiveness of S/S systems as applied to specific sites even if the technology is not selected in the final analysis of remedial techniques. Costs and engineering considerations are critical to these evaluations.

The performance expected from S/S waste must also be assessed as accurately as possible. Cost estimates must take into consideration future expenditures needed to maintain the final waste disposal site after response work is complete. The NCP emphasizes the selection of proven remediation technologies. Examples of successful applications are an important part of any technical evaluation.

Overall guidance on remedial action technologies including a survey of S/S is provided in a Technology Transfer Handbook by the EPA.[1] The decision to implement the S/S option must be preceded by the detailed investigation of many variables. Both waste and site characteristics must be evaluated to ensure that the S/S alternative is cost-effective and environmentally acceptable. The USEPA has provided general guidance on the procedure to be followed in selecting the most appropriate remedial actions.[2,3]

With regard to radioactive waste, 10 CFR Part 61 establishes a waste classification system based on the radionuclide concentrations in the wastes. Class B and C wastes are required to be stabilized, but Class A wastes have lower concentrations and do not have to be stabilized. Liquid waste, including Class A, requires solidification, or absorption into solid media, to meet free liquid requirements. The U.S. Nuclear Regulatory Commission (USNRC) technical position on radioactive waste forms was initially developed in 1983. This technical position was revised in 1991, often referred to as the USNRC Technical Position paper of 1991, with an appendix specifically addressing cement-stabilized radioactive waste forms. This Technical Position paper gives the overall objectives of stabilizing radioactive wastes as well as a list of specific tests and limiting values. These tests are discussed in Chapter 11.

Mixed waste is a hazardous waste as defined by RCRA that contains radionuclides. Such wastes must meet the requirements of both USEPA and USNRC for disposal. Most treatment, storage, or disposal facilities (TSDFs) are licensed for hazardous waste by USEPA and are not equipped to handle radioactive or mixed waste. The USNRC licenses facilities for disposal of commercial radioactive waste, and the U.S. Department of Energy (USDOE) establishes standards and procedures for disposal of radioactive waste on USDOE sites, usually compatible with USNRC guidelines. Some overlap occurs among these three government agencies requiring cooperative agreements for disposing of mixed waste and radioactive waste generated at U.S. government facilities, not only among the agencies, but also among state and local governments. For disposing of hazardous waste from these facilities, the environmental rules and regulations of USEPA, the state, and the local governments take precedence.

2.2 SCIENTIFIC BASIS OF S/S

S/S is typically a process that involves the mixing of a waste with a binder to reduce contaminant leachability by both physical and chemical means and to convert the hazardous waste into an environmentally acceptable waste form, which goes to a landfill or is used in construction. Stabilization processes and solidification processes have different goals. Stabilization attempts to reduce the solubility or chemical reactivity of a waste by changing its chemical state or by physical entrapment (microencapsulation). Solidification systems attempt to convert the waste into an easily handled solid with reduced hazards from volatilization, leaching, or spillage. The two are often discussed together because they have the common purpose of improving the containment of potential pollutants in treated wastes. Combined processes are often termed "waste fixation" or "encapsulation."

Solidification of waste materials is widely used for the disposal of radioactive waste. Many developments relating to solidification originated from low-level radioactive waste disposal. Regulations pertaining to disposal of radioactive waste require that the wastes be converted into a free-standing solid with a minor amount of free water. Most processes used for nuclear waste include a step in which granular ion exchange waste and liquids are incorporated in a solid matrix using a cementing or binding agent (for example, portland cement, organic polymers, or asphalt). The resulting block of waste, with relatively low permeability, reduces the surface area across which the transfer of pollutants can occur.

In hazardous waste disposal and site remediation, treated material must meet certain standards for safe land disposal, by removing the hazardous characteristic for characteristically hazardous waste or by formal delisting for listed waste. For the toxic characteristic, this usually requires passing concentration-based standards using the USEPA TCLP test.[4] To accomplish this goal, a variety of strategies may be used to prevent contaminant leaching, including neutralization, oxidation/reduction, physical entrapment, chemical stabilization, and binding of the stabilized solid into a monolith. Appropriate treatment strategies should (1) treat waste, or contaminated sites, to be chemically inert and nonleachable, to the extent possible and (2) be economical.

A binder is often used to stabilize the contaminants in the waste or contaminated sites and to remove the free liquid. In cases where the waste is extremely soluble or no suitable chemical binder can be found, the waste may be contained by encapsulation in some hydrophobic medium, such as asphalt or polyethylene. This may be done either by incorporating the waste directly in the partially molten material or by forming jackets of polymeric material around blocks of waste.

Many types of binder have been developed for S/S of wastes, as discussed, but not all have been employed in remedial action on uncontrolled waste sites. The most commonly used binders are discussed in Chapters 4, 5, and 6. Portland cement is most commonly used because of its availability and low cost. Supplementary cementing materials such as coal fly ash and ground blast furnace slag are often used to partially replace portland cement, to improve the performance of the treated wastes, and to reduce the cost of the binder.

2.3 SELECTION OF S/S TECHNOLOGY

The process of technology selection, evaluation, and optimization is frequently referred to as "technology screening." The first step in the technology screening process is to identify potentially available technologies. Many treatment/remediation technologies are only suitable for certain types of wastes. A treatment technology that has been properly screened prior to full-scale implementation has the highest probability of success in the field. The book by LaGrega et al. discusses a variety of treatment/remediation technologies.[5] Figure 2.1 is a flowchart for the initial technology screening process.[6] Based on the characterization of the sample and the

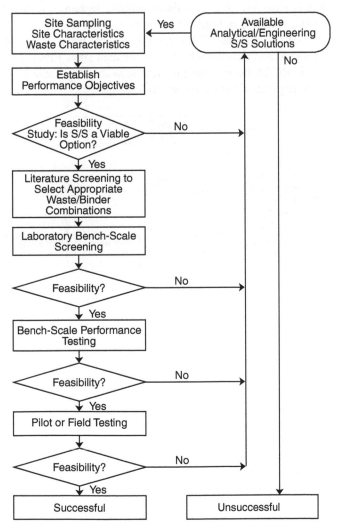

FIGURE 2.1 General technology screening procedure.[6]

site characteristics, performance objectives are established. Based on the performance objectives, a technology is selected mainly based on the following criteria:

- Overall effectiveness in protection of human health and the environment
- Compliance with applicable or relevant regulations and requirements
- Implementability
- Cost-effectiveness

In the broadest sense, most wastes or contaminated sites are potentially treatable with S/S. In some cases, it may require pretreatment or coupling with other technologies. If S/S is regarded as a potentially suitable technology, then the general decision flowpath illustrated in Figure 2.2 can be followed.[6,7]

During the past two decades, S/S technologies have been widely used for hazardous waste disposal and remediation of contaminated sites. EPA has identified S/S as the BDAT for 57 RCRA-listed hazardous wastes, as shown in Table 2.1.[6] About 30% of the Superfund remediation sites used S/S technologies.[8]

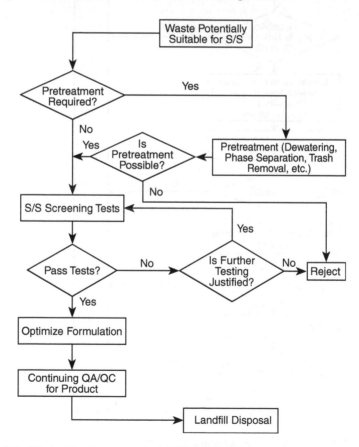

FIGURE 2.2 S/S decision flowchart at an RCRA TSDF.[6,7]

TABLE 2.1
RCRA Wastes for Which S/S is Identified as Best Demonstrated Available Technology (BDAT)[6]

Code	Waste Description	BDAT Treatment/ Treatment Train	Reference
D001	Ignitable (40 CFR 261.21 (a)(2))	S/S (one alternative)	55 FR 22714
D002	Other corrosives (40 CFR 261.22 (a)(2))	S/S (one alternative)	55 FR 22714
D003	Reactive sulfides (40 CFR 261.23 (a)(2))	S/S (one alternative)	55 FR 22714
D005	Barium	S/S (one alternative)	55 FR 22561
D006	Cadmium	S/S (one alternative)	55 FR 22562
D007	Chromium	S/S (one alternative)	55 FR 22563
D008	Lead	S/S	55 FR 22565
D009	Mercury (subclass)	S/S (< 260 mg/kg total Hg)	55 FR 22572
D010	Selenium	S/S	55 FR 22574
D011	Silver	S/S	55 FR 22575
F006	Some wastewater treatment sludges	Alkaline chlorination + precipitation + S/S	55 FR 26600
F007	Spent cyanide plating bath solutions	Alkaline chlorination + precipitation + S/S	55 FR 26600
F008	Plating sludges from cyanide process	Alkaline chlorination + precipitation + S/S	55 FR 26600
F009	Spent stripping and cleaning solutions from cyanide processes	Alkaline chlorination + precipitation + S/S	55 FR 26600
F011	Spent cyanide solutions from salt bath cleaning	Electrolytic oxidation + alkaline chlorination + precipitation + S/S	55 FR 26600
F012	Quenching wastewater treatment sludges from cyanide processes	Electrolytic oxidation + alkaline chlorination + precipitation + S/S	55 FR 26600
F019	Wastewater treatment sludges from coating of aluminum except for some zirconium phosphating processes	S/S	55 FR 22580
F024	Process wastes for the production of certain chlorinated aliphatic hydrocarbons	Incineration + S/S	55 FR 22589
F039	Leachates from listed wastes	S/S (metals)	55 FR 22607
K001	Bottom sediment sludges from the treatment of wastewater from wood-preserving processes that create creosote or pentachlorophenol		54 FR 31153
K006	Wastewater treatment sludges from the production of chromium oxide green pigments (anhydrous or hydrated)	S/S (hydrated form only)	55 FR 22583
K0015	Still bottom from distillation of benzylchloride	Incineration + S/S	55 FR 22598

(continued)

TABLE 2.1 (CONTINUED)
RCRA Wastes for Which S/S is Identified as Best Demonstrated Available Technology (BDAT)[6]

Code	Waste Description	BDAT Treatment/ Treatment Train	Reference
K022	Distillation bottom tars from the production of the phenol/acetone from cumene	Incineration + S/S	55 FR 31156
K028	Spent catalyst from the hydrochlorinator reactor in the production of 1,1,1-trichloroethane	Incineration + S/S	55 FR 22589
K046	Wastewater treatment sludges from the manufacturing, formula, and loading of lead-based initiating compounds	Reactive-Deactivation Stabilization Nonreactive-Stabilization	55 FR 22593
K048	Dissolved air floating float from the petroleum refining industry	Incineration + S/S	53 FR 31160 55 FR 22595
K049	Slop oil emulsion solids from the petroleum refining industry	Incineration + S/S	53 FR 31160 55 FR 22595
K050	Heat exchanger bundle cleaning sludges from the petroleum refining industry	Incineration + S/S	53 FR 31160 55 FR 22595
K051	API separator sludges from the petroleum refining industry	Incineration + S/S	53 FR 31160 55 FR 22595
K052	Tank bottoms (leaded) from the petroleum refining industry	Incineration + S/S	53 FR 31160 55 FR 22595
K061	Emission control dust/sludge from primary steel production in electric furnaces	S/S (< 15% Zn)	55 FR 22599
K069	Emission control dust/sludge from secondary lead smelting	S/S	55 FR 22568
K083	Distillation bottoms from aniline production	Incineration + S/S	55 FR 22588
K087	Decanter tank tar sludge from coking operation	Incineration + S/S	55 FR 31169
K100	Waste leaching solution from acid leaching of emission control dust/sludge from secondary lead production	Precipitation + S/S	55 FR 22568
K115	Heavy ends from the purification of toluenediamine in the production of toluenediamine via hydrogenation of dinitrotoluene	S/S	55 FR 26601
U051	Creosote	Incineration + S/S	55 FR 22482
U144	Lead acetate	S/S	55 FR 22565
U145	Lead phosphate	S/S	55 FR 22565
0146	Lead subacetate	S/S	55 FR 22565

(continued)

TABLE 2.1 (CONTINUED)
RCRA Wastes for Which S/S is Identified as Best Demonstrated Available Technology (BDAT)[6]

Code	Waste Description	BDAT Treatment/ Treatment Train	Reference
U204	Selenious acid	S/S	55 FR 22574
U205	Selenium disulfide	S/S	55 FR 22574
U214	Thallium (1) acetate	S/S or Thermal Recovery	55 FR 3891
U215	Thallium (1) carbonate	S/S or Thermal Recovery	55 FR 3891
U216	Thallium (1) chloride	S/S or Thermal Recovery	55 FR 3891
U217	Thallium (1) nitrate	S/S or Thermal Recovery	55 FR 3891
P074	Nickel cyanide	Electrolytic oxidation + alkaline chlorination + precipitation + S/S	55 FR 26600
P099	Argenate (1-), bis(cyano-C)-potassium	Electrolytic oxidation + alkaline chlorination + precipitation + S/S	55 FR 26600

2.4 DESIGN OF S/S FORMULATIONS

Once S/S technology is selected, the design of a proper S/S formulation is critical in successful treatment of wastes or remediation of contaminated sites. S/S of contaminants by cements includes the following three aspects: (a) chemical fixation of contaminants (chemical interaction between the hydration products of the cement and the contaminants); (b) physical adsorption of the contaminants on the surface of hydration products of the cements; and (c) physical encapsulation of contaminated waste or soil (low permeability of the hardened pastes).[9,10] The first two aspects depend on the nature of the hydration products, and the third aspect relies on both the nature of the hydration products and the density and physical structure of the pastes. Thus, from all aspects, the selection of cementing materials for S/S of wastes may have to consider the following aspects based on the characteristics of the wastes: (1) compatibility between cement and waste materials; (2) chemical fixation of contaminants; (3) physical encapsulation of contaminated waste and soils; (4) leachability of contaminants from the treated waste or soil; (5) durability of the treated waste or contaminated materials; and (6) cost-effectiveness of S/S. These aspects are discussed in the subsections that follow.

2.4.1 COMPATIBILITY BETWEEN CEMENT AND WASTE MATERIALS

Many substances can significantly affect the hydration of cement.[11,12] The intent of using admixtures, such as retarders, accelerators, or superplasticizers, is to affect one or more properties of cements and concrete in a manner that is beneficial for a given application. The interferences between contaminants and binders are discussed in Chapter 7. The same contaminant may have different interference effects on different types of cementing materials, or different interference effects on the same binder may result from different concentrations of a given contaminant. Thus, matching a cementing material with the waste or contaminant is important to avoid interference effects that may compromise treatment. Because of differences in interaction, regulatory agencies such as the USEPA have published regulations or guidance concerning the degree of interaction between a contaminant and binder, because of the potential re-release of the compound under various disposal conditions.[13] Physical forces such as gravity or mechanical compression may cause release. Also, pH and redox changes might affect future releases. The degree of interaction may be evaluated by physical methods such as gravity (paint filter test)[14] and mechanical compression (liquids release test).[15] Chemical extraction (solvents) or analytical techniques such as Fourier transform infrared (FTIR) spectroscopy, energy-dispersive x-ray analysis (EDXA), transmission electron microscopy (TEM), and x-ray diffraction (XRD) may be used to assess the degree of interaction.[16-18]

2.4.2 CHEMICAL FIXATION OF CONTAMINANTS

Hardened cement and concrete are porous materials that allow free ions to transport within and external to the binder matrix. In general, precipitation, absorption, or adsorption fix or stabilize contaminants chemically, primarily by the following three methods: (1) pH control of the system, (2) chemical reaction between contaminants and hydration products of cement, and (3) use of special additives. A variety of sorption processes have been evaluated for reducing the leachability of inorganic or organic compounds in waste material. Applications and evaluations of commercial processes based upon sorption are available in the literature.[19-21]

S/S for inorganic contaminants has been well accepted by the regulatory community, but not for organic contaminants. Water-soluble organic compounds such as phenol and ethylene glycol have been shown to chemically interact with cement, but were not effectively immobilized.[16] Less-soluble organics such as naphthalene were shown to be bound to quaternary ammonium compounds by stronger sorption reactions.[22] In general, hydrophobic organic material is not compatible with inorganic material such as cement. By substituting quaternary ammonium ions for group IA and IIA metal ions in clays, however, the interplanar distance between aluminum and silica can be increased, allowing clays to sorb organic compounds. For example, quaternary ammonium ions (R_4N^+) can be substituted for group IA and IIA metal ions (such as Li^+, Na^+, K^+, Mg^{2+}, Ca), yielding clays that have both organic and inorganic interactive properties, enabling the clay to also sorb organic compounds.[22] An evaluation of the organophilic clays conducted with four polyaromatic compounds, each with a concentration of less than 20 mg/kg, showed significant reductions in an organic extraction test and possible chemical reaction between the organophilic clay and organic compounds.[18]

Surfactants may be manufactured so as to have different chemical compatibilities on each end of the molecule. In simple terms, this configuration allows organic waste material to be sorbed on one end, while the other end is compatible with water mixed with the inorganic cement. With the use of a surfactant to lower surface tension, organic waste material can be dispersed with water in which the continuous phase is aqueous. This suspension can then be mixed with cement for solidification. Surfactants or emulsifiers can be used also to disperse organic waste material in an aqueous phase and then combined with cement for solidification.

2.4.3 PHYSICAL ENCAPSULATION OF CONTAMINATED WASTE AND SOILS

Some contaminants remain soluble despite best efforts at chemical fixation. In these cases, the mechanical barrier of the physical cement matrix plays the dominant role in preventing leaching of these contaminants and their subsequent release to the environment. As discussed above, the permeability of a hardened solidified waste depends on its pore structure. Thus, a dense matrix ensures a good mechanical barrier to migration of dissolved constituents. The initial porosity is determined by the water to cement ratio (W/C). Taylor discusses the effect of W/C on the structure and properties of portland cement paste and equations to calculate volumes (total paste, hydration product, unreacted cement, and nonevaporable water) and porosity (total water, free water, and capillary) as a function of W/C.[23] As cement hydrates, hydration products fill the voids between cement particles. According to Powers, the volume of hydration products is about 1.6 times the volume of its constituents.[24] Two types of pores form within the paste: gel pores and capillary pores. Gel pores constitute about 28% of the total C-S-H gel volume with a size of 1.5 to 2.0 nm (too small to permit the flow of water, with its molecular diameter of about 0.3 nm combined with the strong hydrogen bonding between water molecules). The capillary pores depend on the initial W/C and determine the movement of water, as shown in Figure 2.3.[25] Figure 2.4 shows the effect of the cement paste age (W/C = 0.51) on the d'Arcy's permeability coefficient,[26] and Table 2.2 shows the approximate age required for discontinuity of the capillary pores in cement pastes as a function of W/C.[25] At a W/C of 0.7, the continuous capillary pores within cement pastes are never segmented.

2.4.4 LEACHABILITY OF CONTAMINANTS FROM THE TREATED WASTE OR CONTAMINATED MATERIALS

Leaching happens when the contaminants in a treated waste form come in contact with a leachant. The manner in which this contact occurs is determined by the properties of the waste form such as water content, pore structure, homogeneity, and d'Arcy's permeability or hydraulic conductivity. Leaching can take place through convection or diffusion and depends on the physical and chemical fixation of contaminants. Chapter 10 discusses leaching mechanisms. One of the main goals of S/S formulation design is a leach-resistant waste form.

2.4.5 DURABILITY OF THE TREATED WASTE OR CONTAMINATED MATERIALS

Direct durability concerns for treated hazardous wastes or contaminated materials include stability after immersion in water, resistance to wetting/drying cycles, resistance

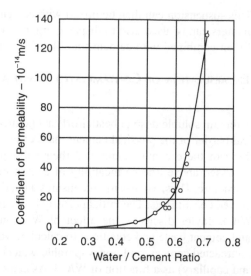

FIGURE 2.3 Effect of water/cement ratio on coefficient of permeability of mature cement pastes.[25]

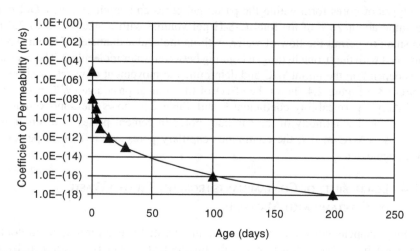

FIGURE 2.4 Effect of age on coefficient of permeability of cement paste with W/C = 0.51 (data from Reference 26).

to freezing/thawing cycles, and biodegradability. Most so-called "durability tests" address the physical stability of the waste form over a period of time that is long for laboratory testing, but short for the projected disposal life. Typically, some accelerated tests are used to evaluate the stability of treated wastes over 90 days or shorter, although some tests can be found spanning years.

The durability requirements for treated radioactive waste are slightly different from treated hazardous wastes due to the differences in the nature of the wastes. The recommended testing in the 1991 Technical Position paper of the USNRC summarized some durability requirements for Class B and C wastes:

TABLE 2.2
Approximate Time Until
Capillary Pore Discontinuity[24]

Water/Cement Ratio (by mass)	Time Required (days)
0.40	3
0.45	7
0.50	14
0.60	180
0.70	365
Over 0.70	Infinite

---

1. The waste should be a solid form or in a container or structure that provides stability after disposal.
2. The waste should not contain free-standing and corrosive liquids. That is, the wastes should contain only trace amounts of drainable liquid, and, as required by 10 CFR 61.56(b)(2), in no case may the volume of free liquid exceed 1% of the waste volume when wastes are disposed of in containers designed to provide stability, or 0.5% of the waste volume for solidified wastes.
3. The waste or container should be resistant to degradation caused by radiation effects.
4. The waste or container should be resistant to biodegradation.
5. The waste or container should remain stable under the compressive strength inherent in the disposal environment.
6. The waste or container should remain stable if exposed to moisture or water after disposal.
7. The as-generated waste should be compatible with the solidification medium or container.

The tests to demonstrate stability of cement-solidified Class B and C wastes include an average unconfined compressive strength of 3.4 MPa (500 psi) (ASTM C39), resistance to thermal cycling (ASTM B553), resistance to 1 MGy (100 MRad) dose, resistance to biodegradation (ASTM G21 and G22), resistance to leaching (ANSI/ANS-16.1-2003), resistance to immersion for 90 days, and generate < 0.5 vol% free-standing liquid (ANSI/ANS-55.1-1979). Long-term leaching properties are applicable to the long-lived radionuclides.

2.4.6 COST-EFFECTIVENESS

Cost is always one of the major concerns for selecting waste treatment/disposal technologies. Stabilization is popular in large part because it is cheap compared with many other technologies. A stabilization operation at a remedial site is usually priced in the range of $40 to $100 per ton of waste treated, not including other site costs such as excavation and landfilling. Of this total, about 40 to 50% goes to the cost of the reagents used.[27]

2.5 EVALUATION OF S/S FORMULATIONS

TCLP is the regulatory testing procedure in several countries and is sometimes used as the only test for delisting hazardous wastes. However, the TCLP test has been criticized for a long time, since the maximum amount of acid for the test is two equivalents of acetic acid per dry gram of waste material. People can easily add some alkaline materials to the waste based on the neutralization capacity to obtain a leachate with a pH around 10, where most regulated heavy metals have the lowest solubility. Although such treated wastes or remediated sites are regarded as environmentally acceptable, they may still release contaminants into fresh water at a concentration higher than the concentration limits for TCLP testing.[28,29] The Science Advisory Board of USEPA reviewed the leaching evaluation framework employed by the agency in 1991 and 1999.[30,31] In the 1999 review, the Science Advisory Board stated:

"The current state of the science supports, even encourages, the development and use of different leach tests for different applications. To be most scientifically supportable, a leaching protocol should be both accurate and reasonably related to conditions governing leachability under actual waste disposal conditions."

and:

"The multiple uses of TCLP may require the development of multiple leaching tests. The result may be a more flexible, case-specific, tiered testing scheme or a suite of related tests incorporating the most important parameters affecting leaching. Applying the improved procedure(s) to the worst-case scenario likely to be encountered in the field could ameliorate many problems associated with current procedures. Although the Committee recognizes that these modifications may be more cumbersome to implement, this type of protocol would better predict leachability."

The Science Advisory Board also criticized the TCLP protocol on the basis of several technical considerations, including the test's consideration of leaching kinetics, liquid-to-solid ratio, pH, potential for colloid formation, particle size reduction, aging, volatile losses, and commingling of the tested material with other wastes (i.e., co-disposal).

On March 23, 2000, the USEPA published a guidance document on delisting, entitled *EPA RCRA Delisting Program Guidance Manual for the Petitioner.*[32] Section 6.2.2 of the manual provides guidelines for stabilized wastes. If the petitioned waste is generated from the chemical stabilization of a listed waste, then leachable metal concentrations should be tested using the Multiple Extraction Procedure (MEP), SW-846 Method 1320, as well as by TCLP analyses to assess the long-term stability of the waste.

The purpose of S/S is to maximize the containment of environmental contaminants by both physical and chemical means and to convert hazardous waste into an

environmentally acceptable waste form. Whenever possible, utilization of stabi-
lized/solidified hazardous wastes should be preferred over disposal in order to
decrease the burden of land disposal.[33] If a waste form does require landfilling,
however, the degree of environmental protection provided by stabilization/solidifi-
cation should allow for disposal in less-costly landfilling facilities, by reducing or
eliminating the need for engineered barriers or liner systems. The evaluation protocol
for S/S waste forms developed by the Wastewater Technology Centre (WTC) of
Environment Canada includes three levels of evaluation: Level 0 – Information on
Untreated Waste, S/S Process, and Waste Forms; Level 1 – Chemical Immobilization;
and Level 2 – Physical Entrapment. This protocol can be used as a decision-making
tool for different disposal or utilization scenarios based on the degree of contaminant
containment and physical properties of an S/S waste form. Recently, Kosson et al.[34]
proposed an alternative framework for evaluation of inorganic constituent leaching
from wastes and secondary materials. The framework is based on the measurement
of intrinsic leaching properties of the material in conjunction with mathematical
modeling to estimate release under field management scenarios. A description of
the WTC evaluation protocol and an alternative leaching evaluation framework is
discussed in Chapter 11.

2.6 SUMMARY

This chapter provides a general guideline for the use, design, and evaluation of a
process for S/S of a waste. The selection of an S/S technology should be based on
both the legal and technical foundations. The evaluation of an S/S waste form should
consider not only the intrinsic leaching properties of the material, but also the field
management scenarios.

References

1. USEPA, *Remedial Action at Waste Disposal Sites (revised)*, EPA-625/6-85-006, U.S.
   Environmental Protection Agency, Cincinnati, OH, 1985.
2. USEPA, *Guidance on Feasibility Studies under CERCLA*, EPA-540/G-85-003, U.S.
   Environmental Protection Agency, Washington, DC, 1985.
3. USEPA, *Guidance on Remedial Investigations Under CERCLA*, EPA-540/G-85-002,
   U.S. Environmental Protection Agency, Washington, DC, 1985.
4. U.S. Federal Register, Appendix I, Part 268, *Toxicity Characteristic Leaching Procedure (TCLP)*, Vol. 51, No. 216, November 7, 1986.
5. LaGrega, M.D., Buckingham, P.L., and Evans, J.C., *Hazardous Waste Management*,
   McGraw-Hill, New York, 1994.
6. Means, J.L. et al. *The Application of Solidification/Stabilization to Waste Materials*,
   Lewis Publishers, Boca Raton, FL, 1995.
7. Barth, E.D. et al., *Stabilization and Solidification of Hazardous Wastes*, Noyes Data
   Corporation, Park Ridge, NJ, 1990.
8. USEPA, *Innovative Treatment Technologies: Annual Status Report*, 8th Edition,
   EPA/542/R-96/010, U.S. Environmental Protection Agency, Washington, DC, 1996.

9. Shi, C., Day, R.L., Wu, X., and Tang, M., Uptake of Metal Ions by Autoclaved Cement Pastes, *Proceedings of Materials Research Society*, Vol. 245, pp. 141–149, 1992.
10. Shi, C., Shen, X., Wu, X., and Tang, M., Immobilization of Radioactive Wastes with Portland and Alkali-Slag Cement Pastes, *Il Cemento*, Vol. 91, No. 2, pp. 97–108, 1994.
11. Mattus, C.H. and Gilliam, T.M., *A Literature Review of Mixed Waste Components: Sensitivities and Effects Upon Solidification in Cement-Based Matrices*, ORNL/TM-12656, Oak Ridge National Laboratory, Oak Ridge, TN, 1994.
12. Trussell, S. and Spence, R.D., A Review of Solidification/Stabilization Interferences, *Waste Management*, Vol. 14, No. 6, pp. 507–519, 1994.
13. USEPA, *Test Methods for the Evaluation of Solid Wastes, Physical/Chemical Methods*, Office of Water and Waste Management, Washington, DC, SW-846, 1986.
14. USEPA, *Code of Federal Regulations. Final Rule — Paint Filter Liquids Test*, April 30, 1985.
15. USEPA, *Code of Federal Regulations. Liquids Release Test*, October 29, 1991.
16. Cartledge, F., Eaton, H., and Tittlebaum, M., *Morphology and Microchemistry of Solidified/Stabilized Hazardous Waste Systems*, USEPA Cooperative Agreement CR-812318-01-0, Risk Reduction Engineering Laboratory, Cincinnati, OH, 1990.
17. Tittlebaum, M., Seals, R., Cartledge, F., and Engels, S., State of the Art Stabilization of Hazardous Organic Liquid Wastes and Sludges, *CRC Critical Reviews in Environmental Control*, 15 (2), pp. 179–211, 1985.
18. Soundararajan, R., Barth, E., and Gibbons, J., Using an Organophilic Clay to Chemically Stabilize Waste Containing Organic Compounds, *Hazardous Materials Control Journal*, 2(1), 1990.
19. American Academy of Environmental Engineers, *Innovative Site Remediation Technology: Solidification/Stabilization, Volume 4*, Annapolis, MD, 1994.
20. Barth, E., Summary of Solidification/Stabilization SITE Demonstrations at Uncontrolled Hazardous Waste Sites, *Stabilization and Solidification of Hazardous, Radioactive, and Mixed Wastes*, ASTM STP 1123, American Society for Testing and Materials, Philadelphia, 1992.
21. Bates, E., Akindele, F., and Sprinkle, D., American Creosote Site Case Study: Solidification/Stabilization of Dioxins, PCP, and Creosote for $64 per Cubic Yard, *Environmental Progress*, 21(2), 2002.
22. Gibbons, J.J. and Soundararajan, R., The Nature of Chemical Bonding Between Organic Wastes and Organophilic Binders, Part 2, *American Laboratory*, 21(7), pp. 70–79, 1989.
23. Taylor, H.F.W., *Cement Chemistry*, Academic Press, Ltd., London, 1990, Chap. 8, pp. 243–275.
24. Powers, T.C., Physical properties of cement paste, in *Proceedings of the Fourth International Symposium on the Chemistry of Cement*, Washington, D.C., 2, pp. 577–613, 1960.
25. Powers, T.C., Structure and Physical Properties of Hardened Portland Cement Pastes, *Journal of the American Ceramic Society*, Vol. 41, No. 1, pp. 1–6, 1958.
26. Mindess, S. and Young, J.F., *Concrete*, Prentice-Hall, 1981.
27. Conner, J.R. and Hoeffner, S.L., A Critical Review of Stabilization/Solidification Technology, *Critical Reviews in Environmental Science and Technology*, 28(4):397–462, 1998.
28. Perket, C.L., Krueger, R.J., and Whitehurst, D.A., The Use of Extraction Tests for Deciding Waste Disposal Options, *Trends in Analytical Chemistry*, Vol. 1, No. 14, 1982.

29. Poon, C.S. and Lio, K.W., The Limitation of Toxicity Characteristic Leaching Procedure for Evaluation of Cement-Based Stabilized/Solidified Waste Forms, *Waste Management*, Vol. 17, No. 1, pp. 15–23, 1997.

30. USEPA, *Leachability Phenomena*. EPA-SAB-EEC-92-003. Washington, DC, USEPA Science Advisory Board, 1991.

31. USEPA, *Waste Leachability: The Need for Review of Current Agency Procedures*. EPA-SAB-EEC-COM-99-002. Washington, DC, USEPA Science Advisory Board, 1999.

32. USEPA, *EPA RCRA Delisting Program Guidance Manual for the Petitioner*, Washington, DC, March 23, 2002.

33. Wastewater Technology Centre, Proposed Evaluation Protocol for Cement-based Solidified Waste, Environment Canada Report EPS 3/HA/9, Environment Canada, Ottawa, 1991.

34. Kosson, D.S., van der Sloot, H.A., Sanchez, F., and Garrabrants, A.C., An Integrated Framework for Evaluating Leaching in Waste Management and Utilization of Secondary Materials, *Environmental Engineering Science*, Vol. 19, No. 3, pp. 159–204, 2002.

3 Classification and Characterization of Hazardous, Radioactive, and Mixed Wastes

Xiaosheng S. Qin and Guo H. Huang

CONTENTS

1-56670-444-8/05/$0.00+$1.50
© 2005 by CRC Press

3.1 BACKGROUND AND OVERVIEW

The improper management of hazardous, radioactive, and mixed wastes poses a serious threat to the health of humans and other living organisms and their environment. Even when the above wastes are managed or disposed of in an abortive manner, serious harm is possible, including endangering human health. For instance, toxic leachates that include hazardous wastes from a poorly constructed or improperly maintained hazardous waste landfill could severely contaminate the groundwater and surface waters; incidents of nuclear power plant explosion or leakage can seriously injure or kill surrounding workers or residents. It has become obvious that these waste management projects must be regulated with sufficiently knowledgeable, capable, and unfailingly trustworthy legislations and regulations. Clearly, if a regulatory agency is to regulate something, there should be an unambiguous means of identifying and describing what is to be regulated. Therefore, fundamentals of the management of hazardous, radioactive, and mixed wastes are needed for adequate definition, designation, classification, and characterization to provide bounds to the problem.

The definition, classification, and management of hazardous, radioactive, and mixed wastes vary from one country to another. In the U.S., the above wastes are covered under different agencies and regulations. The regulations contain important guidelines for determining what exactly is the type of waste, how to identify a specific substance to be the regulated waste and provide objective criteria for including other materials in the universe, and what the exclusions and exemptions from the regulations are.

As for hazardous waste, since 1980, the U.S. Environmental Protection Agency (EPA) has developed a comprehensive program to ensure that hazardous waste is managed safely. A "cradle-to-grave" strategy for hazardous wastes from the point of generation to ultimate disposal is established for hazardous waste identification, recycling, storage, and disposal. The Resource Conservation and Recovery Act (RCRA), enacted in 1976, was motivated by concern over improper disposal of solid and hazardous wastes and contains a great deal of prescriptive detail. RCRA was an amendment to the Solid Waste Disposal Act to address the huge volumes of municipal and industrial solid wastes generated nationwide. RCRA Subtitle C gives a comprehensive program regarding identification, generation, transportation, treatment, storage, and disposal relevant to hazardous waste.

Radioactive waste arises from commercial nuclear power generation as well as from other industrial activities and from the use of radioactive materials in several human activities such as medical uses of radioactive materials. The Atomic Energy Act (AEA) of 1954 is the basic law governing production, use, ownership, liability, and disposal of radioactive materials in the United States.[1] A number of laws also specify radioactive waste management procedures and authorities such as the Low-Level Radioactive Waste Policy Act (LLWPA, amended in 1985, LLWPAA) and the Nuclear Waste Policy Act (NWPA, amended in 1987 NWPAA). Radioactive waste is regulated by either the Nuclear Regulatory Commission (NRC) or the U.S. Department of Energy (DOE) under the AEA. The classification and characterization of such wastes are normally specified by relevant laws and regulations.

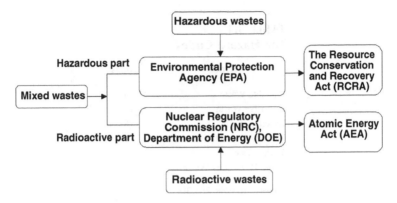

FIGURE 3.1 Regulations of hazardous, radioactive, and mixed wastes.

Many of the waste materials are classified as hazardous waste under RCRA, among which a number of these wastes also exhibit radioactivity, making them "mixed waste." The hazardous components of mixed waste are subject to RCRA regulations by EPA, while the radioactive component is subject to DOE or NRC regulations under AEA of 1954. Relationships between legislations, administrators, and corresponding wastes are shown in Figure 3.1.

3.2 CLASSIFICATION AND CHARACTERIZATION OF HAZARDOUS WASTES

Before classification and characterization of hazardous wastes, the exact definition of what is a hazardous waste is paramount. The RCRA briefly defines hazardous waste as any toxic, corrosive, reactive, or ignitable material that could damage the environment or negatively affect human health.[2] Some examples of hazardous waste include oils, solvents, acids, metals, and pesticides. In another word, a hazardous waste is any waste that has substantial dangers, now or in the future, to human, plant, or animal life, and which therefore cannot be handled or disposed of without special precautions. A more complete and specific definition has been published by the U.S Environmental Protection Agency.[3] In the EPA definition, a waste material either presents on the EPA-developed lists or has evidence that the waste exhibits ignitable, corrosive, reactive, or toxic characteristics as defined to be hazardous waste. Actually, the determination of a hazardous waste is a complex task because the hazardous waste may either come from multiple sources or may exist in many forms such as solids, liquids, and even gases.

3.2.1 HAZARDOUS WASTE CLASSIFICATION

There are two ways that EPA identifies hazardous waste: the waste is either listed, called listed wastes, or the waste has a toxic or dangerous characteristic, called characteristic wastes.

TABLE 3.1
The Hazard Codes

Types	Codes
Ignitable Waste	(I)
Corrosive Waste	(C)
Reactive Waste	(R)
Toxicity Characteristic Waste	(E)
Toxic Waste	(T)
Acute Hazardous Waste	(H)

Source: 40 CFR 261.30

3.2.1.1 Hazard Codes

To indicate its reason for listing a waste, EPA assigns a hazard code to each waste listed on the F, K, P, and U lists (40 CFR 261.30). As shown in Table 3.1, the first four hazard codes apply to wastes that have been listed because they typically exhibit one of the four regulatory characteristics of hazardous waste. The last two hazard codes apply to listed wastes whose constituents pose additional threats to human health and the environment.

The hazard codes assigned to listed wastes affect the regulations that apply to treat the waste. For example, acute hazardous wastes followed by the hazard code (H) are subject to stricter management standards than most other wastes.

3.2.1.2 Listed Wastes

EPA has published lists of wastes in the regulations that are classified as hazardous. The reason they are considered as listed hazardous wastes is that these wastes are dangerous enough to warrant full Subtitle C regulation of RCRA based on their origins.[4] Each listed waste carries its own unique EPA hazardous waste code. The codes begin with F, K, P, or U followed by three numbers. The listed wastes are coming from generic industrial processes, certain sectors of industry, and unused pure chemical products and formulations. The list ranges from spent halogenated and nonhalogenated solvents, heavy ends, light ends, and bottom tars from various distillation processes to some commercial chemical products such as arsenic acid, cyanides, and many pesticides.

At this writing, the EPA has established four categories of listed hazardous wastes. Some criteria developed by EPA are used to decide whether or not a waste will be selected, as follows:[5]

- It exhibits any of the characteristics of hazardous waste: ignitability, corrosivity, reactivity, and toxicity.
- The waste is found to be fatal to humans and animals even in very low doses, or the waste shows in studies to have such dangerous chemicals that it could cause or significantly contribute to an increase in serious

irreversible illness. Wastes in accordance with these criteria are known as Acute Hazardous Wastes.

- When improperly treated, stored, transported, or disposed of, the waste is capable of posing a substantial threat to human health and the environment. Such wastes are known as Toxic Listed Wastes.
- EPA has reason to believe that the waste is typically or frequently hazardous under the statutory definition of hazardous waste.

EPA has grouped hundreds of specific hazardous wastes into three categories located at 40 CFR 261, Subpart D under the listing criteria, as follows:

(a) Nonspecific Source Wastes

The nonspecific source wastes, codified in regulations at 40 CFR 261.31, are also called F list wastes, which means the identification code begin with the letter F. They are commonly produced by manufacturing and industrial processes. In brief, the nonspecific source wastes consist of seven groups:[4]

- Spent solvent wastes (Codes from F001 to F005) that generated from the use of certain common organic solvents widely used in various industries such as dry cleaning, electronics manufacturing, degreasing, and cleaning. They mostly consist of the spent halogenated solvents, including methylene chloride, carbon tetrachloride, and chlorobenzene, and nonhalogenated solvents, including xylene, acetone, ethylbenzene, cyclohexanone, toluene, pyridine, and methanol.
- Electroplating and other metal finishing wastes (Codes from F006 to F012, including F019) generated from electroplating, metal heat treating operations, and aluminum can washing processes. Examples include spent cyanide plating bath solutions, wastewater treatment sludges, and quenching bath residues.
- Dioxin-bearing wastes (Codes from F020 to F023, and from F026 to F028) generated from production of specific pesticides or specific chemicals used in the production of pesticides such as wastes from the discarded unused formulations and residues from incineration or thermal treatment containing tri- or tetrachlorophenol, pentachlorophenol, and hexachlorophene.
- Chlorinated aliphatic hydrocarbon production wastes (F024 and F025) produced from the manufacture of chlorinated aliphatic hydrocarbons having carbon chain lengths ranging from one to five, with varying amounts and positions of chlorine substitution.
- Wood-preserving wastes (F032, F034 and F035) generated from wood-preserving operations such as coating lumber with pentachlorophenol, creosote, or preservatives containing arsenic or chromium.
- Petroleum refinery wastewater treatment sludges (F037 and F038) generated from the gravitational and physical/chemical separations of oil/water/solids during the storage or treatment of process wastewaters and oily cooling wastewaters from petroleum refineries.

- Multisource leachate (F039), normally resulting from a hazardous waste landfill containing high concentrations of chemicals in liquid forms.

(b) Specific Source Wastes

The specific source wastes, also called K list wastes, are codified in regulations at 40 CFR 261.32. They are wastes generated from specially identified industries such as wood preserving, petroleum refining, and organic chemical manufacturing. The waste descriptions in K listed wastes have been presented in detail for waste identification. There are about 13 industries covering the list (there used to be 17; EPA has withdrawn the K waste codes applicable to waste streams in the primary copper, primary lead, primary zinc, and ferroalloys industries [40 CFR 261.32]). These industries are wood preservation; inorganic pigments; organic chemicals; inorganic chemicals; pesticides; explosives; petroleum refining; iron and steel; primary aluminum; secondary lead; veterinary pharmaceuticals; ink formulation; and coking.

(c) Commercial Chemical Products

The commercial chemical products, codified in Regulation 40 CFR 261.33, include specific commercial chemical products or manufacturing chemical intermediates. These wastes are given an EPA hazardous waste number that begins with either the letter P or U. More detailed classification can be applied to commercial chemical products as follows:

- Specific substances identified as acute hazardous waste (with hazardous code H) that are discarded commercial chemical products, off-specification species, container residues, and spill residues. The waste number begins with the letter P.
- Specific substances identified as toxic wastes (with hazardous code T) that are discarded commercial chemical products, off-specification species, container residues, and spill residues. The waste number begins with the letter U.

To identify whether a waste qualifies as P or U listed, it must meet the following criteria (40 CFR 261.33):

- The waste is discarded or is intended to be discarded.
- The waste contains a generic name listed in the P or K list in any commercial chemical product (CCP), or manufacturing chemical intermediate (MCI), any residue remaining in a container that has held any CCP or MCI, and any residue or contaminated soil, water, or other debris of any CCP or MCI.
- The waste contains a generic name listed in the P or K list, if it met specifications, in any off-specification CCP or MCI it previously held, and any residue or contaminated soil, water, or other debris of any off-specification chemical product and MCI.

A waste listed in CCP and MCI can also be understood as a chemical substance that is manufactured or formulated for commercial or manufacturing use, which consists of the commercially pure grade of the chemical (the chemical is the only chemical constituent in the product); any technical grades of the chemical (not 100% pure, but recognized in general usage by chemical industries) that are produced or marketed; and all chemical formulations in which the chemical is the sole active ingredient (the chemical is the only ingredient serving the function of the formulation).

3.2.1.3 Characteristic Wastes

The characteristic wastes can be defined as wastes that are not specifically identified elsewhere, or exhibit properties of ignitability, corrosivity, reactivity, or toxicity. They are not listed hazardous wastes under RCRA but pose sufficient threat to human health and the environment that they deserve regulation as hazardous wastes. Obviously, the characteristic wastes are an essential supplement to the hazardous waste listings.

If a listed waste still exhibits a characteristic that poses an additional hazard to human health and the environment, additional regulatory precautions should be implemented. There are four hazardous waste characteristics. A characteristic waste has its own unique EPA hazardous waste code, beginning with a D followed by three numbers.

3.2.2 HAZARDOUS WASTE CHARACTERISTICS

The EPA has established four characteristics for hazardous waste identification. Two criteria are used in selecting these characteristics as follows:

- The characteristics of the properties of hazardous waste should be detectable by using a standardized test method or by applying general knowledge of the properties of the waste.
- The characteristics must be defined in terms of physical, chemical, or other properties that cause the waste to meet the definition of hazardous waste in the RCRA.

The first criterion was adopted because EPA believed that unless generators were provided with widely available and uncomplicated methods for determining whether their wastes exhibited the characteristics, the identification system would prove unworkable.[6]

Based on the above criteria, four characteristics and their respective rationales described in 40 CFR 261 are summarized as follows.

(a) Ignitability

The ignitability characteristic refers to wastes that can readily catch fire and sustain combustion such as paints, cleaners, and other industrial wastes that pose such a hazard. Most ignitable wastes are in liquid form. EPA used a flash point test as the method for determining whether a liquid waste is combustible enough to deserve regulation as hazardous. Many wastes in solid or nonliquid physical form (e.g., wood, paper) can also readily catch fire and sustain combustion, but they are not all regulated by EPA. A solid waste is said to exhibit the characteristic of ignitability if a representative sample of the waste has any of the following properties:

- It is a liquid other than an aqueous solution containing less than 24% alcohol by volume and has a flash point less than 60°C (140°F), as determined by a Pensky–Martens Closed Cup Tester or a Setaflash Closed Cup Tester, or as determined by an equivalent test method (40 CFR 260.11 and 260.20).
- It is not a liquid and is capable, under standard temperature and pressure, of causing fire through friction, absorption of moisture, or spontaneous chemical changes and, when ignited, burns so vigorously and persistently that it creates a hazard.
- It is an ignitable, compressed gas.
- It is a substance meeting the Department of Transportation's definition of oxidizer.

The ignitable wastes are codified as D001 and are among the most common hazardous wastes.

(b) Corrosivity

The corrosivity characteristic identifies wastes that are acidic or alkaline (basic) that can readily corrode or dissolve flesh, metal, or other materials such as sulfuric acid from automotive batteries and pickle liquor employed to clean steel during its manufacturing. The EPA chose pH as an indicator of corrosivity because wastes with high or low pH can react dangerously with other wastes or cause toxic contaminants to migrate from certain wastes. A solid waste that exhibits any of the following properties is considered a hazardous waste due to its corrosivity:

- It is an aqueous material with pH less than or equal to 2 or greater than or equal to 12.5, as determined by a pH meter using the method described in EPA Publication SW-846 incorporated by reference in 40 CFR 260.11.
- It is a liquid and corrodes steel at a rate greater than 6.35mm (0.25 inch) per year at a test temperature of 55°C (130°F) as determined by the test method in EPA Publication SW-846.

Steel corrosion is used as a criteria because wastes capable of corroding steel can escape from their containers and liberate other wastes. Physically solid, non-aqueous wastes are not evaluated for corrosivity. A waste that exhibits the characteristic of corrosivity is given an EPA hazardous waste number of D002.

(c) Reactivity

The reactivity characteristic refers to wastes that readily explode or undergo violent reactions. Reactivity was chosen as a characteristic to identify unstable wastes that can pose a problem at any stage of the waste management cycle, e.g., an explosion. Examples of reactive wastes include water from TNT manufacturing operations, contaminated industrial gases, and deteriorated explosives. A solid waste exhibits the characteristic of reactivity if a representative sample of the waste has any of the following properties (40 CFR 261.23):

- It is normally unstable and readily undergoes violent change without detonating.
- It reacts violently with water.
- It forms potentially explosive mixtures with water.
- When mixed with water, it generates toxic gases, vapors, or fumes in a quantity sufficient to present a danger to human health or the environment.
- It is a cyanide- or sulfide-bearing waste which, when exposed to pH conditions between 2 and 12.5, can generate toxic gases, vapors, or fumes in a quantity sufficient to present a danger to human health or the environment.
- It is capable of detonation or explosive reaction if it is subjected to a strong initiating source or if heated under confinement.
- It is readily capable of detonation or explosive decomposition or reaction at standard temperature and pressure.
- It is a forbidden explosive as defined in 49 CFR 173.51, or a Class A explosive as defined in 49 CFR 173.53, or a Class B explosive as defined in 49 CFR 173.88.

Wastes exhibiting the characteristic of reactivity are assigned the waste code D003.

(d) Toxicity

The term toxicity refers to both a characteristic of a waste and a test. The test procedure is called Toxicity Characteristics Leaching Procedure (TCLP), which is designed to produce an extract simulating the leachate that may be produced in a land disposal situation. The extract is then analyzed to determine if it includes any of the toxic contaminants listed in 40 CFR 261.24, shown in Table 3.2. If the concentrations of any of the Table 3.2 constituents are greater than or equal to the levels listed in the table, the waste is classified as hazardous. The regulatory levels listed in Table 3.2 are based on groundwater modeling studies and toxicity data that calculate the limit above which these common toxic compounds and elements will threaten human health and the environment by contaminating drinking water. The wastes in this category are often referred to as toxicity characteristic (TC) wastes.

3.2.3 IDENTIFICATION PROCESS OF A HAZARDOUS WASTE

3.2.3.1 Question Determination Process

The characteristics and classification introduced in the above sections, albeit simple and straightforward, still could not give easy hazardous waste identification. Actually, RCRA have adopted an effective question determination process to decide if a waste belongs to regulated hazardous waste. The questions are listed in Figure 3.2.

TABLE 3.2
Toxicity Characteristic Constituents and Regulatory Levels

Types	Contaminants	EPA HW No.	Regulatory Level
Metals	Arsenic	D004	5.0 mg/L
	Barium	D005	100.0 mg/L
	Cadmium	D006	1.0 mg/L
	Chromium	D007	5.0 mg/L
	Lead	D008	5.0 mg/L
	Mercury	D009	0.2 mg/L
	Selenium	D010	1.0 mg/L
	Silver	D011	5.0 mg/L
Volatile Organic Compounds	Benzene	D018	0.5 mg/L
	Carbon tetrachloride	D019	0.5 mg/L
	Chlorobenzene	D021	100.0 mg/L
	Chloroform	D022	6.0 mg/L
	1,2-Dichloroethane	D028	0.5 mg/L
	1,1-Dichloroethylene	D029	0.7 mg/L
	Methyl ethyl ketone	D035	200.0 mg/L
	Tetrachloroethylene	D039	0.7 mg/L
	Trichloroethylene	D040	0.5 mg/L
	Vinyl chloride	D043	0.2 mg/L
Pesticides	Chlordane	D020	0.03 mg/L
	Endrin	D012	0.02 mg/L
	Heptachlor (and its epoxide)	D031	0.008 mg/L
	Lindane	D013	0.4 mg/L
	Methoxychlor	D014	10.0 mg/L
	Toxaphene	D015	0.5 mg/L
Herbicides	2,4-D	D016	10.0 mg/L
	2,4,5-TP (Silvex)	D017	1.0 mg/L
Semivolatile Organic Compounds	o-Cresol	D023	200.0 mg/L (a)
	m-Cresol	D024	200.0 mg/L (a)
	p-Cresol	D025	200.0 mg/L (a)
	Cresol	D026	200.0 mg/L (a)
	1,4-Dichlorobenzene	D027	7.5 mg/L
	2,4-Dinitrotoluene	D030	0.13 mg/L (b)
	Hexachlorobenzene	D032	0.13 mg/L (b)
	Hexachlorobutadiene	D033	0.5 mg/L
	Hexachloroethane	D034	3.0 mg/L
	Nitrobenzene	D036	2.0 mg/L
	Pentachlorophenol	D037	100.0 mg/L
	Pyridine	D038	5.0 mg/L
	2,4,5-Trichlorophenol	D041	400.0 mg/L
	2,4,6-Trichlorophenol	D042	2.0 mg/L

Source: 40 CFR 261.24

(a): If o-, m- and p-cresol concentrations cannot be differentiated, the total cresol (D026) concentration is used. The regulatory level of total cresol is 200 mg/L.

(b): The quantification limit is greater than the calculated regulatory level. The quantification level therefore becomes the regulatory level.

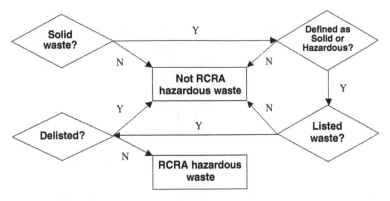

FIGURE 3.2 Questions in determining a hazardous waste.

(a) Is the Material a Solid Waste?

According to 40 CFR 261.2, a material must first be a solid waste before it is classified as a hazardous waste. A solid waste is a waste that is discarded by being either abandoned, inherently waste-like, a certain military munitions, or recycled. They are described in detail as follows:

- Abandoned waste. A material is considered abandoned if it is disposed of, burned, or incinerated; or if it is accumulated, stored, or treated before or in lieu of being abandoned.
- Inherently waste-like waste. Some materials are considered to be inherently waste-like for they are posing such a threat to human health and the environment that they are always taken account of as solid waste. Inherently waste-like waste consists of hazardous waste numbers F020–F023, F026, and F028 when they are recycled in any manner.
- Military munitions. Some ammunition products and components generated from the U.S. Department of Defense (DOD) or U.S. Armed Services for national defense and security are defined as military munitions. Unused or defective munitions are solid wastes when (1) disposed of, burned, incinerated, or treated prior to disposal; (2) nonrecyclable or unusable; or (3) used munitions if collected for storage, recycling, treatment, or disposal.
- Recycled waste. Materials that are recycled are a special subset of solid waste. The definition is that, if a material is used or reused, reclaimed, or used in a manner constituting disposal, burned for energy recovery, or accumulated speculatively, it is recycled waste.[4] Some materials are no longer solid wastes when recycled, while others are solid waste subject to less-stringent regulatory controls. RCRA does not exempt all recycled materials from the definition of solid waste, for some types of recycling also pose threats to human health and the environment. The procedures of how to determine a recycled waste to be solid waste are summarized in Figure 3.3.

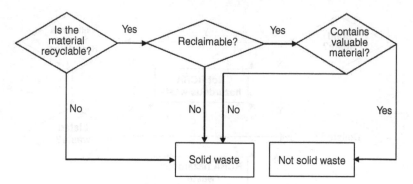

FIGURE 3.3 Questions in determining whether a recycled waste is a solid waste.

From Figure 3.3, it can be seen that a material is defined as a solid waste under the circumstance that it is a secondary material. The determination could be simply made in terms of the yes or no judgment as described in Table 3.3.

TABLE 3.3
Secondary Material Determinations

The materials are solid waste under circumstance that:

	Use Constituting Disposal	Energy Recovery/Fuel	Reclamation	Speculative Accumulation
Spent materials	Yes	Yes	Yes	Yes
Sludges listed in 40 CFR 261.31 or 261.32	Yes	Yes	Yes	Yes
Sludges exhibiting a characteristic of hazardous waste	Yes	Yes	No	Yes
By-products listed in 40 CFR 261.31 or 261.32	Yes	Yes	Yes	Yes
By-products exhibiting a characteristic of hazardous waste	Yes	Yes	No	Yes
Commercial chemical products listed in 40 CFR 261.33	Yes	Yes	No	No
Scrap metal other than excluded scrap metal listed in 261.1	Yes	Yes	Yes	Yes

Source: 40 CFR 261.2
Yes: material is a solid waste; No: material is not a solid waste

(b) Is It an Excluded Waste?

As can be seen from Figure 3.2, the second step to identify a hazardous waste is to determine whether or not the waste is excluded from the definition of solid or hazardous waste. Some special wastes containing dangerous chemicals are impractical or undesirable to be regulated under stringent RCRA waste management regulations, such as household solvents. As a result, they are excluded or exempted by Congress and EPA from the definitions and regulations of hazardous waste.

The exclusions are divided into five categories as follows:[4]

- Solid waste exclusions. If a material does not meet the definition of a solid waste, it cannot be considered as a hazardous waste. Obviously, the exclusions of solid waste will not be subject to RCRA hazardous waste regulation. They are (1) domestic sewage and mixtures of domestic sewage; (2) industrial wastewater discharges; (3) irrigation return flows; (4) radioactive waste; (5) *in situ* mining waste; (6) pulping liquors; (7) spent sulfuric acid; (8) closed-loop recycling; (9) spent wood preservatives; (10) coke by-product wastes; (11) splash condenser dross residue; (12) hazardous oil-bearing secondary materials and recovered oil from petroleum refining operations; (13) condensates from kraft mill steam strippers; (14) processed scrap metal; (15) shredded circuit boards; (16) mineral processing spent materials; (17) petrochemical recovered oil; (18) spent caustic solutions from petroleum refining; (19) glass frit and fluoride-rich baghouse dust generated by the vitrification of K088; and (20) zinc fertilizers made from recycled hazardous secondary materials.
- Hazardous waste exemptions. They are (1) household hazardous waste; (2) agricultural waste; (3) mining overburden; (4) Bevill and Bentsen wastes including fossil fuel combustion wastes, oil, gas, and geothermal wastes, mining and mineral processing wastes, and cement kiln dust; (5) trivalent chromium wastes; (6) arsenically treated wood; (7) petroleum-contaminated media and debris from underground storage tanks; (8) spent chlorofluorocarbon refrigerants; (9) used oil filters; (10) used oil distillation bottoms; (11) landfill leachate or gas condensate derived from K169, K171, and K172 listings; (12) project XL pilot project exclusions.
- Raw material, product storage, and process unit waste exclusions. If a hazardous waste remains in units of tanks, pipelines, vehicles, and vessels used either in the manufacturing process or for storing raw materials or products (surface impoundments are not included), it is exempted from RCRA hazardous waste regulation. When a unit stops operation for over 90 days, or when the waste is removed from the unit, the waste will be considered generated and regulated as hazardous waste under RCRA.
- Sample and treatability study exclusions. Hazardous waste samples are small amounts of waste that are essential to ensure accurate characterization and proper hazardous waste treatment. Samples sent to a lab to determine whether or not a waste is hazardous or to determine if a particular treatment method will be effective on a given waste or what

types of wastes remain after the treatment is complete are exempt from regulation.

- Dredge materials exclusions. Dredge materials subject to the permitting requirements of Section 404 of the Federal Water Pollution Control Act (FWPCA) of section 103 of the Marine Protection, Research, and Sanctuaries Act (MPRSA) of 1972 will not be considered as hazardous wastes.

3.2.3.2 Special Regulatory Conventions for Hazardous Waste

Under certain circumstances, the hazardous waste may be mixed with other non-hazardous waste; generate residues during treatment, storage, and disposal processes; and be spilled on soils or constructions. The Special Regulatory Conventions, including mixture rule, derived-from rule, and contained-in policy are proposed by RCRA to deal with such situations.

(a) Mixture Rule

Hazardous wastes sometimes become mixed with other nonhazardous wastes when generated. The mixture rule is intended to ensure these mixtures are regulated in a manner that minimizes threats to human health and the environment. Regardless of what percentage of the waste mixture is composed of listed hazardous wastes, the mixture bears the same waste code and regulatory status as the original listed component of the mixture, unless the generator obtains a delisting. This is intended to prevent any generators from evading RCRA requirements simply by mixing or diluting the listed wastes with nonhazardous solid waste. Characteristic wastes are hazardous because they possess one of four unique and measurable properties. According to the mixture rule, a mixture involving characteristic waste will no longer be regulated as hazardous if the characteristic waste does not exhibit one of the four dangerous properties.

The mixture rule also contains some exceptions, or exclusions, such as a comparatively small quantity of listed hazardous waste routed to large-volume waste-water treatment systems.

(b) Derived-From Rule

The derived-from rule is applied to residues generated from hazardous waste treatment, storage, and disposal processes that may contain high concentrations of both listed and characteristic hazardous waste. The rule states that any material derived from a listed waste or characteristic waste (only if it exhibits a hazardous characteristic) will also be considered as corresponding hazardous waste. Thus, even if a hazardous waste is burned out, the ash will still bear the same regulations as the original hazardous waste.

(c) Contained-In Policy

The contained-in policy is created to handle the condition when some listed or characteristic wastes spill onto the environmental media such as soil and groundwater, and debris such as dismantled construction materials and discarded personal protective equipment. According to this policy, the environmental media and debris will be regulated as hazardous waste when contaminated by a RCRA listed or characteristic waste.

3.2.4 SPECIAL WASTES

Some wastes do not come under the regulation of RCRA but are still considered to be hazardous. They are polychlorinated biphenyls (PCBs), asbestos, and radionuclides.[7] PCBs and asbestos are regulated under the Toxic Substances Control Act (TSCA).[8] Radionuclides are regulated as radioactive waste under the AEA.

3.3 CLASSIFICATION AND CHARACTERIZATION OF RADIOACTIVE WASTES

Radioactive wastes arise in many forms from a wide range of activities, most notably from plants and processes associated with nuclear power research and production, and with military applications. Large numbers of industrial, research, and medical uses of radioactive materials also give rise to wastes, albeit in total on a relatively small scale and in most cases giving rise to only small volumes, containing small amounts and low concentrations of activity.

3.3.1 RADIOACTIVE WASTE CLASSIFICATION

3.3.1.1 Background

The objectives of classification are to facilitate understanding and simplify management of the multiple elements of a diverse system.[9] Waste classes were based primarily on practical factors of immediate concern, such as exposure rates and proliferation security, or on the process that produced the waste. For the purpose of constructing a reasonable classification system of radioactive waste, the number of classes is desired to be as low as possible, while diversity is retained to promote proper management. There are some different classification schemes based on different purposes. The first official distinction between different kinds of radioactive waste was between high-level waste (aqueous waste from the first-cycle solvent-extraction in reprocessing spent nuclear fuel) and "other than high level" waste.[10] After several years of development, the classification system has become more complicated and diversified.

3.3.1.2 Radioactive Waste Classification

Most nations categorize wastes into different classes for simplifying waste management actions, rules, and regulations while protecting human health. An acceptable classification and characterization system should depend on a reasonable definition of radioactive waste. The DOE defines radioactive waste as solid, liquid, or gaseous material that contains radionuclides regulated under the AEA, and of negligible economic value considering costs of recovery.[11]

Rather than having a common basis for each waste class in its classification system, such as a set of classes based on a combination of the half-life, gas production, and heat generation rate, the U.S. system has some classes that participate on one basis and others that partake another. A number of waste and material classes have been specified in AEA, which does not clearly distinguish what is regulated from what is not. A commonly used source-defined classification system (defined by the source of the waste rather than by measurable quality of the waste) is summarized as follows:

- **High-Level Waste (HLW).** The NRC description of HLW includes (1) the highly radioactive material resulting from the reprocessing of spent nuclear fuel, including liquid waste produced directly in reprocessing and any solid material originating from such liquid waste that contains fission products in sufficient concentrations, and (2) other highly radioactive material that the Commission, in accordance with existing law, determines by rule requires permanent isolation.[12] The Commission has determined that irradiated reactor fuel shall, for the purposes of the repository, be considered as HLW.[13]

- **Spent Nuclear Fuel (SNF).** SNF is the fuel that has been withdrawn from a nuclear reactor following irradiation, the constituent elements of which have not been separated by reprocessing.[13] It is worth noting that spent nuclear fuel is regulated as HLW under 10 CFR 60.

- **Transuranic Waste (TRU).** TRU elements are those having atomic numbers greater than 92 (i.e., having more protons than uranium). In the United States, TRU waste is defined as radioactive waste that is not classified as HLW, but contains an activity of more than 100 nCi/g from alpha-emitting TRU isotopes having half-lives greater than 20 years.[14] Typical waste includes metal tools, lab coats, gloves, equipment, debris, and so on contaminated with plutonium during laboratory and facility operations.[15] Much of the TRU contains sufficiently high concentrations of gamma-emitting nuclides that necessitate remote handling. Most TRU wastes contain primarily alpha-emitters which are safe for contact handling when packaged. As a result of long half-lives of TRU waste, the most suitable method for disposal is isolation in geologic repositories.

- **By-Product Materials.** including uranium mining and mill tailings. The by-product materials include (1) any radioactive material (except special nuclear material) yielded in or made radioactive by exposure to the radiation occurrences, or to the process of producing or utilizing special nuclear material, and (2) the tailings or wastes produced by the extraction or concentration of uranium or thorium from any ore processed primarily for its source material content.[16] The tailings or wastes produced by the extraction or concentration of uranium or thorium from any ore processed primarily for its source material content are also called by-product materials under 42 U.S.C. 2014.

- **Low-Level Waste (LLW).** LLW is defined by the Low-Level Radioactive Waste Policy Act (LLWPA) as radioactive material that (1) is not classified as high-level waste, transuranic waste, spent nuclear fuel, or mill tailings; and (2) the NRC, in accordance with existing law and consistent with part (1).[16] LLW often contains small amounts of radioactivity dispersed in large amounts of material. It is mainly generated by uranium enrichment processes, reactor operations, isotope production, and medical and research activities.[17]

- **Naturally Occurring and Accelerator-Produced Radioactive Materials (NORM/NARM).** Naturally occurring radioactive material and accelerator-produced radioactive material lie outside NRC's regulatory authority and are subject to health and safety regulation by the states and other Federal agencies.[18] The waste is generally subclassified as diffuse (< 2 nCi/g 226 Ra or equivalent) or discrete (> 2 nCi/g 226 Ra or equivalent).[19] They are under review by EPA and may be regulated under TSCA or RCRA.[20]

The above categories, defined according to relevant statutes and regulations, are not the only classification scheme used to specify radioactive wastes. There are a variety of purposes for different systems.

3.3.1.3 LLW Waste Subclassification

As for LLW subclassification, NRC has developed a scheme that is implemented by 10 CFR 61. LLW is divided into two broad categories: waste that qualifies for near-surface burial, and waste that requires deeper disposal (greater than Class C LLW, or greater confinement waste). LLW that is regulated by the NRC and qualifies for near-surface burial is separated into three classes. Subclasses of LLW can be listed as follows:

- **Class A.** Class A waste is waste that is usually segregated from other waste classes at the disposal site. They have low levels of radiation and heat.
- **Class B.** Class B waste is waste that must meet more rigorous requirements on waste form to ensure stability after disposal. The waste has higher concentrations of radioactivity than Class A and requires greater isolation and packaging (and shielding for operations) than Class A waste.
- **Class C.** Class C waste is waste that not only must meet more rigorous requirements on waste form to ensure stability, but also requires additional measures at the disposal facility to protect against inattentive invasion. It requires isolation from the biosphere for 500 years and must be buried at least 5 m below the ground surface and must have an engineered barrier (container and grouting).
- **Greater than Class C.** Greater than Class C waste is the waste that is not generally allowable for near-ground surface disposal, and is the waste for which form and disposal methods must be different, and in general more stringent, than those specified for Class C waste. This is the LLW that does not qualify for near-surface burial. This includes commercial transuranics that have half-lives greater than 5 years and activity greater than 100 nCi/g.

The LLW identification process is related to the types of radionuclides.[20] For classification determined by long-lived radionuclides, if radioactive waste contains only the radionuclides listed in Table 3.4, classification can be determined as (1) Class A if the concentration does not exceed 0.1 times the value in Table 3.4; (2) Class C if the concentration is greater than 0.1 times the value in Table 3.4 but does not exceed the value in Table 3.4; (3) if the concentration is greater than the value in Table 3.4, the waste is not acceptable for near-surface disposal.

For classification is determined by short-lived radionuclides, if radioactive waste does not contain any of the radionuclides listed in Table 3.4, classification shall be determined based on the concentrations shown in Table 3.5. The classification can be determined as (1) Class A if radioactive waste does not contain any nuclides listed in either Table 3.4 or Table 3.5; (2) Class A if the concentration does not exceed the value in Column 1; (3) Class B if the concentration exceeds the value in Column 1, but does not exceed the value in Column 2; (4) Class C if the concentration exceeds the value in Column 2, but does not exceed the value in Column 3; (5) if the concentration exceeds the value in Column 3, the waste is not generally acceptable for near-surface disposal.

If radioactive waste contains a mixture of radionuclides, some of which are listed in Table 3.4, and some of which are listed in Table 3.5, classification shall be determined as follows: (1) if the concentration of a nuclide listed in Table 3.4 does not exceed 0.1 times the value listed in Table 3.4, the class shall be that determined by the concentration of nuclides listed in Table 3.5; (2) if the concentration of a nuclide exceeds 0.1 times the value listed in Table 3.4 but does not exceed the value

TABLE 3.4
Radionuclides and Concentrations for LLW Classification (1)

Radionuclide	Concentration (Ci/m^3)
C-14	8
C-14 in activated metal	80
Ni-59 in activated metal	220
Nb-94 in activated metal	0.2
Tc-99	3
I-129	0.08
Alpha-emitting transuranic nuclides with half-life greater than 5 years	100*
Pu-241	3,500*
Cm-242	20,000*

Source: 10 CFR 61.55
* = Units are nanoCuries per gram (nCi/g)

TABLE 3.5
Radionuclides and Concentrations for LLW
Classification (2)

Radionuclide	Concentration (Ci/m³)		
	Column 1	Column 2	Column 3
Total of all nuclides with less than 5-year half-life	700	(*)	(*)
H-3	40	(*)	(*)
Co-60	700	(*)	(*)
Ni-63	3.5	70	700
Ni-63 in activated metal	35	700	7000
Sr-90	0.04	150	7000
Cs-137	1	44	4600

Source: 10 CFR 61.55

* = There are no limits established for these radionuclides in Class B or C wastes. Practical considerations such as the effects of external radiation and internal heat generation on transportation, handling, and disposal will limit the concentrations for these wastes. These wastes shall be Class B unless the concentrations of other nuclides in Table 3.2 determine the waste is Class C independent of these nuclides.

in Table 3.5, the waste shall be Class C, provided the concentration of nuclides listed in Table 3.5 does not exceed the value shown in Column 3 of Table 3.5.

3.3.2 RADIOACTIVE WASTE CHARACTERIZATION

The definition of Radioactive Waste Characterization can be described as the determination of the radiological, chemical, and physical properties of the waste to establish the need for further adjustment, treatment, conditioning, or its suitability for further handling, processing, storage, or disposal.

Radiological waste characterization involves detecting the presence of individual radionuclides and quantifying their inventories in the waste. This detection and inventory can be done by a variety of techniques, depending on the waste form, radionuclides involved, and level of detail required. For instance, a simple radiation dose rate measurement will give an indication of the total quantity of gamma-emitting radionuclides in a waste package, but will not identify individual radionuclides or their concentrations. Gamma spectroscopy will identify the individual radionuclides and their quantities as well. Other techniques, such as active or passive neutron interrogation, alpha spectroscopy, and liquid scintillation counting, are used for other classes of radionuclides. The preferred methods are often referred to as "nondestructive" or "noninvasive," since they do not involve opening a waste package to take samples. The terms most frequently used are NDA (nondestructive assay), NDE (nondestructive examination), and NDT (nondestructive testing).

The nature of the waste is important in terms of radionuclides present, their half-life, and mode of decay, since the disposal route selected will often depend on the presence of several key radionuclides. A radionuclide inventory will provide the necessary information regarding specific radionuclides present and their activity levels. At a minimum, this information allows a waste stream to be evaluated in terms of maximum concentration of each radionuclide, which is available at any time in the future.

Chemical waste characterization involves the determination of the chemical components and properties of the waste. This is most often done by chemical analysis of a waste sample. All radioactive wastes that contain chemicals should be considered potentially regulated under federal or state regulations until the chemical portion has been determined to be nonhazardous.

Radiological and chemical waste characterization can also be inferred from process knowledge. For example, if you are a medical researcher who only uses a few particular radionuclides under controlled experimental conditions, or a manufacturer who uses a particular chemical, then you can determine from your own knowledge process which radionuclides or chemicals, or their combinations, exist in your waste. To use and justify process knowledge for characterization of radioactive waste, you should (1) be able to estimate, as precisely as possible, the radioactive content of a unit of waste; (2) understand the radioactive decay processes that may result in daughter isotopes that are not in secular equilibrium; (3) know whether the chemicals used in the process were hazardous; (4) have a thorough understanding of how the chemicals were used; (5) understand the chemistry of the reaction to determine if hazardous chemicals were produced where none existed before; (6) know whether the process converted unlabeled chemicals to radio-labeled ones; and (7) be able to report the origin and purity of isotope preparations if such information is needed by the Waste Management Group.

Physical characterization involves inspection of the waste to determine its physical form and strength. Closed waste packages can be inspected using a variety of techniques, such as radiography (x-ray) and sonar.

Waste acceptance criteria (WAC) are the conditions imposed on a waste producer by the regulator or operator of a waste handling, transportation, storage, processing, or disposal service. The WAC usually specify such things as the required physical form of the waste, maximum levels of radioactivity, and packaging requirements as well as what wastes are excluded from their service. The WAC from radioactive waste disposal sites require that all waste generators be able to validate the chemical and radioactive constituents of their waste by referring to pertinent written procedures, logs of activities, and results of analyses conducted in the course of their experiments.

3.4 CLASSIFICATION AND CHARACTERIZATION OF MIXED WASTES

Radioactive waste that also contains a "conventional" hazard, such as chemical toxicity, or material that contains both radioactive material and RCRA hazardous waste, is referred to as mixed waste under RCRA. An example would be radioactive lead. If it were not radioactive, the lead would still be considered to be an environmental hazard in most countries. In some countries, notably the United States, the radioactive properties are regulated by one set of rules under the NRC, while the conventional hazard is regulated by another set of rules under the jurisdiction of the EPA. This dual regulation greatly increases the complexity of mixed waste management, especially if the regulations impose contradictory requirements. In other countries, the radioactive hazard takes precedence over the nonradioactive hazard, and a single government body regulates the waste.

Mixed waste contains radioactive and hazardous waste. A dual regulatory framework exists for mixed waste, with the EPA or authorized states regulating the hazardous waste and the NRC, NRC agreement states, or the DOE regulating the radioactive waste. NRC generally regulates commercial and non-DOE federal facilities. DOE is currently self-regulating and its orders apply to DOE sites and contractors.

Under AEA, NRC and DOE regulate mixed waste with regard to radiation safety. Using the RCRA authority, EPA regulates mixed waste with regard to hazardous waste safety. NRC is authorized by the AEA to issue licenses to commercial users of radioactive materials. RCRA gives EPA the authority to control hazardous waste from "cradle-to-grave." Once a waste is determined to be a mixed waste, the waste handlers must comply with both AEA and RCRA statutes and regulations. The requirements of RCRA and AEA are generally consistent and compatible. However, the provisions in Section 1006(a) of RCRA allow the AEA to take precedence in the event provisions of requirements of the two acts are found to be inconsistent.

3.4.1 MIXED WASTE CLASSIFICATION

The most common type of mixed waste is a scintillation vial that contains flammable (toluene-based) scintillation cocktail and a small amount of radioactive isotope. The classification of a mixed waste is based on the containment of toxic or corrosive materials, transuranic elements, or high levels of radioactivity. With reference to Radioactive Waste classification, the Mixed Waste can be divided into two basic categories: DOE mixed waste and Non-DOE mixed waste.

3.4.1.1 DOE Mixed Waste

There are three main types of mixed waste being produced or stored at DOE facilities: Low Level, High Level, and Transuranic.

- DOE Low-Level Mixed Waste (DLLMW) is generated, projected to be generated, or stored as a result of research, development, and production of nuclear weapons. In the U.S., waste management activities will require management of an estimated 226,000 m^3 of DLLMW over the next 20 years.
- DOE High-Level Waste (DHLW) is radioactive waste resulting from reprocessing spent nuclear fuel and irradiated targets from reactors. Some

of its elements will remain radioactive for thousands of years. DHLW is also a mixed waste because it has highly corrosive components or has organics or heavy metals that are regulated under RCRA. In the U.S., DOE has about 399,000 m³ of HLW stored in large tanks at some locations such as Hanford, Washington; Idaho National Engineering Laboratories (INEL), Idaho; Savannah River Site (SRS), South Carolina; etc. DOE is proceeding with plans to treat DHLW by processing it into a solid form (e.g., borosilicate glass) that would not be readily diffusive into the air or leak into the ground or surface water. In the U.S., this treatment process is called vitrification that will generate approximately 29,000 canisters to be disposed of in a geologic repository.

- DOE Mixed Transuranic Waste (DMTRU) is waste that has a hazardous component and radioactive elements heavier than uranium. The radioactivity in the DMTRU must be greater than 100 nCi/g and co-mingled with RCRA hazardous constituents. The principle hazard from DMTRU is alpha-particle radiation through inhalation or ingestion. DMTRU is primarily generated from nuclear weapons fabrication, plutonium-bearing reactor fuel fabrication, and spent fuel reprocessing. The percentage of non-DOE MTRU is negligible. Approximately 55% of DOE's TRU waste is MTRU. In the U.S., DMTRU is currently being treated and stored at some sites such as Los Alamos National Laboratories, New Mexico (8,000 m³); Rocky Flats, Colorado (1,500 m³); Oak Ridge National Laboratory, Tennessee (1,500 m³).

DOE's 1995 Baseline Environmental Management Report roughly estimates that the life-cycle costs for DHLW, DMTRU, and DLLMW are $34 billion, $13 billion, and $13 billion, respectively, over a 75-year period.

3.4.1.2 Non-DOE Mixed Waste

Almost all of the commercially generated (non-DOE) mixed waste is composed of Low-Level Radioactive Waste (LLRW) and Hazardous Waste and is called Low-Level Mixed Waste (LLMW). Commercially generated LLMW is produced in all 50 states of the U.S. at industrial, hospital, and nuclear power plant facilities. Radioactive and hazardous materials are used in a number of processes such as medical diagnostic testing and research, pharmaceutical and biotechnology development, pesticide research, as well as nuclear power plant operations. Based on the results of a survey conducted by NRC and EPA, approximately 4,000 m³ of LLMW were generated in the U.S. in 1990. Of this amount, approximately 2,840 m³ was liquid scintillation cocktail (LSC). Organic solvents such as chlorofluorocarbons (CFCs), corrosive organics, and waste oil made up 18%, toxic metals made up 3%, and "other" waste made up the remaining 8%.

Under the 1984 Amendments to RCRA, Land Disposal Restriction (LDR) regulations prohibit disposal of most mixed waste including LLMW until it meets specific treatment standards. While most of the commercial mixed waste that is generated and stored can be treated to meet the LDRs by commercially available treatment technology, there still exists a small percentage of commercial mixed waste for which no treatment or disposal capacity is available. Commercial mixed waste volumes are very small (approximately 2%) compared to the total volume of mixed waste being generated or stored by DOE.

As mandated by the Federal Facilities Compliance Act (FFCA), which was signed into law on October 6, 1992, DOE has developed Site Treatment Plans to handle its mixed wastes under the review of EPA or its authorized States. These are being implemented by orders issued by EPA or the state regulatory authority. It will take the cooperation of DOE, EPA, NRC, and the states to manage treatment, storage and disposal of mixed waste.

3.4.2 Mixed Waste Characterization

Proper characterization of mixed waste is extremely important and encompasses all of the challenges of both radioactive waste characterization and hazardous waste characterization.

Mixed waste is a waste material that contains radioisotopes and possesses other hazardous properties; i.e., the waste is (1) ignitable or explosive; (2) toxic; (3) corrosive (pH greater than 12.5 or less than 2); (4) reactive; (5) persistent (halogenated hydrocarbons and polycyclic aromatic hydrocarbons with more than three and less than seven rings); or (6) carcinogenic.

A waste is mixed if it is classified by federal agencies as being both radioactive and hazardous. At Berkeley Lab, a waste is characterized as radioactive if either process knowledge or monitoring and sampling show that radioactive material has been added to the waste through DOE research or support activities.

- Hazardous wastes containing naturally occurring radioactive material as the sole radioactive constituent are disposed of as mixed waste only if they have been isotopically enriched during research or support activities. For example, photographic fixers and other hazardous materials containing only ^{40}K in its naturally occurring isotopic abundance are managed for hazardous characteristics.
- Hazardous wastes containing only isotopes with half-lives shorter than 7 hours (e.g., ^{18}F) may be decayed for 10 or more half-lives in generator areas, surveyed to confirm activity level indistinguishable from background, and managed as hazardous waste.
- Hazardous wastes containing isotopes with half-lives greater than 7 hours and no more than 90 days (e.g., ^{32}P, ^{35}S, ^{125}I, etc.) must be managed as mixed waste in generator areas. After transfer to the Hazardous Waste Handling Facility (HWHF), such waste is decayed through 10 or more half-lives, surveyed or sampled to confirm low activity, and disposed of as hazardous waste.
- Hazardous scintillation fluids with activities less than 0.05 µCi/g due solely to ^3H and/or ^{14}C are managed as mixed waste in generator areas. After the waste is transferred to the HWHF and the composition is confirmed, the waste may be managed as nonradioactive.

Both hazardous waste and radioactive characterizations have been discussed in Sections 3.2.2 and 3.3.2, respectively.

REFERENCES

1. *United States Code*, title 42, sections 2011 et seq. (42 U.S.C. § 2011 et seq.).
2. *The National Biennial RCRA Hazardous Waste Report*, U.S. Environmental Protection Agency, August 1995.
3. *Code of Federal Regulations*, 40 CFR 260, 1 July 1996.
4. U.S. Environmental Protection Agency, *RCRA Orientation Manual*, The U.S. EPA Office of Solid Waste/Communications, Information, and Resources Management Division, September 2002, EPA530-R-02-016.
5. *Code of Federal Regulations*, 40 CFR 261, 1 July 1996.
6. U.S. Environmental Protection Agency, *RCRA Orientation Manual*, 1990 Edition. Office of Solid Waste, Washington, D.C. 1990.
7. Freeman, H.M., *Standard Handbook of Hazardous Waste Treatment and Disposal*, Second Edition, McGraw-Hill, 1998.
8. *Toxic Substances Control Act (TSCA)*, 40 CFR 700, U.S. EPA, 1976.
9. International Atomic Energy Agency, *Classification of Radioactive Waste: a Safety Guide*, Safety Series No. 111-G-1.1, Vienna, 1994.
10. Berlin, R.E. and Stanton, C.C., *Radioactive Waste Management*, John Wiley & Sons, New York, 1989, Chapter 3.
11. Attachment 2, DOE Order 5820.2 (1984) and replaced by 5820.2A in 1988, both titled "Radioactive Waste Management."
12. *United States Code,* title 42, section 10101 (42 U.S.C. 10101).
13. *U.S. Code of Federal Regulations*, title 10, Part 60.2 (10 C.F.R. 60.2).
14. *United States Code*, title 42, section 2014 (42 U.S.C. 2014).
15. U.S. Department of Energy, *Environmental Fact Sheets*, Office of Environmental Management, Washington, D.C., 1994.
16. *United States Code*, title 42, section 2021 (42 U.S.C. 2021).
17. Tang, Y.S. and Saling, J.H., *Radioactive Waste Management*, Hemisphere Publishing Corporation, New York, 1990.
18. Oak Ridge National Laboratory, *Integrated Data Base Report*, 1995: U.S. Spent Fuel and Radioactive Waste Inventories, Projections, and Characteristics, DOE/RW-006, Rev. 12, Oak Ridge, TN, December 1996.
19. The League of Women Voters Education Fund, *The Nuclear Waste Primer 1993 Revised Edition*, Lyons & Burford, New York, 1993, PP25-26.
20. *U.S. Code of Federal Regulations*, Title 10, Part 61 (10 C.F.R. 61).

4 Hydraulic Cement Systems for Stabilization/ Solidification

Caijun Shi

CONTENTS

4.1 INTRODUCTION

Hydraulic cement is widely used for stabilization/solidification (S/S) of low-level radioactive, hazardous wastes, mixed wastes, and remediation of contaminated sites because it has many advantages.[1,2] The S/S of contaminants by cements includes the following three aspects: (a) chemical fixation of contaminants — chemical interactions between the hydration products of the cement and the contaminants; (b) physical adsorption of the contaminants on the surface of hydration products of the cements; and (c) physical encapsulation of contaminated waste or soil (low permeability of the hardened pastes).[3,4] The first two aspects depend on the nature of the hydration products and the contaminants, and the third aspect relates to both the nature of the hydration products and the density and physical structure of the paste. If the leaching and other performance of the cement-solidified waste forms meet appropriate environmental criteria, the cement-based waste forms may be suitable for controlled construction applications.

The selection of cementing materials for S/S of waste must consider the following aspects, based on the characteristics of the waste: (1) compatibility of the cement and the waste; (2) chemical fixation of contaminants; (3) physical encapsulation of contaminated waste and soil; (4) durability of final waste forms; (5) waste form leachability; and (6) cost-effectiveness of S/S. In practice, a variety of additives are often used with cementing materials to address all of these aspects. Some commonly used additives are described in detail in Chapter 8 as well as in previous publications.[1,5] This chapter focuses mainly on the various hydraulic cement systems used for S/S of wastes.

4.2   PORTLAND CEMENT-BASED MATERIALS

Portland cement is the most widely used cement because of its commercial availability and low cost. However, for construction use, supplementary cementing materials, such as granulated/pelletized blast furnace slag, coal fly ash, volcanic ashes, condensed silica fume, rice husk ash, and natural pozzolans, are often used to replace portland cement to reduce the cost or to improve the performance of concrete. For the same reasons, these supplementary cementing materials are very often used to replace portland cement in S/S of wastes. This section will discuss portland cement-based cementing materials used for S/S of wastes, which include portland cement, portland blast furnace slag cement, portland pozzolan cement, portland cement–sodium silicate mixture, polymer-modified portland cement, and masonry cement.

4.2.1   PORTLAND CEMENT

4.2.1.1   Characterization and Hydration Chemistry

Portland cement is a hydraulic cement produced by pulverizing clinker and calcium sulfate (usually gypsum — $CaSO_4.2H_2O$) as an interground addition. Cement clinker consists mainly of tricalcium silicate ($3CaO.SiO_2$), dicalcium silicate ($2CaO.SiO_2$), tricalcium aluminate ($3CaO.Al_2O_3$), and tetracalcium aluminoferrite ($4CaO.Al_2O_3.Fe_2O_3$). Different types of portland cement are manufactured to meet various normal physical and chemical requirements for specific purposes. ASTM C150 specifies five types of common portland cement produced by adjusting the proportions of their minerals and finenesses as shown in Table 4.1.[6] Canadian CAS-A5 cement Types 10, 20, 30, 40, and 50 are essentially the same as ASTM C150

TABLE 4.1
Composition and Finenesses of Different Types of Portland Cement[6]

Type of Portland Cement	Chemical Composition (%)							Mineral Composition (%)				Blaine Fineness (m²/kg)
	SiO_2	Al_2O_3	Fe_2O_3	CaO	MgO	SO_3	Na_2O_{eq}	C_3S	C_2S	C_3A	C_4AF	
I (min–max)	18.7–22.0	4.7–6.3	1.6–4.4	60.6–66.3	0.7–4.2	1.8–4.6	0.11–1.2	40–63	9–11	6–14	5–13	300–421
I (mean)	20.5	5.4	2.6	63.9	2.1	3.0	0.61	54	18	10	8	369
II (min–max)	20.0–23.2	3.4–5.5	2.4–4.8	60.2–65.9	0.6–4.8	2.1–4.0	0.05–1.12	37–68	6–32	2–8	7–15	318–480
II (mean)	21.2	4.6	3.5	63.8	2.1	2.7	0.51	55	19	6	11	377
III (min–max)	18.6–22.2	2.8–6.3	1.3–4.9	60.6–65.9	0.6–4.6	2.5–4.6	0.14–1.20	46–71	4–27	0–13	4–14	390–664
III (mean)	20.6	4.9	2.8	63.4	2.2	3.5	0.56	55	17	9	8	548
IV (min–max)	21.5–22.8	3.5–5.3	3.7–5.9	62.0–63.4	1.0–3.8	1.7–2.5	0.29–0.42	37–49	27–36	3–4	11–18	319–362
IV (mean)	22.2	4.6	5.0	62.5	1.9	2.2	0.36	42	32	4	15	340
V (min–max)	20.3–23.4	2.4–5.5	3.2–6.1	61.8–66.3	0.6–4.6	1.8–3.6	0.24–0.76	43–70	11–31	0–5	10–19	275–430
V (mean)	2.9	3.9	4.2	63.8	2.2	2.3	0.48	54	22	4	13	373
White (min–max)	22.0–24.4	2.2–5.0	0.2–0.6	63.9–68.7	0.3–1.4	2.3–3.1	0.09–0.38	51–72	9–25	5–13	1–2	384–564
White (mean)	22.7	4.1	0.3	66.7	0.9	2.7	0.18	63	18	10	1	482

Types I through V, respectively, except for the allowance of up to 5% limestone in Type 10 and Type 30 cements. The European cement standard EN197 is very different from ASTM C150 or CAS-A5, and classifies cement into CEM I, CEM II, CEM III, CEM IV, and CEM V. These do not correspond to the cement types in ASTM C150. CEM I is portland cement and CEM II through V are blended cements. EN197 also has strength classes and ranges.

In the presence of water, C_3S and C_2S in cement hydrate to form calcium silicate hydrate gel (C-S-H gel) and $Ca(OH)_2$. In the presence of calcium sulfate, C_3A hydrates to form calcium trisulfoaluminate hydrate ($3CaO.Al_2O_3.3CaSO_4.32H_2O$ - AFt or ettringite), or calcium monosulfoaluminate hydrate ($3CaO.Al_2O_3.CaSO_4.12H_2O$ — AFm or monosulfate). In the absence of calcium sulfate, C_3A reacts with water and calcium hydroxide to form tetracalcium aluminate hydrate [$3CaO.Al_2O_3.Ca(OH)_2.12H_2O$]. C_4AF reacts with water to form calcium aluminoferrite hydrates ($6CaO.Al_2O_3.Fe_2O_3.12H_2O$). These hydration reactions can be expressed as follows:

$$3(3CaO.SiO_2) + 6H_2O \rightarrow 3CaO.2SiO_2.3H_2O + 3Ca(OH)_2 \qquad (4.1)$$

$$2(2CaO.SiO_2) + 4H_2O \rightarrow 3CaO.2SiO_2.3H_2O + Ca(OH)_2 \qquad (4.2)$$

$$3CaO.Al_2O_3 + 3CaSO_4.2H_2O + 26H_2O \rightarrow 3CaO.Al_2O_3.3CaSO_4.32H_2O \quad (4.3)$$

$$3CaO.Al_2O_3.3CaSO_4.32H_2O + 2(3CaO.Al_2O_3) + 4H_2O \rightarrow$$
$$3(3CaO.Al_2O_3.CaSO_4.12H_2O) \qquad (4.4)$$

$$3CaO.Al_2O_3 + 12H_2O + Ca(OH)_2 \rightarrow 3CaO.Al_2O_3.Ca(OH)_2.12H_2O \quad (4.5)$$

$$4CaO.Al_2O_3.Fe_2O_3 + 10H_2O + 2Ca(OH)_2 \rightarrow 6CaO.Al_2O_3.Fe_2O_3.12H_2O \quad (4.6)$$

A hardened cement paste is a heterogeneous multiphase system. At room temperature, a fully hydrated portland cement paste consists of 50 to 60% C-S-H gel, 20 to 25% $Ca(OH)_2$, 15 to 20% ettringite (or AFt) and AFm by volume. These minor hydration products, such as $Ca(OH)_2$, $3CaO·Al_2O_3·6H_2O$, and AFt, form in small quantities depending on the composition of the cementing material and hydration conditions. The detailed description on microstructural characteristics of hardened cement pastes can be found in many books,[7a,7b] and is summarized in Chapter 9.

4.2.1.2 S/S with Portland Cement

Portland cement has been used alone for S/S of many wastes.[1] C-S-H gel is the main binding component and responsible for the mechanical strength of hardened pastes. It also plays an important role in S/S of contaminants. Komarneni et al.[8] found that poorly crystallized C-S-H formed below 100°C has cation ion-exchange and ion-absorption properties. The lower the C/S ratio of C-S-H is, the more amount of cation the C-S-H can retain. Bhatty[9] proposed several mechanisms for the immobilization of metal ions by C-S-H, as discussed in Chapter 7. Successful treatment of metal-bearing wastes by S/S may involve the following:

1. Control of excess acidity by neutralization
2. Destruction of metal complexes
3. Control of oxidation state
4. Conversion to insoluble species
5. Formation of a solid with solidification reagents

The portland cement system is a high-alkaline and porous material and can address requirements 1, 4, and 5. In many cases, these requirements are enough to stabilize/solidify some hazardous wastes. Compared with other binders, portland cement has advantages of availability, low cost, and simple operation. However, it also has some disadvantages that need to be addressed in some cases.

One concern is the final pH of the system. Chapter 7 provides detailed descriptions of how pH affects the solubility and leachability of some heavy metals in waste forms. The solubility of amphoteric metals, such as Cd, Cr, Cu, Pb, Ni, and Zn, varies with pH, and the optimum pH range to precipitate them is about 10. Thus, it seems that the pH of portland cement is not ideal for precipitating heavy metals; the pH of the waste forms solidified with portland cement containing very low alkali content and very high alkali content can be very different, which in turn can have a significant effect on the leachability of heavy metals.

Portland cement contains some alkali; the amount is dependent on the raw materials and production process used. Usually, the pH value of the pore solution of a hardened portland cement is over 12.5, maybe up to 13.5 due to the presence of alkalis. A survey on characteristics of North American portland cements in 1994 indicates that the alkali content of portland cement can be as low as 0.05% by mass and as high as 1.2% by mass.[10] Cement can release almost all its alkalis to the pore solution, and the pore solution of concrete consists mainly of alkali hydroxide after 28 days of hydration.[11] Thus, the alkalis in cements have an important effect on the pH of pore solution of solidified waste forms.

The relationship between the C/S ratio of C-S-H and equilibrium pH is shown in Figure 4.1. There are two plateaus for the equilibrium pH. One corresponds to around pH 12, where the C/S ratio is greater than 1.0. The other plateau corresponds to pH 10, where the C/S ratio of C-S-H varies from 0.05 to 0.6. Typically, the C/S ratio of C-S-H from the hydration of portland cement ranges from 1.4 to 1.7, which is associated with a pH that is not ideal for the retention of heavy metals. To produce a porewater with a pH value that minimizes solubility of metals, the C/S ratio of C-S-H in a hydrated cement paste should be lower than 1.

Another potential disadvantage of using pure portland cement is that many inorganic and organic contaminants interfere with the hydration and setting of portland cement, as discussed in Chapter 7. Furthermore, portland cement cannot destroy metal complexes or control the oxidation state of metals as needed. Thus, portland cement cannot be directly used as an S/S agent alone when these issues are concerns. The following sections discuss some portland cement-based cementing systems containing various additives, which have also been used for S/S of wastes.

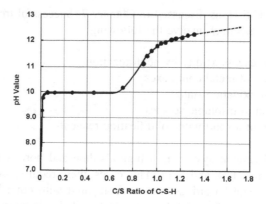

FIGURE 4.1 Relationship between C/S ratio and equilibrium ph (based on reference 12).

4.2.2 PORTLAND BLAST FURNACE SLAG CEMENT

Blast furnace slag is a by-product from iron production. To control the iron quality, the raw materials are carefully controlled and the range of chemical composition of slag is fairly narrow for a specific ore and furnace operating conditions. However, the chemical compositions of slag vary with different ores and furnace operating conditions. It can be represented in the quaternary diagram CaO-SiO_2-Al_2O_3-MgO and may vary within a wide range: CaO — 30 to 50%, SiO_2 — 28 to 38%, Al_2O_3 — 5 to 24%, MgO — 1 to 18%, S — 0.4 to 2.5%, Fe_2O_3 — 0.3 to 3%, MnO — 0.2 to 3%, TiO_2 - < 4%, Na_2O+K_2O - < 2%.

Ground granulated blast furnace slag (GBFS) has been widely used as a portland cement replacement due to its low cost and beneficial effect on some properties of the corresponding concrete. The use of GBFS for S/S of wastes may also have several beneficial functions: (1) it decreases the pH value of the initial pore solution to ~11, which increases the precipitation of some heavy metals; (2) it lowers the oxidation-reduction potential, which reduces the solubility of most radionuclides and the corrosion of steel containers;[2] (3) it precipitates some metals as sulfides, which are even more insoluble than the corresponding hydroxides; and (4) it reduces the permeability of the waste form.

The standard oxidation potential, E_h, measured for cement–slag mixes drops precipitously to become reducing at slag contents above 60% by mass.[13-16] Slag-based cement effectively stabilizes chromates even when the waste loading approaches 90% by mass or higher,[17] which suggests that Cr(VI) is reduced to Cr(III). In addition to providing a reducing environment, slag may precipitate metals as sulfides that are even more insoluble than the corresponding hydroxides. Soluble inorganic mercury compounds are not stabilized effectively in a portland cement matrix. However, the use of more than 10% slag in the binder can greatly decrease the extracted Hg during TCLP testing.[17]

A slag-based waste form was designed to solidify/stabilize low-level radioactive alkaline salt solution at the Savannah River Plant.[18] The use of blast furnace slag together with portland cement and fly ash significantly reduced the release of chromium, technetium, and nitrate in the waste due to three mechanisms: by decreasing the permeability of the waste form, reducing Cr^{6+} and Tc^{7+} to Cr^{3+} and Tc^{4+}, and precipitating them in the form of $Cr(OH)_3$ and TcO_2.

Akhter et al.[19] used various combinations of Type I portland cement, Class F fly ash, blast furnace slag, lime, and silica fume as additives to stabilize/solidify soils containing heavy metals at concentrations ranging from 10,000 to 12,200 mg/kg. A blended cement consisting of 85% GBFS and 15% portland cement was used to stabilize steelmaking flue dust in two consecutive pelletization stages.[20] The first involves pelletizing the steelmaking flue dust and the second involves coating these pellets with the blended cement. The results showed that a 2-mm thin coating could stabilize the dust classified as a hazardous product.

4.2.3 PORTLAND POZZOLAN CEMENT

4.2.3.1 Introduction

According to ASTM Standard Specification C618, a pozzolan is defined as "siliceous or siliceous and aluminous materials which in themselves possess little or no cementitious value but will, in finely divided form and in the presence of moisture, chemically react with calcium hydroxides at ordinary temperatures to form compounds possessing cementitious properties."

4.2.3.2 Hydration and Microstructure of Portland Pozzolan Cements

When portland pozzolan cement contacts water, portland cement hydrates first; the lime released from the hydration of portland cement reacts with pozzolan to form new products depending on the composition of the pozzolan and hydration conditions. The main hydration product in a lime–pozzolan mixture is C-S-H with a C/S ratio less than 1.5, depending on the local concentration of reactants.[23] The aluminate in the pozzolan may yield a variety of hydrates: calcium aluminate hydrate (C_4AH_{19}), gehlenite hydrate (C_2ASH_8), AFt ($C_3A.3CaSO_4.32H_2O$), and AFm ($C_3A.CaSO_4.12H_2O$). Carbon dioxide (CO_2) can be combined in tetracalcium aluminate hydrate ($C_3A.CaCO_3.12H_2O$).[24] The pozzolanic reactions can be generally expressed as follows:

$$Ca(OH)_2 + SiO_2 + (n-1)H_2O \rightarrow xCaO.SiO_2.nH_2O \qquad (4.7)$$

$$Ca(OH)_2 + Al_2O_3 + (n-1)H_2O \rightarrow xCaO.Al_2O_3.nH_2O \qquad (4.8)$$

$$Ca(OH)_2 + Al_2O_3 + SiO_2 + (n-1)H_2O \rightarrow 3CaO.Al_2O_3.2SiO_2.nH_2O \qquad (4.9)$$

$$(1.5\text{-}2.0)CaO.SiO_2.aq + SiO_2 \rightarrow (0.8\text{-}1.5)CaO.SiO_2.aq \qquad (4.10)$$

Because of the chemical nature of pozzolans, they can be very effective sorbents for some contaminants. Pozzolans may be used in two different ways for S/S of wastes. In the first way, pozzolan is blended with cement, then mixed with the waste and water, if applicable, to stabilize/solidify the waste. In the second way, a pozzolanic material is mixed with the waste first to absorb certain contaminants, then cement is added to stabilize/solidify the waste material. The use of some pozzolanic materials as additives for S/S enhancement is discussed in Chapter 8. However, pozzolanic reactions will happen in the same way as described above. As pozzolanic reaction takes place, the contaminants absorbed on the pozzolan particles may be partially released and leached out, depending on the nature of the pozzolan and contaminants.[25]

4.2.3.3 Portland Natural Pozzolan Cements

Several natural pozzolans such as zeolites, calcined clays, and volcanic ashes are used for production of portland natural pozzolan cements.[26] No publication can be identified on the direct use of blended or interground portland natural pozzolan cements for S/S. However, zeolites and several clay minerals have been widely used as sorbents for certain contaminants before portland cement is added. Chapter 7 provides detailed discussions on the use of these natural pozzolans as sorbents.

4.2.3.4 Portland Fly Ash Cement

Fly ash is a by-product of the combustion of pulverized coal in thermal power plants. Fly ashes are heterogeneous fine powders consisting mostly of rounded or spherical glassy particles. The composition of fly ash depends on the coal used, but also on the various substances injected into the coal or gas stream to reduce gaseous pollutants or to improve efficiency of particulate collectors. When limestone and dolomite are used for desulfurization of the exit gases, CaO and MgO content in fly ash will be increased. Conditioning agents such as sulfur trioxide, sodium carbonate and bicarbonate, sodium sulfate, phosphorus, magnesium oxide, water, ammonia, and triethylamine are often used to improve the collection efficiency.

Fly ash consists mainly of SiO_2, Al_2O_3, Fe_2O_3, and CaO. According to ASTM C618,[27] fly ash belongs to Type F if the $(SiO_2+Al_2O_3+Fe_2O_3) > 70\%$, and belongs to Type C if $70\% > (SiO_2+Al_2O_3+Fe_2O_3) > 50\%$. The most abundant phase in fly ashes is glass. Crystalline compounds usually account for 5 to 50% and include quartz, mullite, hematite, spinel, magnetite, melilite, gehlenite, kalsilite, calcium sulphate, and alkali sulfate.[28] High-calcium fly ash may contain appreciative amounts of free CaO, C_3A, C_2S, $CaSO_4$, MgO, and $C_3A.CaSO_4$. The x-ray diffraction (XRD) technique is very useful in identifying these crystallized substances in fly ashes. A broad diffraction halo, which is attributed to the glassy phase, always appears on the XRD patterns of fly ashes.

Fly ashes are widely used to partially replace portland cement in concrete to reduce materials cost and to improve some properties of the concrete.[29] The introduction of fly ash will decrease the C/S ratio of C-S-H and increase the retention of cationic contaminants in C-S-H. The pozzolanic reactions between lime and fly ashes can be described using Equations 4.7 to 4.10.

TABLE 4.2
Coal Combustion By-Product Used in
Waste S/S (short tons)[30]

Year	Fly Ash	Bottom Ash	FDG* Material
1999	1,930,000	61,000	15,600
2000	2,000,000	35,400	20,900
2001	1,439,407	68,930	47,258
2002	3,187,773	19,091	67,053

* Flue gas desulfurization gypsum.

Fly ashes react with lime much more slowly than the hydration of portland cement. Rod-like ettringite grows on the surface of fly ash at the age of about 1 day and the surface of fly ash is thickly coated with hydration products at 3 days.[29] The presence of fly ash decreases or eliminates the free lime content in hardened cement pastes. The products resulting from the pozzolanic reactions between lime and fly ash refine the pore structure of hardened pastes and reduce the permeability of hardened pastes.

Coal fly ash and other coal combustion by-products have been used extensively for waste stabilization. Table 4.2 shows the amounts of these materials used for waste S/S.[30] Fly ashes are used in three ways: (1) fly ash alone (mainly Class C fly ash), (2) lime–fly ash mixture, and (3) portland cement–fly ash mixture. A previous report reviewed the use of coal fly ash for waste S/S.[31]

Many laboratory research projects and commercial applications have used portland fly ash cement to stabilize/solidify hazardous, radioactive, and mixed wastes.[31,32] A portland cement/Class F fly ash binder was used to solidify a heavy metal sludge containing cadmium, chromium, mercury, and nickel.[31] Results indicate a wide variability in the composition of partially dewatered sludges. Such variability can lead to localized differences in the chemical composition of the solidified material. Microanalyses of the cement and fly ash mixtures indicate that fly ash spheres reacted with the portland cement component to form a variety of reaction products, including ettringite. To a minor degree, fly ash was involved in the chemical fixation of the waste elements. Lead nitrate up to a concentration of 10% (by the mass of the binder) was solidified with Type I portland cement and a cement–fly ash (equal proportion) mixture.[33] The quantity of divalent lead leached depended on the initial lead nitrate concentration and the binder systems adopted. Lead solidified with a cement–fly ash mixture showed slightly less leaching compared to the cement binder.

The use of fly ash can enhance the sorption of Cs in a cement-based system.[34] Weng and Huang[35] employed fly ash to adsorb metals in industrial water, then used cement to fix the metal-contaminated adsorbent. It was found that fly ash could provide an acceptable level of metal adsorption for zinc and cadmium in dilute wastewater streams, with adsorption capacities of 0.27 and 0.05 milligrams per gram, respectively. Tests of leachates derived from the cement-fixed metal-laden fly ash indicated that concentrations of the metals in the leachates were lower than the existing drinking water standards. The liberation of Na^+ and K^+ during the hydration of fly ashes will compete with the potential sorption sites for other metals. For this reason, the fly ashes used for cation sorption should have a low alkali content. Also, the pozzolanic reaction between pozzolans and $Ca(OH)_2$ released from the hydration of cement or high-calcium silicate hydrate or both may reduce the usefulness of the adsorption capability of pozzolanic materials.[25]

In a slag-based waste form for S/S of low-level radioactive alkaline salt solution at the Savannah River Plant,[17] Class F fly ash was used to reduce the heat of hydration and the permeability of the waste form. Both portland cement and portland fly ash cement were used to solidify ion-exchange resin wastes, loaded with radionuclides, transition metals, and organic chelating agents.[36] The solidified waste forms were tested following the ANSI/ANS-16.1-2003 procedure for release rates, effective diffusivities, and leachability indexes. Portland fly ash cement–solidified waste forms exhibited lower leachability than portland cement–solidified waste forms. The author attributed this to the decreased permeability by introducing fly ash. This is confirmed by another study.[37]

For over 20 years, the Oak Ridge National Laboratory has disposed of low-level radioactive waste by mixing it with portland cement, Class F fly ash, and clay minerals and injecting the grout into an impermeable shale formation at a depth of approximately 1000 feet.[38] In Florida, a mixture of portland cement and fly ash was used to treat 62,000 cubic yards of polychlorinated biphenyl (PCB)-contaminated soil in a flowable fill mixture.[39] The contaminated soil was excavated, crushed, mixed, and replaced in the trenches from where it had been excavated.

However, some publications also report that the use of fly ash is not helpful during S/S of waste materials. Akhter et al.[19] used various combinations of Type I portland cement, Class F fly ash, blast furnace slag, lime, and silica fume as additives to immobilize soils containing heavy metals at concentrations in the range of 10,000 to 12,200 ppm. They concluded that fly ash, used together with lime, blast furnace slag, or portland cement, does not serve to improve performance when used as an additive.

4.2.3.5 Portland Silica Fume Cement

Condensed silica fume, also known as volatilized silica, microsilica, or simply silica fume, is a by-product of the manufacture of silicon or of various silicon alloys by reducing quartz to silicon in an induction arc furnace at temperatures up to 2000°C. Gasified SiO at high temperatures condenses in the low-temperature zone to tiny spherical particles consisting of noncrystalline silica.

The chemical composition of silica fume depends not only upon the raw materials used, but also upon the quality of the electrodes and the purity of the silicon product. Generally speaking, the impurities in condensed silica fume decrease as the amount of silicon increases in the final products. The by-products from the silicon metal and the ferrosilicon alloy industries, producing alloys with 75% or higher silicon content, contain 85 to 95% noncrystalline silica; the by-product from the production of ferrosilicon alloy with 50% silicon contains a much lower silica content. Minor components in silica fume are 0.1 to 0.5% Al_2O_3, 0.1 to 5% Fe_2O_3,

2 to 5%, carbon, 0.1 to 0.2% S, less than 0.12% CaO, less than 0.1% TiO_2, less than 0.07% P_2O_5, and less than 1% alkalis.[40]

It is well known that the use of silica fume in concrete decreases or eliminates the free lime content, increases the strength, and decreases the permeability of concrete. Silica fume showed an effect on the released lime 8 hours after the hydration of portland cement,[41] which is significantly faster than slag or fly ashes due to its small particle size.

Laboratory results[37,42] have indicated that the use of silica fume decreases the diffusion coefficient of contaminants very significantly. Shin and Jun[43] used silica fume as an admixture to solidify a waste containing a high concentration of organic contaminants and chromium in a monolithic mass with high strength and very low leachability for these contaminants. They noticed that silica fume was highly effective in achieving high compressive strength and low permeability. In another study, both silica fume and fly ash were blended with portland cement to solidify K061 hazardous waste (electric arc furnace dust [AFD]).[32] Some short-term results indicated that the use of silica fume is more effective than fly ash based on TCLP results.

4.2.4 PORTLAND CEMENT–SODIUM SILICATE SYSTEM

Sodium silicate is the generic name for a series of compounds with a formula $Na_2O.nSiO_2$. Theoretically, the ratio n can be any number. Commercial liquid sodium silicates have a ratio from 1.60 to 3.85. Based on ^{29}Si-NMR (nuclear magnetic resonance) testing, the SiO_4 structural units can be classified into seven types. Q^0, Q^1, Q^2_{cy-3}, Q^2, Q^3_{cy-3}, Q^3, and Q^4. The superscript on the Q represents the number of linkages between the given Si atom and neighboring Si atoms by = Si-O-Si = bonds. The symbols Q^2_{cy-3} and Q^3_{cy-3} designate intermediate or branched SiO_4 structural units in cyclo-tristructure (six-membered rings).[44] The percentages of structural units in a sodium silicate solution depend on the Na_2O/SiO_2 ratio, concentration, temperature, and age. Sodium silicate was widely used as an accelerator of concrete. Today, it is still a widely used accelerator for shotcrete. Sodium silicate is also a very effective activator for ground GBFS. The Na_2O/SiO_2 ratio and concentration have great effect on the properties of concrete.

Conner[1] first discovered using liquid sodium silicate as an additive for cement-based S/S in the late 1960s. It was demonstrated in 1970 and applied on full-scale S/S in 1971 by Chemfix Technologies, Inc. under the trade name of Chemfix®. The use of sodium silicate adsorbs extra water in liquid wastes, accelerates the setting and hardening of the system, and decreases the leachability of heavy metals. The adsorption of water is attributed to the gelation of sodium silicate in the presence of portland cement. Thus, a portland cement–sodium silicate system is especially suitable for liquid wastes from both technical and economical aspects.[1,45] Later on, a number of vendors have marketed the S/S process using sodium silicates in different ways and forms, which include Fujimasu Process, Lopat process, SolidTek, and enviroGuard/ProTek/ProFix. Conner[1] gives more details about the sodium silicate technologies.

It is reported that many investigators have difficulty reproducing in the laboratory the results that are routinely obtained in the field.[1] The most important reason is that these investigators do not realize that the properties of liquid silicates vary significantly with concentration, temperature, Na_2O/SiO_2 ratio, and storage time. Very limited information can be found on how the Na_2O/SiO_2 ratio and concentration of sodium silicate affect the properties of stabilized waste forms.

Butler et al.[46] used Type I portland cement and Type N sodium silicate solution to solidify Cd, Pb, and phenols. They used a combination of conduction calorimetry, XRD, and solid-state NMR to obtain information about immobilization mechanisms of heavy metals and organics in cement matrices. The results from conduction calorimetry showed a greater rate of heat evolution for cement/sodium silicate compared to blank (control) cement samples, although the total heat output was greater for the blank cement. Following the course of silicate transformation beginning with the orthosilicates (SiO_4^{4-}) through tenninal silicates (SiO_3), internal silicates (SiO_2), and branching silicates (SiO) with the aid of NMR showed an excellent correlation with the calorimetry results. Fourier transform infrared spectroscopy (FTIR) and thermogravimetry analysis (TGA) methods have also been used to investigate the solidification of Cr, Pb, Ba, Hg, Cd, and Zn in portland cement. The analyses have been found to support the leaching and mechanical property test results.

Caldwell et al.[46] attempted the solidification of a largely inorganic residue containing trace amounts of organic compounds using several solidification processes: cement, cement–fly ash, cement-activated carbon, cement–bentonite, and cement-soluble silicates. Twenty organic compounds at three contaminant levels (10, 100, and 1000 mg/g of solidified waste) were studied. The compounds were chosen to represent a wide array of organics: aliphatic/aromatic, volatile/nonvolatile, monocyclic/polycyclic, halogenated/nonhalogenated compounds, pesticides, and PCBs. Results based on a distilled water extraction of the solidified wastes indicated that the performance of the solidified forms was highly dependent on the compound in question. Nonvolatile and insoluble organic compounds were contained better than soluble/nonvolatile contaminants. For the processes studied, the cement-soluble silicates system showed better performance than cement, cement–fly ash, cement-activated carbon, and cement–bentonite systems.

A site in Florida, which belonged to General Electric Co. and was contaminated with PCBs (up to 950 mg/kg soil), volatile organic compounds (up to 1500 mg total VQC/kg soil), chromium (up to 400 mg/kg), copper (up to 900 mg/kg), lead (up to 2500 mg/kg), and zinc (up to 1000 mg/kg), was solidified using a proprietary commercial mix with sodium silicate as an additive in April 1988. Microstructural analyses of the treated waste showed that the process produced a dense homogeneous product with low porosity. The treated soil had a compressive strength of up to 5 MPa and permeability up to 21×10^{-7} cm/s. Results from TCLP tests showed that PCBs were successfully immobilized. The viral quality control (VQC) concentrations in the leachates ranged between 320 and 605 mg/L. Metal concentrations ranged between 120 and 210 mg/L. The detected concentrations were below regulatory levels for all the contaminants.

4.2.5 POLYMER-MODIFIED PORTLAND CEMENT

A replacement of 10 to 15% (by mass) of the cement binder by some synthetic organic polymer can greatly improve the flexural strength and impermeability of hydraulic cements. The polymer includes a monomer, prepolymer–monomer mixture, or a dispersed polymer (latex). To effect the polymerization of the monomer or prepolymer–monomer, a catalyst is added to the mixture. Curing of latex polymer cement concrete is generally shorter than for conventional concrete because the polymer forms a film on the surface of the product and retains some of the internal moisture needed for continuous cement hydration. Therefore, polymer-modified cement concrete can be cast in place in field applications.

Polymer-modified cement can exhibit good ductility and excellent resistance to penetration of water, aqueous salt solutions, and freeze/thaw cycles. The use of an improper polymer may have negative effects on the properties of the cement due to the incompatibility between the cement and the polymers. To achieve a substantial improvement over unmodified concrete, fairly large proportions of these polymers are required. The improvement does not always justify the additional cost.

Polymers are sometimes added into hydraulic cements to improve the effectiveness of S/S of some contaminants, especially organic wastes. The USEPA even acknowledges that organics interfere with the cement stabilization process particularly when the organic concentration exceeds 1% total organic carbon by mass.[48] In one study,[49] latex-modified cement systems were used for S/S of inorganic wastes containing lead and chromium. Results of freeze/thaw durability tests conducted on the synthetic wastes show little or no weight loss after 50 cycles. In addition, unconfined compressive strength tests show no loss of strength due to the presence of lead and chromium. Preliminary TCLP and extraction procedure tests indicated latex-modified cement systems had considerable improvement over regular portland cement. In another study, it is noticed that a mercury-containing sludge solidified by polymer latex-modified cement shows much higher strength and lower leachability than that by regular portland cement.[50] The author attributed this to the formation of a cross-linked structure in the solidified monoliths, which enhances the cohesive or chemical bonding between the contaminants and binder.

A latex polymer was added to portland cement to stabilize several pure polynuclear aromatic compounds present in the K051 waste, which include naphthalene, phenanthrene, and pyrene.[51] Leach test results indicate that the retention levels were very high for all these compounds and was over 99% for pyrene in the polymer-modified cement-stabilized waste form. FTIR analysis indicated the formation of chemical bonding between the additives and the organic contaminants. It was reported that the polymer enhances the encapsulation and penetration of the cement system into the interstitial spaces of the waste.[52]

4.2.6 MASONRY CEMENT

In North America, masonry cement is usually an interground cement of 50% Type I portland cement and 50% limestone. The replacement of 50% portland cement with limestone powder decreases the initial pH of the system. On the other hand, limestone has a significantly high acid-neutralization capacity. A comparison between Type I portland cement and masonry cement indicated, in the early leaching stages, masonry cement-solidified waste forms showed lower pH and lower metal leachability. In a more acidic solution, portland cement outperforms the masonry cement.[53] In an early study, masonry cement was also used for solidification of low-level radioactive wastes.[54]

4.3 LIME–POZZOLAN CEMENT

Lime–pozzolan mortars have been used by humans for centuries.[55] These materials were widely used in the masonry construction of aqueducts, arch bridges, retaining walls, and buildings during Roman times and have earned a good reputation of durability.[56] The invention of portland cement in the 19th century resulted in a reduction in the use of lime–pozzolan cements because portland cement shows a faster setting time and higher early strength. In the past 50 years, lime–pozzolan cements have again come under investigation, especially in some developing countries, because of their low cost. The hydration of lime–pozzolan cements are the same as those described for portland pozzolan cements.

Lime–pozzolan cements have been successfully used for treatment and delisting of several hazardous wastes. In one case,[58] lime–fly ash cement was used to treat electric AFD, which contains hexavalent chromium, lead, and cadmium and is listed by USEPA as hazardous waste designation K061. TCLP test indicated that contaminants in the leachate from treated waste were all brought into compliance with regulatory criteria. The process, with a trade name of Super Detox Process, as illustrated in Figure 4.2, entered commercial operation in Illinois in 1989 and had processed 120,000 tons of electric AFD when reported by Smith.[58]

In another case, a steel manufacturer generates about 240 wet tons per year of filter cake from its chromium wastewater treatment operations.[59] This filter cake is classified as EPA Hazardous Waste Code F006. The concerns for F006 are cadmium, hexavalent chromium, nickel, and complexed cyanide. The waste generated contains chromium (III) produced from the reduction of spent chromium passivation and plating solutions. The steel manufacturer used a lime–fly ash mixture to adsorb or

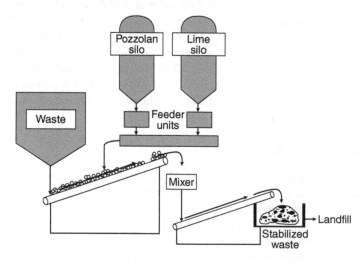

FIGURE 4.2 Illustration of lime–pozzolan cement stabilization process.[58]

encapsulate the heavy metals present in the chromium filter cake into a calcium–alumino-silicate matrix, thereby rendering them essentially immobile.[59] The product of this reaction is identified as a chemically stabilized filter cake (CSFC). Several leaching tests on the treated waste indicated the concentrations of Cr in leachates are well below the corresponding levels of regulatory concern. Based on the test results, it can be concluded that the chromium CSFC did not meet the criteria for which EPA hazardous waste code F006 was listed in 40 CFR Part 261.32, nor did it meet any other hazardous criteria/characteristics that might cause the waste to be considered as hazardous. In addition, the materials used in the treatment process are not expected to introduce any additional constituents of concern. The manufacturer submitted a delisting petition for excluding the CSFC from the list of hazardous wastes.

Debroy and Dara[60] evaluated the immobilization of zinc and lead present in waste sludges by chemical fixation and encapsulation methods using lime–pozzolan and fiber-reinforced lime–pozzolan admixtures. Fixation and coating with sodium silicate solution produced good results for immobilizing lead and zinc ions, but encapsulation of heavy metal hydroxide sludge in simple and fiber-reinforced lime–fly ash mixtures was more efficient. Fiber additives to the admixture may increase the structural strength of the monolith. Coating of an encapsulated admixture may provide an additional immobilization barrier.

Mining tailings are waste materials left over after mineral ores have been milled and the valuable minerals extracted. Lime is often mixed with mining tailings for backfill, or with acid-generating tailings to increase the pH to a level where the solubility of most metals is at its lowest.[61,62] Pozzolanic reactions happen between lime and the minerals in the tailings, whose rate and products are dependent on the size and the nature of the minerals in the tailings.

Lime is often used to treat contaminated soils alone. Of course, lime provides an alkaline environment for precipitation of heavy metals. The addition of lime to a fine-grained soil initiates several reactions, which can be cataloged as physical adsorption, ion exchange, and pozzolanic reaction. Physical adsorption and ion exchange take place rapidly and produce immediate changes in soil plasticity, workability, and the immediate uncured strength and load-deformation properties.[20]

In most cases, the silica and aluminum in soils are reactive, and a soil–lime pozzolanic reaction may happen. The pozzolanic reaction results in the formation of various cementing compounds, which improve the strength and durability of the mixture. Since the pozzolanic reactions are very slow and time dependent, the strength development is gradual but continuous for long periods of time amounting to several years in some instances. An increase in temperature accelerates the pozzolanic reaction between lime and soil, and the strength development, significantly. Laboratory results have confirmed that heavy metals, such as lead and chromium, were included and immobilized by the soil–lime pozzolanic reaction products.[63] Thus, lime treatment of contaminated soil, to some extent, belongs to the use of lime–pozzolan cements.

TABLE 4.3
Chemical Composition Ranges for High-Alumina Cements[64]

Grade	Composition (%)			
	Al_2O_3	CaO	SiO_2	Fe_2O_3+FeO
Standard/Low alumina	36–42	36–42	3–8	12–20
Low alumina/Low Iron	48–60	36–42	3–8	1–3
Medium alumina	65–75	25–35	< 0.5	< 0.5
High alumina	> 80	< 20	< 0.2	< 0.2

4.4 HIGH-ALUMINA CEMENTS

High-alumina cements are different from portland cements in that they are composed of calcium aluminates rather than calcium silicates. There are a number of different high-alumina cements with their alumina content from 36 to 80%. Scrivener and Capmas[64] divide them into four types, whose composition is summarized in Table 4.3. The main mineral phase of high-alumina cements is CA. Some possible minor phases are $C_3(A,F)$, CA_2, $C_{12}A_7$, C_2AS, and C_2S. High-alumina cements are mainly used in refractory concrete, since they can withstand high temperatures, but are often used in combination with other minerals and admixtures in concrete for construction to give a wide range of properties, including rapid setting and drying, and controlled expansion or shrinkage compensation.[64]

CA hydrates differently at low and high temperature. The hydration product is CAH_{10} at low temperature (< 20°C) and C_3AH_6 and AH_3 at high temperature (> 30°C). The hydration product of $C_3(A,F)$ is $C_3(A,F)H_6$ at low and high temperatures. The hydration reaction can be expressed as follows:

At low temperature:

$$CA + 10\ H_2O \rightarrow CAH_{10} \tag{4.11}$$

$$C_3(A,F) + 6\ H_2O \rightarrow C_3(A,F)H_6 \tag{4.12}$$

At high temperature:

$$CA + 12\ H_2O \rightarrow CAH_6 + 2AH_3 \tag{4.13}$$

CAH_{10} will convert to CAH_6 and AH_3 at high temperature:

$$CAH_{10} \rightarrow CAH_6 + 2AH_3 + 18H_2O \tag{4.14}$$

The conversion is accompanied by the increase of porosity and the decrease of strength of hardened cement paste and concrete. The hydration chemistry, as described above, has indicated that C_3AH_6 is the most stable hydration product of aluminous cement.

There are many laboratory research projects using high-alumina cements for S/S of wastes.[65-67] Shen[67] compared three cements — portland cements, high-alumina cement, and alkali-activated slag cements — for immobilization of cesium. After 28 days of curing at room temperature, the three waste forms were tested for Cs leachability at 25 and 70°C. At 25°C, high-alumina cement showed the lowest, and portland cement the highest leachability, which may be attributed to the lower solubility of $Al(OH)_3$ compared with that of $Ca(OH)_2$. As the temperature was increased to 70°C, the cesium leachability increased drastically due to the conversion of hydration produced. Actually, it was found that the Sr leachability from high-alumina cement–solidified wastes is lower than that from portland cement–solidified waste forms.[68] However, high-alumina cement–solidified waste forms showed better thermal integrity and stability than portland cement–solidified waste forms at high temperatures, especially at above 600°C.[68] High-alumina cements are highly recommended for incorporating wastes by hot pressing.

The addition of some additives into high-alumina cement can change its hydration products and increase the efficacy of immobilization of contaminants. Amorphous silica was added to calcium aluminate cement to give a composition similar to zeolites and clay minerals.[69] Laboratory test results indicated that silica-adjusted high-alumina cement gave lower leachability for Ca, Sr, and the rare earths, but is higher for Na, Mg, and U. Another study noticed that the addition of silica fume to high-alumina cement resulted in the formation of zeolitic phases, which drastically reduces the leaching of cesium.[70,71] Analysis of ^{27}Al and ^{29}Si using solid NMR and microanalysis with an electronic microscope indicated the formation of chabazite with the following composition: $Ca_{1.38}Cs_{1.38}Al_{5.24}Si_{6.76}O_{24}.nH_2O$.[72] The lab research also confirmed that an increase of curing temperature accelerates the formation of chabazite and decreases the leaching of Cs very significantly.

With the adjustment of chemical composition of high-alumina cement, trisulfoaluminate — $C_3A.3CaSO_4.32H_2O$ (AFt or ettringite), which is a member of an isostructural group of compounds of prismatic or needle morphology — can form. Ettringite can be a material tolerant of many substitutions,[73-79] as discussed in detail in Chapter 7.

Calcium aluminate was used to stabilize municipal solid waste incineration fly ash containing 10.5% chlorides (in alkali chlorides), 8.3% SO_3 (mainly as anhydrite and some as gypsum), and 19.3% CaO.[80] The mixing proportion was 84.5% fly ash and 15.5% calcium aluminate at a water-to-cement ratio of 2.7. After 4 hours of hydration, most anhydrite was consumed and ettringite reached the maximum. After complete consumption of sulfate, monochlorocalcium aluminate starts to form, until the calcium aluminate is completely consumed, after about 2 days.

The main concern with the use of ettringites for waste management is their stability. Thermodynamic calculation indicates that AFt is stable only below 65°C.[81] In the laboratory, it was noticed that AFt was destroyed after a heat treatment of 16 hours at 80°C.[82] After the heat treatment, the aluminate and sulfate were generally present in the C-S-H gel, but some monosulfate and very small amounts of hydrogarnet were also present; some sulfate was found in the pore solution.

The stability of AFt is also dependent on the pH value of the environment. Ettringite was found to be stable up to 60°C in the resulting solution of pH 11.2.[83,84] In another study, Gabrisovil et al.[85] demonstrated that, under nonequilibrium conditions, the boundary for the disappearance of AFt is pH = 10.7 and for the monosulfate, pH = 11.6. Therefore, only gypsum and aluminate sulfate are stable at pH values below 10. A laboratory study has indicated that AFt-based waste form is dissolved very quickly in acidic environments.[86] Although no published information is available on the stability of those nonsulfate ettringites, it can be expected that they behave very similarly. The conversion or decomposition of ettringites will release the immobilized heavy metals. Precautions should be taken when ettringites are considered for fixation of contaminants.

4.5 ALKALI-ACTIVATED SLAG CEMENTS

Alkali-activated slag cements usually consist of two basic components: a ground granulated slag and an alkaline activator(s). The former shows little or no cementitious behavior under normal conditions, but may give very high strengths in the presence of an alkaline activator(s). Typical examples are GBFS and granulated phosphorus slag (GPS). NaOH, Na_2CO_3, $Na_2O.nSiO_2$, and Na_2SO_4 are the most common and economical activators.

Although many research papers have been published on the hydration, microstructure, and properties of alkali-activated slag cements, the hydration mechanism of alkali-activated slag cements is still not very clear. The main hydration product of alkali-activated slag cement is C-S-H with a low C/S ratio. The minor hydration products such as C_4AH_{13}, C_2ASH_8, and $C_3A.CaCO_3.11H_2O$ may appear depending on the nature of slag and alkaline activators used. The detailed information on microstructure and properties of alkali-activated slag cements can be found in several publications.[3,88-90] When a proper alkaline activator is used, alkali-activated slag cement can show much higher early and later strength than even ASTM Type III portland cement. It was noticed that, for a given water-to-cement or water-to-slag ratio, alkali-activated slag cement exhibits a much less porous structure than portland cement.[4] Since the main hydration product of alkali-activated slag cement is C-S-H with a low C/S ratio, and no $Ca(OH)_2$ forms, alkali-activated slag cements show much better resistance in aggressive environments than portland cement. The former also has much better stability at high temperatures in dry and wet conditions than the latter. This means that alkali-activated slag cement is a better cement for S/S of wastes than portland cement.

Figure 4.3 represents the mercury intrusion pore structure of portland cement and alkali-blast furnace slag cement pastes cured 28 days at 25°C. It can be seen that the alkali slag cement paste has not only a lower porosity, but also a finer pore structure than portland cement paste. The alkali slag cement pastes contain mainly pores with r < 100 Å, which can restrict the flow of liquid or diffusion of ions in the pastes. This means that hydrated alkali-activated slag can have much finer pore structure and lower porosity than portland cement pastes, which indicates that the former is a better physical barrier than the latter.[91,92]

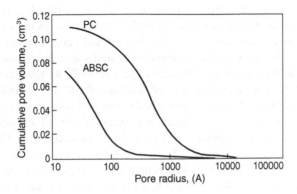

FIGURE 4.3 Cumulative pore volume of portland cement (PC) and alkali-activated slag cement (ABSC) pastes at 28 days.[4]

Cs is a difficult radionuclide to stabilize in radioactive wastes. Cesium leachability from portland cement and alkali-activated slag cements were measured using a standard semidynamic test in which monolithic samples are suspended in deionized water in Teflon containers at the testing temperature, these samples are transferred to other containers with fresh dionized water at specified intervals and the leachate cesium concentrations are measured.[4] The cesium leachability, L_t (cm^{-1}), at time t was calculated using the following equation:

$$L_t = (a_t/A) \cdot (F/V)$$

where:

a_t – mass of leached Cs$^+$ at time t (g);
A – total original mass of Cs$^+$ in the specimens (g);
F – total surface area of the monolithic specimens (cm^2);
V – volume of the monolithic specimens (cm^3).

Figure 4.4 shows a higher cesium L_t(cm^{-1}) from portland cement than from alkali-activated slag cement for 0.5 wt% CsNO$_3$ loading after 28 days of moist curing at 25°C.

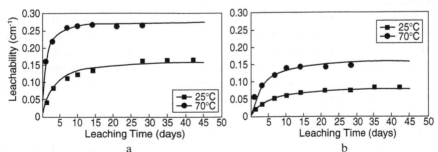

FIGURE 4.4 Leached fraction of Cs$^+$ in hardened portland and alkali-activated slag cement pastes.[4] (a) portland cement pastes, (b) alkali-activated slag cement pastes.

Figure 4.4 shows the leached fraction of Cs^+ in portland cement and alkali-activated slag cement pastes containing 0.5% $CsNO_3$ after 28 days of moist curing at 25°C. The results indicate that the Cs^+ in portland cement pastes shows a much higher leached fraction than that in alkali-activated slag cement pastes at the same temperature. As the temperature increases from 25 to 70°C, the leached fraction of Cs^+ in both pastes escalates. The leached fraction of Cs^+ in portland cement pastes at 25°C is even higher than that from alkali-activated slag cement pastes at 70°C. The calculation using Arrhenius' Equation indicated that the Cs^+ leaching activation energy of portland cement pastes is 18.96 kJ/mol compared with 25.19 kJ/mol for alkali-activated slag cement pastes. The lower leached fraction and higher leaching activation energy of Cs^+ in alkali-activated slag cement pastes than in portland cement pastes can be attributed to the less porous structure and lower C/S ratio in C-S-H. A partial replacement of slag with metakaolin will increase the porosity of the hardened cement pastes. However, it decreases the leached fraction of Cs^+ and Sr^{2+} in the hardened pastes.[93] The authors attribute the decrease in the leached fraction to the formation and adsorption properties of (Al+Na) substituted C-S-H and the self-generated zeolite precursor.

Many substances can significantly interfere with the hydration of cement, as discussed in Chapter 7. This is the basic principle for the use of different cement chemical admixtures such as retarders, accelerators, and superplasticizers to obtain some special properties of cements and concrete. It has been reported that heavy metals show much less interference with the hydration of alkali-activated slag cements than with portland cement. Shi et al.[94,95] investigated the S/S of electrical AFD with portland cement and an alkali-activated slag cement using an adiabatic calorimeter. The AFD is from the production of a specialty steel, which contains a high concentration of a variety of heavy metals. When 30% AFD is added, it retarded the hydration of the cement, but did not show an obvious effect on the hydration of cement at later ages. As the AFD content is increased from 30% to 60%, it retarded the hydration of portland cement very significantly, and the solidified waste forms did not show measurable strength after 6 months of hydration. When AFD is mixed with alkali-activated slag-based cement, the early hydration of the cement was retarded more obviously as the AFD content increased. However, the cement continued to hydrate with time-released more heat as the AFD content increased. It seems that the presence of AFD retarded the early hydration, but was beneficial to the later hydration of the slag.[14,15,96]

Several heavy metals, such as Zn^{2+}, Pb^{2+}, Cd^{2+}, and Cr^{6+}, were stabilized in $NaOH-$, Na_2CO_3-, and sodium silicate-activated slag cements.[97,98] These alkali-activated slag cements could immobilize these heavy metals very well regardless of activator. Cho et al.[99] investigated the leachability of Pb^{2+} and Cr^{6+} immobilized in NaOH and sodium silicate-activated slag cement pastes. They noticed that the leachability of Pb^{2+} and Cr^{6+} in alkali-activated slag cement pastes varied with curing conditions, but was very small. There is a very good relationship between the diffusion coefficient of Cr^{6+} and the pore volume with a radius less than 5 nm.

The other advantage of alkali-activated slag cement is its dense structure and stability at high temperatures. Under hydrothermal conditions, the main hydration products of portland cement are $C_2SH(A)$ and $Ca(OH)_2$, which give the cement paste a very porous and unstable structure. However, the main hydration products of alkali-activated slag cement are C-S-H (B) and tobermorite or xonotlite, which give the hardened pastes a much less porous and stable structure.[100] They also have an obvious cation ion exchange capacity and enhance the chemical fixation of contaminants.[101-104]

Alkali-activated calcined metakaoline/fly ash cements have also been proven as effective binders for S/S of wastes.[105-111] They are sometimes called geopolymer or hydroceramics. Some details can also be found in Chapter 6.

4.6 CEMENT KILN DUSTS

Cement plants generate cement kiln dust (CKD) as a means of removing alkalis, chlorides, and sulfates from the kiln system. There are two types of cement kiln processes: wet-process kilns, which accept feed materials in a slurry form; and dry-process kilns, which accept feed materials in a dry, ground form. In each type of process the dust can be collected in two ways: (1) a portion of the dust can be separated and returned to the kiln from the dust collection system (e.g., cyclone) closest to the kiln, or (2) the total quantity of dust produced can be recycled or discarded.

CKD is composed primarily of fine ground particles of limestone, clay, or shale, lime, sodium, potassium chlorides, sulfates, metal oxides, calcium silicates, and other salts. It is enriched in sodium and potassium chlorides and sulfates, as well as volatile metal compounds. The lime content can be up to 15%. Trace constituents in CKD (including certain trace metals such as cadmium, lead, and selenium, and radionuclides) are generally found in concentrations less than 0.05% by mass. Because some of these constituents are potentially toxic at low concentrations, it is important to assess their levels (and mobility or leachability) in CKD before considering their use. CKD is a fine powdery material similar in appearance to portland cement.

Approximately 12.9 million metric tons of CKD are produced annually. About 64% of the total CKD generated (or about 8.3 million metric tons) is reused within the cement plant. Approximately 6% of the total CKD generated is utilized off-site. The chemical and physical characteristics of CKD that is collected for use outside of the cement production facility will depend in great part on the method of dust collection employed at the facility. The most common beneficial use of CKD is its use as a stabilizing agent for wastes, where its absorptive capacity and alkaline properties can reduce the moisture content, increase the bearing capacity, and provide an alkaline environment for waste materials. Chemical Waste Management conducted a great deal of work on CKD-based S/S process and made a specification as summarized in Table 4.4.[1] There is also a patented process using CKD and acid to produce a neutral, environmentally stable and hardened solid product.[112]

Calcium silicates in CKD can act as binders. Reactive silica from original or heated clay in a CKD can react with lime to form a weak binder directly. Thus, a CKD with high lime and calcium silicate contents is the most desirable for S/S purposes.

In August 1999, the U.S. Environmental Protection Agency proposed new regulations for management of CKD, which was designated as "high-volume, low-toxicity" special wastes requiring individualized treatment under the Resource Conservation

TABLE 4.4
Specification on CKD for S/S Purposes[1]

Catalog	Property	Criteria
Visual Observation	Free-flowing powder	
	No lumps larger than 3/8 in.	
	No impurities (rocks, plant fragments, etc.)	
Chemical Property	pH of 10% slurry	> 10
	Alkalinity as CaO	> 15%
	Moisture Content at 105°C	< 1%
	Loss of ignition at 1100°C	< 30%
	Temperature rise	20°C
Physical Property	Apparent bulk density (loose)	35–60 lb/ft^3
	Apparent bulk density (packed)	35–60 lb/ft^3
	Fineness (+8 mesh)	< 1
	Fineness (–325 mesh)	< 90

and Recovery Act. Although the proposed rule does not limit the beneficial use of CKD in a commercial landfill, the handling and transportation regulations could potentially pose difficulties for customers and haulers.

4.7 FLY ASHES

As discussed above, since Class C fly ash has a high content of lime and exhibits some cementitious properties, it can be used alone for S/S purposes. Redd et al.[113] conducted a series of experiments to stabilize the phenolics in foundry sands from Kansas using four different types of binders: portland cement, Class C fly ash, kaolinite, and bentonite. Strength and leachability of stabilized mixes of foundry sand were analyzed to assess their feasibility in construction and geotechnical applications. The results suggest that compressive strength was acquired relatively faster in Class C fly ash than in cement and, in general, varied inversely with the proportion of foundry sand in the stabilized mix. Lesser amounts of phenolic compounds leached from Class C fly ash-stabilized mixes than from cement-stabilized mixes. The leachate analyses for both total phenolics and 2,4,6-trichlorophenol indicate that increasing percentage replacement of foundry sands enhances stabilization. SEM observations confirmed that the cement-stabilized wastes were more porous than the Class C fly ash-stabilized wastes.[113]

REFERENCES

1. Conner, J. R., *Chemical Fixation and Solidification of Hazardous Wastes*, Van Nostrand Reinhold, New York, 1990.
2. IAEA, *Improved Cement Solidification of Low and Intermediate Level Radioactive Wastes*, Technical Report Series No. 350, International Atomic Energy Agency, Vienna, 1993.

3. Shi, C., Day, R. L., Wu, X., and Tang, M., Uptake of metal ions by autoclaved cement pastes, in *Proceedings of Materials Research Society*, Vol. 245, MRS, Boston, 1992, 141.

4. Shi, C., Shen, X., Wu, X., and Tang, M., Immobilization of radioactive wastes with Portland and alkali-slag cement pastes, *Il Cemento*, 91, 97, 1994.

5. Conner, J. R., *Guide to Improving the Effectiveness of Cement-Based Stabilization/Solidification*, Portland Cement Association, Skokie, IL, 1997.

6. Kosmatka, S. et al. *Design and Control of Concrete Mixtures*, Portland Cement Association, Skokie, IL, 2002.

7a. Mindess, S. and Young, J. F., *Concrete*, Prentice-Hall, New York, 1981.

7b. Odler, I., Hydration, setting and hardening of Portland cement, in *Lea's Chemistry of Cement and Concrete*, Shewlett, P. C., Ed., Edward Arnold, London, 1998, Chapter 6.

8. Komarneni, S., Roy, D. M., and Kumar, A., Cation exchange properties of hydrated cements, in *Advances in Ceramics*, Vol. 8, Nuclear Waste Management, Wicks, G. A. and Ross, W. A., Eds., American Ceramic Society, Columbus, OH, 1984, 441.

9. Bhatty, M. S. Y., Fixation of metallic ions in Portland cement, *Proceedings of 4th National Conference on Hazardous Wastes and Hazardous Materials*, 1987, 140.

10. Gebhardt, R., Survey of North American Portland cements: 1994, *Cement, Concrete and Aggregates,* 17, 145, 1995.

11. Duchesene, J. and Berube, M. A., Relationships between Portlandite depletion, available alkalis and expansion of concrete made with mineral admixtures, *Proceedings of 9th International Conference on AAR in Concrete*, The Concrete Society, London, 1992, 287.

12. Beaudion, J. J. and Brown, P. W., The Structure of Hardened Cement Paste, Proceedings of 9th International Congress on the Chemistry of Cement, New Delhi, 1992, Vol. I, pp. 485–525.

13. Angus, M. J. and Glasser, F. P., The chemical environment in cement matrixes, *Materials Research Society Symposium*, Vol. 50, MRS, Boston, 1986, 547.

14. Spence, R. D. et al. Immobilization of technetium in blast furnace slag grouts, *Proceedings of 3rd International Conference on the Use of Fly Ash, Silica Fume, Slag and Natural Pozzolans in Concrete*, American Concrete Institute, Farmington Hills, MI, 1989, 1579.

15. Glasser, F. P., Chemistry of cement-solidified waste forms, in *Chemistry and Microstructure of Solidified Waste Forms*, Spence, R. D., Ed., Lewis Publishers, Boca Raton, FL, 1993, Chapter 1.

16. Caldwell, R., Stegemann, J. A., and Shi, C., Effect of curing on field-solidified waste properties, part ii: chemical properties, *Waste Management and Research,* 17, 37, 1999.

17. Spence, R. D. et al. *Grout and Glass Performance Maximizing the Loading of ORNL Tank Sludges*, ORNL/TM-13712, Oak Ridge National Laboratory, 1999.

18. Langton, C. A., Slag-based materials for toxic metal and radioactive waste stabilization, *Proceedings of 3rd International Conference on the Use of Fly Ash, Silica Fume, Slag and Natural Pozzolans in Concrete*, ACI SP-114, American Concrete Institute, Farmington Hills, MI, 1989, 1697.

19. Akhter, H., Butler, L. G., Branz, S., Cartledge, F. K., and Tittlebaum, M. E., Immobilization of As, Cd, Cr and Pb-containing soils using cement or pozzolanic fixing agents, *Journal of Hazardous Materials,* 24, 145, 1990.

20. Lopez, F. A., Sainz, E., and Formoso, A., Use of granulated blast furnace slag for stabilization of steelmaking flue dust, *Ironmaking and Steelmaking*, 20, 293, 1993.

21. Diamond, S. and Kinter, E. B., Mechanism of soil-lime stabilization — an interpretive review, *Highway Research Record*, 92, 83, 1965.

22. PCA, *Soil-Cement Construction Handbook*, Portland Cement Association, Skokie, IL, 1995.
23. Ogawa, K., Uchikawa, H., Takemoto, K., and Yasui, I., The mechanism of hydration in the system C_3S-pozzolana, *Cement and Concrete Research*, 10, 683, 1980.
24. Helmuth, R., Some questions concerning ASTM standards and methods of testing fly ash for use with Portland cement, *Cement, Concrete and Aggregate*, 5, 103, 1983.
25. Angus, M. J., McCulloch, C. E., Crawford, R. W., and Glasser, F. P., Kinetics and mechanism of the reaction between Portland cement and clinoptilolite, in *Advances in Ceramics, Nuclear Waste Management*, Wicks, G.G. and Rose, W. A., Eds., American Ceramic Society, Columbus, OH, 1983, 413.
26. Kasai, Y., Tobinai, K., Asakura, E., and Feng, N., Comparative study on natural zeolites and other inorganic admixtures in terms of characterization and properties of mortars, *Proceedings of the 4th International Congress on the Use of Fly Ash, Silica Fume, Slag and Natural Pozzolans in Concrete*, SP-132, American Concrete Institute, Farmington Hills, MI, 1992, 615.
27. ASATM C618-03, Standard specification for coal fly ash and raw or calcined natural pozzolan for use as a mineral admixture in concrete, *Annual Book of ASTM Standards*, Vol. 04.01, Cement; Lime; Gypsum, American Society for Testing & Materials, Philadelphia, 2003.
28. Hemmings, R. T. and Berry, E. E., On the glass in coal fly ashes: recent advances, *Materials Research Society Proceedings*, Vol. 113, MRS, Boston, 1988, 3–38.
29. Helmuth, R., *Fly Ash in Cement and Concrete*, Portland Cement Association, Skokie, IL, 1987.
30. American Coal Ash Association, *Coal Combustion Product Production and Uses*, ACAA, Alexandria, VA, 2003.
31. Eylands, K. E., *Solidification and Stabilization of Wastes Using Coal Fly Ash: Current Status and Direction*, American Coal Ash Association, Alexandria, VA, 1995.
32. Fuessle, R. W. and Taylor, M. A., Comparison of fly ash versus silica fume stabilization: short-term results, *Hazardous Waste and Hazardous Materials*, 9, 335, 1992.
33. Wang, S.Y. and Vipulanandan, C., Leachability of lead from solidified cement-fly ash binders, *Cement and Concrete Research*, 26, 895, 1996.
34. McCulloch, C. E., Rahman, A. A., Angus, M. J., and Glasser, F. P., Immobilization of cesium in cement containing reactive silica and pozzolans, in *Advances in Ceramics, Nuclear Waste Management*, Wicks, G. G. and Rose, W. A., Eds., American Ceramic Society, Columbus, OH, 1983, 413.
35. Weng, C. H. and Huang, C. P., Treatment of metal industrial wastewater by flyash and cement fixation, *Journal of Environmental Engineering*, 120, 1470, 1994.
36. McIsaac, C. V., Leachability of chelated ion-exchange resins solidified in cement or cement and fly ash, *Waste Management*, 13, 41, 1993.
37. Lange, L. C., Hills, C. D., and Poole, A. B., Effect of carbonation on properties of blended and non-blended cement solidified waste forms, *Journal of Hazardous Materials*, 52, 193, 1997.
38. Gilliam, T. M., Use of coal fly ash in radioactive waste disposal, in *Proceedings of the 9th International Ash Use Symposium*, Vol. 2, Stabilization and Aquatic Uses; EPRI GS7162, Electric Power Research Institute, Palo Alto, CA, 1991.
39. Fazzinin, A. J., Collins, R. J., and Colussi, J. J., The role of fly ash in waste stabilization and solidification from a contractor's view point, *Proceedings of the 9th International Ash Use Symposium*, Vol. 2, Stabilization and Aquatic Uses; EPRI GS7162, Palo Alto, CA, 1991.

40. Uchikawa, H., Blended and special cements: effect of blending component on hydration and structure formation, *Proceedings of the 8th International Congress on the Chemistry of Cements*, Principal Report, Vol. 1, Apoio Financeiro Da, Rio de Janeiro, Brazil, 1986, 249.

41. Huang, C. Y. and Feldman, R. F., Hydration reaction in Portland silica fume blends, *Cement and Concrete Research*, 15, 585, 1985.

42. Quaresim, R., Scoccia, G., Volpe, R., Medici, F., and Merli, C., Influence of silica fume on the immobilization properties of cementitious mortars exposed to freeze-thaw cycles, in *Stabilization and Solidification of Hazardous, Radioactive, and Mixed Wastes, 3rd Volume*, Gilliam, T. M. and Wiles, C. C., Eds., ASTM STP 1240, American Society for Testing and Materials, Philadelphia, 1996, 135.

43. Shin, H. S. and Jun, K. S., Cement-based stabilization/solidification of organic contaminated hazardous wastes using na-bentonite and silica fume, *Journal of Environmental Science and Health*, 30, 651, 1995.

44. Hoebbel, D. and Ebert, R., Sodium silicate solution — structure, properties and problems, *Zeitschrift fur Chemie*, 28, 41, 1988.

45. Conner, J. R., Soluble silicates in solidification and fixation technology, *Second International Conference on New Frontiers for Hazardous Wastes Management*, Pittsburgh, 1987, 295.

46. Caldwell, R. J., Cote, P. L., and Chao, C. C., Investigation of Solidification for the Immobilization of Trace Organic Contaminants, Hazardous Wastes and Hazardous Materials, Vol. 7, No. 3, pp. 273–282, 1990.

47. U.S. EPA, *Technology Evaluation Report: SITE Program Demonstration*, Test International Waste Technologies in situ Stabilization/Solidification, Hialeah, FL, Report EPA/540/5-89/004a, June 1989.

48. U.S. EPA, 40 CFR, Part 268 — Land Disposal Restrictions, June 1, 1990.

49. Daniali, S., Solidification/stabilization of heavy metals in latex modified Portland cement matrices, *Journal of Hazardous Materials*, 24, 225, 1990.

50. Yang, G. C. C., Lee, C.-H., and Hsiue, G.-H., Properties of a mercury-containing sludge solidified by polymer latex modified cementitious materials, *Hazardous Waste and Hazardous Materials*, 10, 453, 1993.

51. Frost, D. J. and Carandang, C. M., Stabilization of organic wastes, *Proceedings of the 12th National Conference on Hazardous Materials Control/Superfund '91*, Washington, 1991, 389.

52. Bourgeois, J. C. et al., Designing a better matrix for solidification/stabilization of hazardous waste with the aid of bagasse (lignin) as a polymer additive to cement, *Proceedings of the Spring National Meeting of the American Chemical Society*, New Orleans, LA, 1996, 416.

53. Bhatty, J., Miller, F., West, P., and Ost, B., Stabilization of heavy metals in Portland cement, silica fume/Portland cement and masonry cement matrix, Portland Cement Association, RP 348, Skokie, IL, 1999.

54. Zhou, H. and Colombo, P., *Solidification of Low-level Radioactive Wastes in Masonry Cement*, Report No. BNL-52074, Brookhaven National Laboratory, Upton, NY, March 1987.

55. Malinowski, R. and Garfinkel, Y., Prehistory of concrete, *Concrete International*, 13, 62, 1991.

56. Hazra, P. C. and Krishnaswamy, V. S., Natural pozzolans in India, their utility, distribution and petrography, *Records of the Geological Survey of India*, 87, Part 4, 675, 1987.

57. Shi, C. and Day, R. L., Comparison of different methods for enhancing reactivity of pozzolans, *Cement and Concrete Research*, 31, 813, 2001.

58. Smith, C. L., Pozzolanic delisting — stabilization of electric arc furnace dust, *12th PTD Conference Proceedings, Ion and Steel Society*, 1993, 375.
59. MacGregor, A. and Kraeski, M., Stabilization and delisting of hazardous wastes: an effective approach for reducing high sludge disposal costs, *Remediation*, 455, 1994.
60. Debroy, M. and Dara, S. S., Immobilization of zinc and lead from wastes using simple and fiber-reinforced lime pozzolana admixtures, *Journal of Environmental Science and Health*, A29, 339, 1994.
61. Mohamed, A. O. M., Boily, J. F., Hossein, M., and Hassani, F. P., Ettringite formation in lime-remediated mine tailings: I. Thermodynamic modeling, *CIM Bulletin*, 88, 69, 1995.
62. Hossein, M., Mohamed, A. O. M., Hassani, F. P., and Elbadri, H., Ettringite formation in lime-remediated mine tailings: II. Experimental study, *CIM Bulletin*, 92, 75, April 1999.
63. Dermatas, D. and Meng, X., Stabilization/solidification (S/S) of heavy metal contaminated soils by means of a quicklime-based treatment approach, in *Stabilization and Solidification of Hazardous, Radioactive, and Mixed Wastes, 3rd Volume*, Gilliam, T. M. and Wiles, C. C., Eds., ASTM STP 1240, American Society for Testing and Materials, Philadelphia, 1996, 499.
64. Scrivener, K. L. and Capmas, A. Calcium aluminate cements, in *Lea's Chemistry of Cement and Concrete*, Hewlett, P. C., Ed., Edward Arnold, London, 1998.
65. Roy, D. M. and Gouda, G. R., High-level radioactive waste incorporation into (special) cements, *Nuclear Technology*, 40, 214, 1978.
66. Roy, D. M., Scheetz, B. E., Wakeley, L. D., and Barnes, M. W., Leach characterization of cement encapsulated wastes, *Nuclear and Chemical Waste Management*, 3, 35, 1982.
67. Shen, X., *Stabilization/Solidification of High-Level Radioactive Wastes with Cements*, M. Eng thesis, Nanjing University of Chemical Technology, Nanjing, China, 1988.
68. Stone, J. A., Studies of concrete as a host for Savannah River Plant radioactive waste, in *Scientific Basis for Nuclear Waste Management, Vol. 1*, McCarthy, G. J., Ed., 1978, 443.
69. Barnes, M. W., Scheetz, B. E., and Roy, D. M., The effect of chemically adjusting cement compositions on leachability of waste ions, in *Advances in Ceramics: Nuclear Waste Management*, Vol. 20, The American Ceramic Society, Columbus, OH, 1986, 311.
70. Fryda, H., Vetter, G., Ollitrault-Fichet, R., Boch, P., and Capmas, A., Formation of chabazite in mixes of calcium aluminate cement and silica fume used for cesium immobilization, *Advances in Cement Research*, 8, 29, 1996.
71. Fryda, H., Boch, P., and Scrivener, K. L., Formation of zeolite in mixes of CAC and silica fume used for ceasium immobilization, in *Mechanisms of Chemical Degradation of Cement-Based Systems*, Scrivener, K. and Young, J. F., Eds., E& FN SPON, London, 1997.
72. Helard, L., Letourneux, J. P., and Fryda, H., Examples of chemical interactions between pollutants and hydraulic binders, in *Proceedings of the International Congress on Waste Solidification-Stabilization Processes*, Cases, J. M. and Thomas, F., Eds., Nancy, France, 1995:143.
73. Brown, D. R. and Grutzeck, M. W., Iodine waste forms: calcium aluminate hydrate analogues, in *Scientific Basis for Nuclear Waste Management VIII*, Materials Research Society Symposia Proceedings, Vol. 44, 1985a, 911.
74. Brown, D. R. and Grutzeck, M. W., Synthesis and characterization of calcium aluminate monoiodide, *Cement and Concrete Research*, 15, 1068, 1985b.

75. Hassett, D. J., Pflughoeff-Hassett, D. F., Kumarathasan, P., and McCarthy, G. J., Ettringite as an agent for the fixation of hazardous oxyanions, *Proceedings of the Twelfth Annual Madison Waste Conference on Municipal and Industrial Waste*, Madison, WI, 1989, 20.

76. McCarthy, G. J., Hassett, D. J., and Bender, J. A., Synthesis, crystal chemistry and stability of ettringite, a material with potential applications in hazardous waste immobilisation, *Materials Research Society Symposium Proceedings*, Vol. 245, MRS, Boston, 1992, 129.

77. Poellmann, H. et al., Solid solution of ettringites: Part II: Incorporation of $Ba(OH)_4^-$ and CrO_4^{2-} in $3CaO.Al_2O_3.3CaSO_4.32H_2O$, *Cement and Concrete Research*, 23, 422, 1993.

78. Kindness, A., Macias, A., and Glasser, F. P., Immobilization of chromium in cement matrices, *Waste Management*, 14, 3, 1994.

79. Klemn, W. A., *Ettringite and Oxyanion-Substituted Ettringites — Their Characterization and Applications in the Fixation of Heavy Metals: a Synthesis of the Literature, Research and Development Bulletin RD116W*, Portland Cement Association, Skokie, IL, 1998.

80. Auer, S., Kuzel, H.-J., Poellmann, H., and Sorrentino, F., Investigation on MSW fly ash treatment by reactive calcium aluminates and phases formed, *Cement and Concrete Research*, 25, 1347, 1995.

81. Babushkin, V. I., Matveyev, G. M., and Mchedlov-Petrossyan, O. P., *Thermodynamics of Silicates*, Springer-Verlag, New York, 1985.

82. Scrivener, K. L. and Taylor, H. F. W., Delayed ettringite formation: a microstructural and microanalytical study, *Advances in Cement Research*, 5, 139, 1993.

83. Ghorab, H. Y. and Kishar, E. A., Studies on the stability of the calcium sulfoaluminate hydrates. Part 1: Effect of temperature on the stability of ettringite in pure water, *Cement and Concrete Research*, 15, 93, 1985.

84. Ghorab, H. Y. and Kishar, E. A., The stability of the calcium sulfoaluminate hydrates in aqueous solutions, *Proceedings of the 8th International Congress on the Chemistry of Cement, Vol. V*, Rio de Janeiro, Brazil, 1986, 104.

85. Gabrisovii, A., Havlica, J., and Sahu, S., Stability of calcium sulphoaluminate hydrates in water solutions with various ph values, *Cement and Concrete Research*, 21, 1023, 1991.

86. Stegemann, J. A. and Shi, C., Acid resistance of different monolithic binders and solidified wastes, in *Waste Materials in Construction: Putting Theory in Practice, Studies in Environmental Science 71*, Elsevier Science B.V., Amsterdam, 1997, 803.

87. Glukhovsky, V. D., Rostovkaja, G. S., and Rumyna, G. V., High strength slag-alkaline cements, *Proceedings of 7th International Congress on the Chemistry of Cement, Vol. III*, Paris, 1980, pp. v–164.

88. Krivenko, P. V., Alkaline cements, *Proceedings of the 1st International Conference on Alkaline Cements and Concretes*, VIPOL Stock Company, Kiev, Ukraine, 1994, 11.

89. Krivenko, P. V., Alkaline cements and concretes: problems of durability, *Proceedings of the 2nd International Conference on Alkaline Cements and Concretes*, VIPOL Stock Company, Kiev, Ukraine, 1999, 3.

90. Talling, B. and Brandstetr, J., Present and future of alkali-activated slag concrete, *Proceedings of the 3rd International Conference on the Use of Fly Ash, Silica Fume, Slag and Natural Pozzolans in Concrete*, SP-114, American Concrete Institute, Farmington Hills, MI, 1989, 1519.

91. Wu, X. et al., Alkali-activated cement based radioactive waste forms, *Cement and Concrete Research*, 21, 16, 1991.

92. Shi, C. and Day, R. L., Alkali-slag cements for the solidification of radioactive wastes, in *Stabilization and Solidification of Hazardous, Radioactive, and Mixed Wastes*, ASTM STP 1240, Gilliam, T. M. and Wiles, C. C., Eds., American Society for Testing and Materials, Philadelphia, 1996, 163.

93. Qian, G., Li, Y., Yi, F., and Shi, R., Improvement of metakaoline on radioactive sr and cs immobilization of alkali-activated slag matrix, *Journal of Hazardous Materials*, B92, 289, 2002.

94. Shi, C., Stegemann, J. A., and Caldwell, R., An examination of interference in waste solidification through measurement of heat signature, *Waste Management*, 17, 249, 1997a.

95. Shi, C., Stegemann, J. A., and Caldwell, R., Use of heat signature in solidification treatability studies, *Proceedings of 10th International Congress on the Chemistry of Cement*, Amarkai AB and Congrex Goteborg, Goteborg, Sweden, 1997b.

96. Caldwell, R., Shi, C., and Stegemann, J. A., Solidification formulation development for a specialty steel electric arc furnace dust, *Proceedings of Congrès International sur les Procèdes de Stabilization et de Solidification*, Nancy, France, 1995.

97. Malolepszy, J. and Deja, J., Effect of heavy metals immobilization on properties of alkali-activated slag mortars, *Proceedings of the 5th International Conference on the Use of Fly Ash, Silica Fume, Slag and Natural Pozzolans in Concrete*, SP-153, Farmington Hills, MI, 1995, 1087.

98. Deja, J., Immobilization of Cr^{6+}, Cd^{2+}, Zn^{2+} and Pb^{2+} in alkali-activated slag binders, *Cement and Concrete Research*, 32, 1971, 2002.

99. Cho, J. W., Ioku, K., and Goto, S., Effect of Pb^{II} and Cr^{VI} on the hydration of slag-alkaline cement and the immobilization of these heavy metal ions, *Advances in Cement Research*, 11, 111, 1999.

100. Shi, C., Wu, X., and Tang, M., Hydration of alkali-slag cements at 150°C, *Cement and Concrete Research*, 21, 91, 1991.

101. Komarneni, S. and Roy, D. M., New tobermorite cation exchangers, *Journal of Materials Science*, 20, 2930, 1985.

102. Shrivastava, O. P. and Glasser, F. P., Ion-exchange properties of $Ca_5Si_6O_{18}H_2.4H_2O$, *Journal of Materials Science Letter*, 4, 1122, 1985.

103. Komarneni, S., Roy, R., and Roy, D. M., Pseudomorphism in xonotlite and tobermorite with Co^{2+} and Ni^{2+} exchange for Ca^{2+} at 25°C, *Cement and Concrete Research*, 16, 47, 1986.

104. Komarneni, S., Braval, E., Roy, D. M., and Roy, R., Reactions of some calcium silicates with metal ions, *Cement and Concrete Research*, 18, 204, 1988.

105. Xu, H. and van Deventer, J. S. J., The geopolymerisation of alumino-silicate minerals, *International Journal of Mineral Processing*, 59, 247, 2000.

106. Comrie, D. C. and Davidovits, J., Long term durability of hazardous toxic and nuclear waste disposals, *Geopolymer '88 Proceedings*, First European Conference on Soft Mineralurgy, Compiegne, France, 1988, 125.

107. Comrie, D. C., Paterson, J. H., and Ritcey, D. J., Applications of geopolymer technology to waste stabilisation, *3rd International Conference New Frontier Hazardous Waste Management*, 1989, 161

108. Khalil, M. Y. and Merz, E., Immobilization of intermediate-level wastes in geopolymers, *Journal of Nuclear Materials*, 211, 141, 1994.

109. Davidovits, J., Properties of geopolymer cements, *Proceedings of 1st International Conference on Alkaline Cements and Concretes*, VIPOL Stock Company, Kiev, Ukraine, 1994a, 131.

110. Hermann, E., Kunze, C., Gatzweiler, R., Kießig, G., and Davidovits, J., Solidification of various radioactive residues by géopolymère with special emphasis on long-term-stability, *Géopolymère '99 Proceedings*, Saint-Quentin, France, 1999, 1.

111. Van Jaarsveld, J. G. S., van Deventer, J. S. J., and Lorenzen, L., The potential use of geopolymeric materials to immobilise toxic metals, Part I. Theory and Applications, *Minerals Engineering*, 10, 659, 1996.

112. CoCozza, E., U.S. Patent 4,049,462, Sept 20, 1977.

113. Redd, L. N., Riecka, G. P., Schwabb, A. P., Chouc, S. T., and Fan, L. T., Stabilization of phenolics in foundry waste using cementitious materials, *Journal of Hazardous Materials*, 45, 89, 1996.

5 Organic Polymers for Stabilization/ Solidification

Paul D. Kalb

CONTENTS

5.1 INTRODUCTION

Chapter 4 discussed conventional hydraulic cements used for S/S purposes. Conventional hydraulic cement-based (i.e., grout) S/S processes rely on the chemistry of cement hydration to convert the waste, usually in liquid form, to a solid monolithic product. For some types of waste, e.g., toxic metals, the hydration reaction and associated alkaline pH convert ionic metal contaminants to less-soluble hydroxide forms and thus chemically stabilize the waste. In addition, conversion of the waste into a solid monolithic form, with reduced permeability and surface area in contact with potential groundwater leachant, helps to immobilize both radioactive and hazardous constituents from the environment.

In some cases, the effectiveness of hydraulic cement-based waste forms is limited due to interactions between waste constituents and the binder materials that impede the chemistry of hydration.[1] These interactions, discussed in more detail in Chapter 7, can lead to:

- Accelerated or retarded set times or failure to set

- Free-standing water
- Low-strength final waste form products
- Poor durability

Some constituents common in radioactive, hazardous, and mixed wastes such as chlorides and nitrates are known "poisons" to the hydration reaction.[2] Higher concentrations of waste containing these constituents are more likely to result in process or performance failures. This phenomenon is exacerbated by the need to optimize waste loading efficiency, driven by concerns over treatment cost and limited disposal capacity for treated wastes.

Long-term performance in disposal is dependent on many factors and may degrade over time due to changes in the chemical and physical environment. For example, stability of contaminants is dependent on high solution pH, and while waste forms contain reserve alkalinity, exposure to neutral or acidic groundwater leachates will eventually lower the pH, resulting in higher solubility and enhanced leaching. Solidified cement grout waste forms reduce the surface area of contaminant exposed to percolating groundwater, reducing the net leach rate. However, compared with alternative S/S treatment techniques, cement-based waste forms are more porous, allowing enhanced leaching. Cement grout waste forms containing evaporator salt concentrates or ion exchange resins have been shown to swell and crack upon exposure to saturated conditions typical in shallow land disposal at many sites.[3,4] Cracking or disintegration of the monolithic waste form structure rapidly increases the available surface area exposed to leachant, with similar consequences for the mobilization of contaminants. Swelling and cracking can occur in organic polymer waste forms as well, especially in soft materials with low strength (e.g., bitumen). Polymer waste forms with higher tensile and compressive strength (e.g., polyester styrene, epoxy, polyethylene) tend to resist swelling and maintain mechanical integrity even with relatively high (e.g., 40 to 60 wt%) salt loadings.

Investigation and development of organic polymers for SS of wastes have been conducted over the past 25 years to provide improvements in waste loading efficiency and performance, as well as process economics, compared with conventional cement grout technologies. This chapter discusses some common organic polymers and their use in S/S. In general, organic polymer encapsulation technologies can be divided into two main categories: thermosetting and thermoplastic polymers.

Since organic thermosetting and thermoplastic polymers are chemically inert, they do not react chemically with inorganic or radioactive waste constituents to chemically stabilize the waste. However, with the inclusion of additives that react with waste constituents and reduce contaminant solubility, organic polymers can be considered true S/S technologies.

Encapsulation of small solid waste particles (< 60 mm) distributed homogenously throughout the organic polymer matrix is known as microencapsulation. In this case, individual waste particles are fully surrounded and encapsulated by the polymer matrix. For wastes containing larger particles (> 60 mm) such as debris or contaminated lead, clean polymer can be placed around the waste to reduce leachability in a process known as macroencapsulation. Typically, thermoplastic polymers such as high-density or low-density polyethylene are used for macroencapsulation applications. For mixed waste lead and debris, the Environmental Protection Agency (EPA) has identified macroencapsulation as the best demonstrated available technology (BDAT). Macroencapsulation may be deployed by extruding a plastic layer around consolidated waste or by hermetically sealing the waste within plastic sleeves or containers.

Organic polymers can be applied for *in situ* remediation or *ex situ* treatment of waste. However, those polymers with lower viscosity at ambient temperature, e.g., thermosetting resins, are generally more suitable for *in situ* applications than thermoplastic resins. For example, polyester styrene can be readily pumped into subsurface soil via jet grouting techniques, penetrating and filling voids. In these cases, the monomer is pumped in two separate batches; one premixed with catalyst and the other with promoter, so the polymerization reaction is triggered in a controlled fashion within the soil and no solidified product is left to solidify within process piping. The use of thermoplastic polymers for *in situ* application would require heating large masses of soil to maintain the liquid state of the polymers during injection and mixing, and thus is not practical or cost-effective. The following subsections discuss in more detail three thermosetting polymer systems used for S/S (urea-formaldehyde [UF], polybutadiene, and polyester resins) and two thermoplastics (bitumen and polyethylene).

5.2 THERMOSETS

Thermosetting polymers are formed when a liquid monomeric resin is reacted in the presence of a catalyst and promoter to form polymerized organic chains that intermesh and form a stable, monolithic solid. Once polymerized to a solid, thermosetting polymers cannot be reformed. They can be adapted for encapsulation of waste by mixing waste constituents with the monomer prior to polymerization. Thermosetting polymers can be used to treat solid waste or liquids (if mixed as an emulsion with the monomer). Thermosets do require a chemical reaction to initiate polymerization, and as such, like hydraulic cement grouts, are susceptible to interactions between the waste and the binder. However, thermosets are more prone to difficulty in the presence of organic contaminants and are less likely to be affected by inorganic constituents commonly found in radioactive wastes. Polybutadiene, and polyester are all examples of thermosetting polymers.

5.2.1 UREA-FORMALDEHYDE

UF is a solidification process based on the condensation polymerization of UF thermosetting resin that occurs upon the addition of an acidic catalyst.[5] It was a popular commercial technique marketed by several solidification vendors in the 1970s to treat both solid and aqueous low-level radioactive wastes from nuclear power plants. Liquid waste was treated by forming an emulsion in the low-viscosity UF liquid and then adding the catalyst to induce solidification.

The UF resin binder consists of an aqueous emulsion of partially polymerized monomethylol and dimethylol urea with a small amount (less than 3 wt%) of formaldehyde, which is reacted under neutral or alkaline conditions. Partial polymerization is then carried out under slightly acidic conditions, and the reaction is terminated by adjustment of the pH to 7 or 8. Final polymerization is carried out after incorporation of the waste and introduction of a weak acid such as sodium bisulfate or phosphoric acid (approximately 2 vol%) to catalyze the reaction. Typical of condensation polymers, water is produced during the polymerization reaction, which sometimes results in free-standing water in the final waste form product. In addition, interactions between the waste and binder constituents occasionally result in solidification failures. Since the catalytic reaction occurs at low pH, wastes containing highly alkaline or buffered constituents present a technical challenge. For example, the acidic by-product water sometimes leached contaminants, enhancing their release to the environment. These process difficulties ultimately led to discontinuing use of UF for treatment of low-level radioactive wastes by the mid-1970s.

5.2.2 POLYBUTADIENE

Polybutadiene is a thermosetting polymer that is widely used in the production of adhesives and sealants. It has been applied in combination with thermoplastic polyethylene in a two-part process in which polybutadiene is used to agglomerate waste particles that are then macroencapsulated into a final waste form product. The process, originally developed in the early 1970s under sponsorship of the U.S. EPA for treatment of hazardous wastes, is applicable for dry solids and sludge wastes.[6,7] Aqueous liquid wastes require pretreatment prior to processing.

Stabilization of contaminants using polybutadiene is sometimes accomplished by pretreating the waste with some additives such as lime, kiln dust, or portland cement. These additives increase the pH of the waste to reduce solubility of toxic metal constituents. In addition, if the waste contains residual excess moisture (e.g., sludge waste) it is sorbed by these reagents. Next, the waste is mixed with a small quantity of thermosetting polybutadiene (typically 10 wt% or less), which has a viscosity similar to molasses at ambient temperature, to coat the waste particles and agglomerate the mixture. The viscosity of the polybutadiene can be lowered by heating to temperatures between 50 and 93°C if needed for better mixing with the waste. It is then placed in a cylindrical mold and heated under pressure above 176°C to initiate the polymerization chain reaction and form a solid monolith. Once polymerization begins, the reaction is exothermic, reducing the heat load required. In contrast to polar functional group resins (e.g., epoxies, UF) where polymerization is initiated chemically, polymerization of polybutadiene is relatively gradual as the temperature increases. Moisture contained in the waste or sorbed by the additives is evaporated at this stage and collected by an off-gas condensate system. Several types of polybutadiene monomer are available, but 1,2 polybutadiene is the preferred variety of polymer.

Small quantities of polybutadiene used for agglomeration do not have sufficient coverage of the waste particles to provide true microencapsulation. Thus, to further reduce leachability and provide mechanical integrity for the final waste form, the agglomerated waste/polybutadiene mixture is then macroencapsulated in low- or high-density polyethylene. Polyethylene is also available in pellet or powdered form, a wide variety of melt viscosities, and either as virgin feedstock or recycled material.

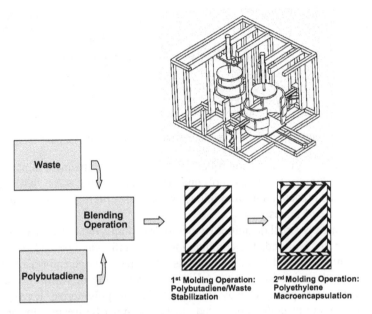

FIGURE 5.1 The conceptual design of a polybutadiene encapsulation system. Based on reference 7.

Additional details on polyethylene macroencapsulation are provided in Section 5.3.2. Rather than conventional macroencapsulation by extrusion, this process uses a jacketed clamshell heater to melt the polyethylene and form the macroencapsulated layer. The conceptual design of a polybutadiene encapsulation system is pictured in Figure 5.1. This process has yet to achieve wide-scale acceptance or use in the industry.

5.2.3 POLYESTER RESINS

Polyesters are a class of thermosetting resins that can be combined with a cross-linking agent (e.g., styrene monomer) and reacted to form a monolithic solid waste form. Unlike condensation polymerization, cross-linking of the polyester with styrene is accomplished by the breaking of double bonds, which form free radicals connecting individual chains. Several forms of polyester resin are available including orthophthalic (the largest group of polyester resins), isopthalic, vinyl ester, and water-extendible polyester.[8] Typically, a mixture of thermosetting resin and styrene monomer is reacted in the presence of a catalytic agent to form a cross-linked polymer, as shown in Figure 5.2.[9]

The catalyst is usually organic peroxide such as methyl ethyl ketone peroxide (MEKP) or benzoyl peroxide, which decompose to form free radicals that drive the cross-linking process. Decomposition of the catalyst can be accomplished by heating, but since the reaction is exothermic, it is difficult to control and is more commonly achieved by using a chemical promoter such as metallic salts (e.g., cobalt naphthenate), anilines (diethyl or dimethyl aniline), and mercaptane (i.e., dodecyl mercaptan). Conversely, premature gelling and setting of the monomers caused by heat,

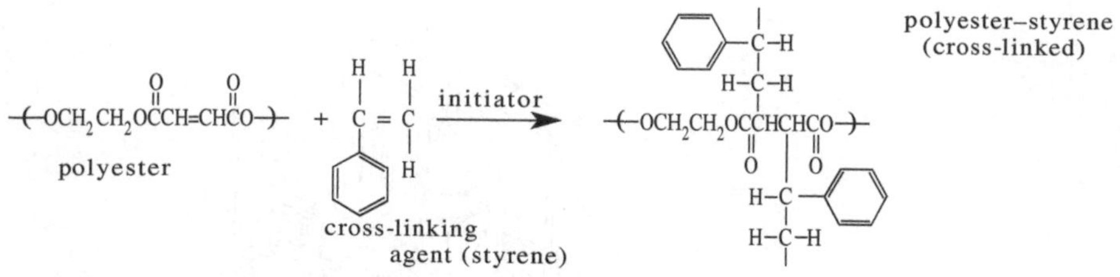

FIGURE 5.2 Polymer structure of a thermoplastic polyester styrene matrix.

light, or contamination can be controlled by the addition of chemical inhibitors (e.g., hydroquinone, tertiary butyl catechol, quaternary ammonium salts) that react with free radicals and prevent them from initiating the cross-linking process, thus extending the shelf life of resins.[10]

While relatively small quantities (typically less than a total of 1 wt%) of catalyst and promoter are required, the combination of specific catalyst and promoter, as well as the amounts and ratios of reagents used, has a large impact on the time required for polymerization and the successful processing of waste. For example, too little catalyst or promoter may result in failure to polymerize, leaving a more difficult and costly hazardous or mixed waste that must be treated. In the case of LLW, failure to solidify creates a mixed waste. Too much of the catalyst/promoter components may result in premature setting of the mixture, resulting in inhomogeneities of the waste within the final waste form or uncontrolled exotherms that can volatilize waste constituents. In the case of sludges or other wastes containing emulsified liquids, exotherms in excess of 100°C result in foaming that can overflow the container and result in a higher waste volume. Typically, polymerization to a solid monolithic final waste form occurs in 1 hour or less.

Thermosetting resins can be applied to treat dry active wastes (e.g., incinerator ash, personnel protective equipment), ion exchange resins, and debris. Water-extendible polyester resins are available that form an emulsion with aqueous solutions so that wet wastes including sludges or evaporator concentrates can be solidified. In systems where the emulsion is encapsulated within closed pores, small droplets of solution are individually surrounded by polymer.

Polyester resins are not highly viscous and can be mixed by relatively simple high-speed stirrer systems that can produce a homogeneous mixture of waste and binder material. For liquid waste applications where an emulsion is created, the mixer must deliver sufficient torque and shear to obtain and maintain the emulsion. Often, polymer encapsulation systems are designed as in-drum mixers to avoid the need to clean the process vessel and thereby saving cost and personnel exposure to contaminants. In some applications the mixing blades are left in the drum and sacrificed, rather than cleaning between runs, which requires solvents and would generate additional secondary waste.

5.2.4 SUMMARY FOR THERMOSETS

Advantages of thermosetting polymer systems include:

- Processing is usually accomplished within waste drums, eliminating the need for and maintenance of a mixing vessel. Viscosity of the thermosetting monomers is relatively low, so that standard, in-drum mixing equipment is generally sufficient to get a homogeneous mixture prior to polymerization.
- Relatively high waste loadings (typically up to 50 to 70 wt% for dry salt wastes).
- Ability to treat both dry solid waste and wet waste without pretreatment.
- Relatively low process temperatures. Since the polymerization reaction can be initiated at ambient temperature, the only heating that occurs is due to reaction exotherms that can be controlled by adjusting the formulation recipe. Typically, the peak exothermic temperature at the centerline of a 200-l waste form during curing reaches a maximum of 50°C.[11]
- Good mechanical integrity of final waste forms (compressive strengths typically ≥ 20 MPa).
- Resistance to radiation degradation (in doses up to 10^8 rad).
- Resistance to microbial degradation.
- Relatively low leachability. As in other solidification systems that rely on physical encapsulation of the waste, leachability is dependent on the amount of waste incorporated. Typically, for soluble salt wastes that are readily leachable, the ANSI/ANS-16.1-2003 Leach Index for thermosetting polymers ranges from 7.5 to 9.3 for waste forms containing 25 to 70 wt% waste, respectively. [12]
- Also amenable to *in situ* treatment of wastes. Low-viscosity monomers can be pressure grouted into soil to encapsulate wastes in place, or form impermeable barriers to contain contaminants.

Process limitations for thermosetting polymers include:

- Sensitive process chemistry and potential interactions with waste constituents
- Relatively high cost for polymers
- Inability to reprocess waste forms at a later date

Similar to conventional cement treatment systems, the most significant process limitation for thermosetting polymers is the potential for interactions between waste and binder that can retard or inhibit the polymerization reaction. Wastes containing natural or synthetic ion exchange media or carbon compounds can preferentially react or sorb the catalyst/promoter, resulting in process failures. Potential problems due to waste–binder interactions are exacerbated by the fact that process waste chemistry is not constant over time. Changes in waste chemistry that go undetected can result in process failures. Thus, operators at a treatment facility must monitor the contents of the waste on a consistent basis and repeat treatability studies once the waste chemistry varies beyond original parameters. Thermosetting polymers tend to be expensive (~\$1.20/lb), which can limit economic feasibility. However, higher binder costs compared with cement grout (~\$0.05/lb) are partially or completely offset by higher waste loading efficiencies for some types of waste. For example, one study comparing the cost of polyester encapsulation versus conventional cement process for dry nitrate salt wastes concluded that the polyester was 28% less expensive ($11.50/kg waste for the polyester system compared with $16.09/kg waste for hydraulic cement process).[9] Thus, cheaper material costs do not guarantee cheaper processing costs. Actually, total cost should be compared, if possible, including processing costs, the effects of volume change, storage costs awaiting final disposal, transport costs, and cost for final disposal. A cost comparison at Oak Ridge National Laboratory (ORNL) indicated that the processing costs for stabilizing transuranic (TRU) waste was dwarfed by the cost of storing the canisters of treated TRU waste on-site, transporting these canisters to the Waste Isolation Pilot Plant (WIPP), and disposing them in the WIPP. Unlike thermoplastic binders (e.g., bitumen and polyethylene), which can be remelted after solidification, once thermosetting resins polymerize they can no longer be processed again.

Thermosetting resins including polyester styrene and vinylester styrene have been used commercially to treat low-level radioactive waste at commercial nuclear power plants and at some DOE facilities. Due primarily to the relatively high cost of materials, the technology has not been widely adapted for waste treatment, but still maintains a niche for the treatment of some types of "problem" wastes, e.g., evaporator concentrates.

5.3 THERMOPLASTICS

The organic chains in thermoplastic polymers are not usually cross-linked; thus, they are solid at ambient temperature but can be heated, melted, and reformed upon cooling without impacting the material structure or properties. These polymers can be applied for waste encapsulation by mixing the waste with the molten polymer and allowing it to cool to a monolithic solid waste form. Since no chemical reaction is required to form the solid, thermoplastics are not susceptible to the failures associated with cement grout and thermosetting polymers. Maximum waste loading is limited by processability and the performance requirements of the final waste form. For example, above maximum loading limits, the viscosity of the waste–binder mixture precludes homogeneous mixing. Likewise, with very high waste loadings, waste particles may not be fully encapsulated, leading to interconnected pores and higher leachability.[13] Processing temperatures for organic polymers are higher than 100°C, so aqueous wastes cannot usually be processed directly due to product foaming as moisture evaporates. Some process techniques are available (e.g., kinetic mixing, vented extrusion, wiped film evaporation) to alleviate or minimize these effects, but if the waste contains significant concentrations of moisture (e.g., > 5 wt%), pretreatment to remove residual moisture is often the most efficient approach.

5.3.1 BITUMEN

Bitumen is a high-molecular weight thermoplastic organic polymer consisting of a mixture of organic solids and organic liquid or oil that is derived from the distillation of petroleum. Its main applications are for asphalt highway paving and roofing, but it has a number of attractive properties for waste solidification, including:

- Low permeability to water
- Resistance to acids, alkalines, and salts
- Low viscosity at relatively low process temperatures that allow good mixing with solid waste particles

Bitumen was first introduced for treatment of radioactive waste in Belgium in the early 1960s and is still in use today. Bitumen is available in several grades based on the method of production, such as:

- Distilled bitumen; softening point 34 to 65°C
- Oxidized or blown bitumen (air is pumped through petroleum during manufacturing); softening point 70 to 140°C
- Cracked bitumen (pyrogenic breakdown of heavy molecules); softening point 77 to 85°C
- Emulsified bitumen (emulsification of bitumen in soapy water)[14]

Direct distillation or oxidized bitumens are most commonly used for waste solidification processes. Composition and physical properties can vary considerably from batch to batch, but typically bitumen specified for waste encapsulation has a density of 1.01 g/cm^3 at ambient temperature, a softening point of 90°C, and a flash point of > 290°C.[5] Bitumen has been used to treat a broad range of low-level radioactive wastes including sludge, evaporator concentrates, incinerator ash, solvents, and debris. It has been deployed in Belgium, France, England, Poland, Russia and other former Soviet countries, Korea, and Japan. Several bitumen systems were installed in the U.S., but due to poor efficiency and safety issues, they are not commonly used.

Several methods have been applied for bituminization of radioactive waste. One method uses preheated molten bitumen in a wiped film evaporator, in which injected liquid wastes are vaporized. The resulting particulate salts and other waste solids are then mixed with the bitumen and discharged to a drum to cool and solidify. The steam is condensed and captured, along with any other volatiles and carryover particulates. A typical wiped film evaporator system installed at the waste treatment facility at the Korean Atomic Energy Research Institute (KAERI) is shown in Figure 5.3. A second method uses a twin-screw extruder, similar to systems used in the plastics industry. Solid or liquid waste and molten bitumen are fed into the extruder and are mixed continuously and conveyed by the action of the intermeshing screws. The heated barrel keeps the bitumen molten and, in the case of aqueous waste, vaporizes the liquid, which is then drawn off of the mixture through several vacuum ports and condensed. The condensate is recycled through the process as long as it remains radioactive. Like the thin film evaporator, the combined mixture of waste solids and bitumen is discharged to a drum for cooling and solidification. A typical extrusion process for bitumen encapsulation is shown in Figure 5.4.[12] A production-scale twin-screw bitumen processing system was installed at the Palisades Nuclear

FIGURE 5.3 Bitumen wiped film evaporator system installed within a hot cell at the Korean Atomic Energy Research Institute.

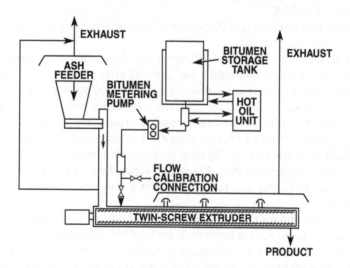

FIGURE 5.4 Process flow diagram for bitumen encapsulation twin-screw extrusion process.

Power Station in Michigan to treat aqueous borated salts in the early 1980s, but was abandoned due to poor process efficiency. Advantages of the bituminization process include:

- Relatively low cost of binder material.
- Relatively low process temperatures compared with vitrification processes. Bitumen is processed at 150 to 230°C, whereas vitrification requires temperatures in excess of 800°C.
- Relatively low leachability due to the impermeable nature of the binder. The Leachability Index for ^{137}Cs contained in soluble salts incorporated in bitumen, determined using the ANSI/ANS-16.1-2003 leach test, was

13, compared with 7.3 for cement waste forms.[12] This represents an improvement of almost six orders of magnitude in the effective diffusion coefficient over cement. Irradiation to 10^8 rad and thermal cycling of bitumen waste forms can increase leachability by up to 1.5 orders of magnitude, but releases are still low with respect to cement waste forms that do not use special sorptive agents to retard ^{137}Cs release.

Process limitations for bituminization of wastes include:

* Flammability
* Poor thermal conductivity
* Relatively low compressive and tensile strength
* Potential radiation damage of the organic matrix that can limit the quantity/activity of the treated waste

Fires and explosions in bitumen processing systems are relatively common and have significantly limited the application of this technology worldwide. This problem is exacerbated by the presence of nitrate and other oxidizing agents in the waste, which can lower the flash point of bitumen and accelerate combustion.[14] For example, a fire and ensuing explosion occurred at the Low-Level Radioactive Waste Bituminization Facility installed at the Tokai Works Nuclear Fuel Reprocessing Plant in Japan, where nitrate-based aqueous concentrates were being encapsulated in bitumen.[15] Poor thermal conductivity requires large heat transfer surfaces and limits the throughput of the process.

The semisolid nature of bitumen results in low strengths and creeping of the material under stress or unconfined compression. It will slowly flow over time under its own weight to the lowest contained point, hence, the use of aggregate or gravel for asphalt to provide structural stability while the asphalt fills the interstitial space to bind and seal the composite. Low compressive strength (0.9 MPa, or 130 psi)[12] is indicative of poor mechanical integrity and can impact long-term durability. Thus, secondary containment may be required for acceptable performance under landfill disposal conditions.[16] Low tensile strength results in significant swelling of solidified salt waste concentrates when subjected to saturated conditions as the salts rehydrate. In testing at Brookhaven National Laboratory (BNL), bitumen waste forms containing soluble sulfate and borate salts swelled to almost ten times their original size following long-term immersion in water. The swelling results from the osmotic pressure exerted as water diffuses through the organic to the salt crystals, first dissolving then diluting the salt as concentrated pockets of salty solution grow around the encapsulated soluble salts. The bitumen creeps under the force of this osmotic pressure, allowing these pockets to grow in size. The driving force is the difference in salt concentration between the bulk immersion solution and these pockets. In general, the organic layer thins, then breaks, allowing a physical pathway to equalize concentrations/pressure, before the concentrations can equilibrate by diffusion through the organic to slow and stop the growth of these pockets.

Radiation damage in bitumen has been shown to be quite low in total doses up to the U.S. Nuclear Regulatory Commission-recommended 10^8 rad, but increases significantly at a dose of 10^9 rad. For example, in tests at BNL, the total gas produced as a result of irradiation (G value given in mol/100eV) was ten times lower than cement for doses up to 10^8 rad (0.029 compared with 0.24 for cement), but more than 2.5 times higher than cement at 10^9 rad (0.43 compared with 0.16 for cement).[5]

5.3.2 POLYETHYLENE

Polyethylene is a thermoplastic polymer that melts to a viscous molten liquid, which can be applied for either microencapsulation or macroencapsulation of waste. If mixed with small particles of waste to form a homogeneous molten mixture and then cooled back to a solid, the waste particles are microencapsulated within the polymer matrix. For larger particles (> 60 mm) such as lead or other debris, polyethylene can be used to form an impermeable macroencapsulation envelope around the waste particles, thereby minimizing leaching in the disposal environment.

Polyethylene is produced by polymerization of ethylene gas, and the structure of the plastic can be tailored to yield products with a wide variety of physical and mechanical properties. Unlike thermosetting polymers, the polymer chains in thermoplastic polymers like polyethylene are not normally cross-linked. It is available in two distinct forms, low-density polyethylene (LDPE) and high-density polyethylene (HDPE). LDPE, with densities ranging between 0.910 and 0.925 g/cm^3, is formed by creating a high degree of chain branching, which effectively keeps layers of polymer apart and reduces the density. HDPE, with densities ranging between 0.941 and 0.959 g/cm^3, is produced from long polymer chains with relatively little branching so that layers pack more tightly and thus increase the density. Each type is available in a wide variety of melt viscosities (specified as melt index), defined by ASTM D-1238 as the quantity of material that can flow through a given orifice at specified temperatures in units of g/10 min.[17] Higher melt indices reflect lower melt viscosity plastics (i.e., flow more readily) and vice versa. LDPE has a lower melt temperature (120°C) than HDPE (180°C) and lower melt viscosities (typically 1 to 55 g/10 min) and thus is more widely used for waste encapsulation.

Typically, polyethylene is processed by extrusion, in which the material is fed through a heated barrel by either a single helical screw or twin screws meshed together. For polyethylene microencapsulation, waste and binder are metered together using volumetric or loss-in-weight feeders to maintain a constant ratio of materials. For macroencapsulation, clear plastic is extruded around compacted waste to form a low-permeability barrier. A production-scale single-screw extruder system for polyethylene microencapsulation is shown in Figure 5.5. Alternatively, polyethylene can be processed by kinetic mixer, in which the materials are mixed in a chamber by a high-speed blade, and the frictional heat that is created melts the plastic. A kinetic mixer system for polyethylene microencapsulation processing is shown in Figure 5.6.

Although polymers in general, and polyethylene in particular, are relatively newly engineered materials, they can be expected to withstand harsh chemical environments over long periods of time under anticipated disposal conditions, based on favorable results in short-term performance testing. For example, compressive strength for polyethylene-encapsulated waste forms can range between 7.03 MPa

FIGURE 5.5 Production-scale single-screw extrusion process system installed at Brookhaven National Laboratory.

and 16.3 MPa (1020 to 2360 psi), but typically is in excess of 13.8 MPa (2000 psi).[18–20] No loss in mechanical integrity was observed following thermal cycling in accordance with ASTM B-553 or immersion in water for 90 days. Irradiation of polyethylene waste forms at a dose of 3.6×10^6 rad/h to a total dose of 10^8 rad resulted in an increase in compressive strength of about 20%, due to cross-linking of the polymer chains, which occurs on exposure to high radiation doses. Polyethylene is known to be resistant to biodegradation due to its branched structure, relatively high molecular weight, and hydrophobic properties.[21,22] Testing of waste form samples according to ASTM G-21 and ASTM G-22 resulted in no fungal or bacterial growth.[18,23,24] Polyethylene is insoluble in virtually all organic solvents and is resistant to many acids and alkaline solutions.[25]

Leachability of polyethylene-encapsulated waste forms has been extensively examined. In general, releases from polyethylene are controlled by diffusion, and its low permeability results in relatively low leach rates. Leaching of soluble salts (e.g., nitrates) and other wastes from polyethylene is inversely proportional to the quantity of waste encapsulated. In one study, the diffusion coefficient ranged from 3×10^{-9} cm^2/sec for waste forms containing 50 dry wt% salts to 5×10^{-8} cm^2/sec for wastes containing 70 dry wt% salts.[19] Using the ASTM D-1308 Accelerated Leach Test, diffusion was determined to be the predominant leaching mechanism for microencapsulated polyethylene waste forms.[26,27] Based on these laboratory results, 200-l (55-gal) LDPE waste forms containing 50 to 70 dry wt% nitrate salt waste were projected to leach 3.7 to 9.5 wt% of the original contaminants at ambient temperature in 300 years. In contrast, similar cement grout waste forms were projected to release 17 wt% of the contaminants in just 11 years, approximately 25 times more rapidly.[28] In another study, the ANSI/ANS-16.1-2003 Leachability Index

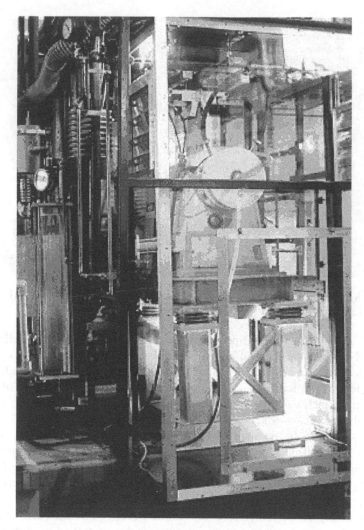

FIGURE 5.6 Kinetic mixer for polyethylene microencapsulation process.

ranged from 11.1 for waste forms containing 30 dry wt% to 7.8 for waste forms containing 70 dry wt% nitrate salt.[18]

In laboratory treatability studies, a wide variety of hazardous and mixed wastes have been treated by polyethylene microencapsulation including evaporator concentrates, incinerator ash, blowdown solution, carbonate and mixed salts, ion exchange resins, sodium-bearing wastes, molten salt oxidation residuals, and sludges. Results varied based on waste loading and toxic metal (see Table 5.1). For example, TCLP leaching of lead was reduced by factors ranging from 750 to 500,000 to concentrations well below the characteristic limit for lead (5 mg/L) and in seven of eight studies was below the allowable concentrations for the Universal Treatment Standard, 0.75 mg/L.[29] Similar results were obtained for leaching of cadmium, chromium, selenium, and arsenic.[29]

TABLE 5.1
Toxicity Characteristic Leaching Procedure Data for Polyethylene-Microencapsulated Waste Forms

Waste Stream	Waste Loading, wt%	Toxic Metal Source Term Concentration[a] (Microencapsulated Waste Form TCLP Concentration, mg/L)					
		Pb	Cr	Cd	Hg	Se	As
DOE Mixed Salts	60	3000 (0.07)	3000 (0.10)	3000 (0.37)			
DOE Incinerator Ash	60	5000 (0.01)	5000 (0.01)	5000 (0.01)			
SRS CIF Blowdown	40	2250 (< 0.05)	500 (0.07)	125 (< 0.05)	250 (< 0.04)		
INEEL Carbonate Salt	50	120 (< 0.14)					
INEEL Ion Exchange Resin	40	1200 (1.6)					
INEEL Sodium-Bearing Waste	40	218 (< 0.14)	351 (2.4)	246 (0.34)	242 (< 0.0002)		
Commercial Incinerator Ash	50					15 (< 0.15)	
Molten Salt Oxidation Residuals	50		2400 (1.6)				
Fernald Silo 1 Sludge	50	(0.712)					
Fernald Silo 3 Sludge	60	(0.058)	(0.029)	(0.002)		(0.088)	(0.245)
Maximum Allowable Concentration, TCLP		5.0	5.0	1.0	0.2	1.0	5.0
Maximum Allowable Concentration, UTS[b]		0.75	0.60	0.11	0.025	5.7	5.0

[a] Total concentration in the undiluted, untreated waste, i.e., source term; the concentration in TCLP extracts of the polyethylene-microencapsulated waste is given in parentheses below the total concentration.
[b] EPA is phasing in new Universal Treatment Standards (UTS) with lower allowable concentrations for toxic metals.

Polyethylene encapsulation shares several advantages with thermosetting resins including:

- Relatively high waste loadings (typically up to 50 to 70 wt% for dry salt wastes). Waste loading may be limited by either physical processing constraints or final waste form performance.
- Relatively low process temperatures minimize generation of secondary waste through collection in the off-gas system. Polyethylene melts at 120°C and is generally processed at approximately 135 to 140°C.
- Compatibility with a wide range of waste types.

- Resistance to microbial degradation.
- Good mechanical integrity of final waste forms (compressive strengths typically ≥ 14 MPa).
- Resistance to radiation degradation (in doses up to 10^8 rad). Compressive strength of polyethylene waste forms improves with radiation doses up to 10^8 rad.
- Relatively low leachability. Leach indices of 8 to 11, depending on type of waste and waste loading.

In addition, other advantages associated with polyethylene encapsulation include:

- One process amenable for both micro- and macroencapsulation. Both technologies can share similar process equipment, so one facility can be equipped for processing both micro- or macroencapsulated waste forms.
- Physical encapsulation process with no chemical reactions required for solidification; i.e., not subject to interactions with waste chemistry and solidification of waste forms is assured.
- Ability to remelt and reform processed waste forms. Waste forms can be melted and recast if needed based on disposal performance and requirements.
- Lower cost of polymer matrix (e.g., $0.50/lb for virgin polyethylene).
- Ability to use recycled plastics. Industrial and consumer recycling programs generate large volumes of plastic materials to reduce disposal loads on municipal landfills and provide feedstock for polyethylene products. However, many manufacturers cannot use recycled feedstock due to product variability issues such as coloration. Availability of these waste products to encapsulate hazardous, radioactive, and mixed wastes provides an ideal market for recycled plastics.

Several vendors, offering services for the treatment of mixed waste lead and debris to both DOE and commercial clients, have successfully commercialized polyethylene macroencapsulation. Each offers a different processing technique. The standard and most commonly used method involves compacting waste within an inner container, placing it within a larger overpack, and then extruding clean polyethylene to fill the space between containers. Millions of pounds of mixed waste lead from across the DOE complex have been treated using this technique. Macroencapsulation can also be conducted by inserting compacted waste into precast polyethylene cylinders and then "welding" plastic caps on each end to make the final seal. This method was recently used to treat metal debris from the Y-12 Plant in Oak Ridge, TN. A third method involves placing the waste in preengineered plastic boxes or containers and then sealing the lid using imbedded electric resistance wires. Polyethylene microencapsulation using both single-screw extrusion and kinetic mixer-based processing has been commercialized, but has not been used to treat large quantities of waste, to date.

REFERENCES

1. Connor, J.R. *Chemical Fixation and Solidification of Hazardous Wastes*, Van Nostrand Publishers, New York, 1990.
2. Lea, F.M. *The Chemistry of Cement and Concrete, Third Edition*, Chemical Publishing Co., Inc., New York, 1971.
3. Neilson, R.M., Jr. and Colombo, P. *Waste Form Development Program Annual Progress Report*, BNL-51614, Brookhaven National Laboratory, Upton, NY, September 1982.
4. Kalb, P.D. and Colombo, P. *Full-Scale Leaching of Commercial Reactor Waste Forms, Final Report*, BNL-35561, Brookhaven National Laboratory, Upton, NY, September 1984.
5. Colombo, P. and Neilson, R.M., Jr. *Properties of Radioactive Wastes and Containers, First Topical Report*, BNL-NUREG 50957, Brookhaven National Laboratory, Upton, NY, August 1979.
6. Lubowitz, H.R., Berham, R.L., Ryan, L.E., and Zakrzewsi, G.A. *Development of a Polymeric Cementing and Encapsulating Process for Managing Hazardous Wastes*, EPA 600/2-77-045,TRW System Group, August 1977.
7. Mechsner, M.J. Process and Equipment Design for Polymer Stabilization of Mixed Low-Level Wastes with No Secondary Containment Required; presented at WM 97 Symposium, Tucson, AZ, February 1997.
8. Biyani, R.K., Agamuthu, P., and Mahalingam, R. Polyester Resin Microencapsulation, in *Hazardous and Radioactive Waste Treatment Technology*, C.H. Oh, Ed., CRC Press, Boca Raton, FL, 2001, Chapter 6.4.
9. Franz, E.M., Heiser, J.H., and Colombo, P. *Immobilization of Sodium Nitrate Wastes with Polymers*, BNL-52081, Brookhaven National Laboratory, Upton, NY, April 1987.
10. Lawrence, J.R. *Polyester Resins, Plastics Applications Series*, Van Nostrand Reinhold Co., New York, 1960.
11. Haighton, A.P. *The Solidification of Ion Exchange Wastes, Part 6, Characterization of a Vinyl Ester Binder*, NW/SSD/RR/75/80, Central Electricity Generating Board, United Kingdom, November 1980.
12. Westick, J.H., Jr. et al. *Characterization of Cement and Bitumen Waste Forms Containing Simulated Low-Level Waste Incinerator Ash*, NUREG/CR-3798 (PNL-5153), Pacific Northwest Laboratory, Richland, WA, August 1984.
13. Fuhrmann, M. and Zhou, H. *Applicability of an Accelerated Leach Test to Different Waste Form Materials*, presented at Spectrum 94, Seattle, WA, August 1994.
14. Dlouhy, Z. *Disposal of Radioactive Wastes, Studies in Environmental Science 15*, Elsevier Scientific Publishing, Amsterdam, 1982.
15. Hasegawa, K. and Li, Y. *Explosion of Asphalt-Salt Mixtures in a Reprocessing Plant*, presented at the AIChE 1999 National Meeting, Houston, TX, 1994.
16. Shin, H.S., Shon, J.S., Yim, S.P., and Kim, K.J. Effect of Additives on the Mechanical Stability of Bitumen-Based Waste Form, *Journal of Environmental Science and Health*, 1998; A33(3): 477–493.
17. American Society of Testing Materials (ASTM), *Standard Test Method for Flow Rates of Thermoplastics by Extrusion Plastometer*, ASTM D-1238-90b, Philadelphia, December 1990.
18. Kalb, P.D., Heiser J.H., and Colombo, P. *Polyethylene Encapsulation of Nitrate Salt Wastes: Waste Form Stability, Process Scale-up, and Economics*, BNL-52293, Brookhaven National Laboratory, Upton, NY, July 1991.

19. Kalb, P.D. and Fuhrmann, M. *Polyethylene Encapsulation of Single-Shell Tank Low-Level Wastes, Annual Progress Report*, BNL-52365, Brookhaven National Laboratory, Upton, NY, September 1992.

20. Lageraaen, P., Patel, B.R., Kalb, P.D., and Adams, J.W., *Treatability Studies for Polyethylene Encapsulation of INEL Low-Level Mixed Wastes*, BNL-62620, Brookhaven National Laboratory, Upton, NY, October 1995.

21. Rodriguez, F. The Prospects for Biodegradable Plastics, *Chemical Technology*, p. 409, July 1971.

22. Barua, P.K. et al. Comparative Utilization of Paraffins by a Trichosporon Species, *Applied Microbiology*, pp. 657–661, November 1970.

23. American Society of Testing Materials (ASTM). *Standard Practice for Determining Resistance to Synthetic Polymeric Materials to Fungi*, ASTM G-21, Philadelphia.

24. American Society of Testing Materials (ASTM) *Standard Practice for Determining Resistance of Plastics to Bacteria*, ASTM G-22, Philadelphia.

25. Raff, R.A.V. and Allison, J.B. *Polyethylene*, Interscience Publishers, New York, 1956.

26. Fuhrmann, M. and Kalb, P.D., Leaching Behavior of Polyethylene Encapsulated Nitrate Waste, in *Stabilization and Solidification of Hazardous, Radioactive and Mixed Wastes*, ASTM Special Technical Publication 1240, T.M. Gilliam and C.C. Wiles, Eds., American Society of Testing and Materials, Philadelphia, 1993.

27. Fuhrmann, M., Heiser, J.H., Pietrzak, R., Franz, E., and Colombo, P. *Users' Guide for the Accelerated Leach Test Computer Program*, BNL-52267, Brookhaven National Laboratory, Upton, NY, November 1990.

28. Fuhrmann, M. et al. *Optimization of the Factors that Accelerate Leaching*, BNL-52204, Brookhaven National Laboratory, Upton, NY, March 1989.

29. Kalb, P.D. Polyethylene Encapsulation, in *Hazardous and Radioactive Waste Treatment Technologies Handbook*, C.H. Oh, Ed., CRC Press, Boca Raton, FL, 2001.

6 Other Binders

Robert C. Moore, Arun Wagh, Paul D. Kalb,
Gerald W. Veazey, Earl W. McDaniel, and
Darryl D. Siemer

Chapter 6 is organized differently than the other chapters. It consists of five sections, each discussing a different matrix: (1) phosphate, (2) Ceramicrete, (3) sulfur polymer cement, (4) gypsum, and (5) hydroceramics. Each section has its own author(s) and references. Thus, they are "mini-chapters."

CONTENTS

6.1   PHOSPHATE STABILIZATION AND IMMOBILIZATION OF HEAVY METAL AND RADIONUCLIDE CONTAMINANTS

Robert C. Moore

Phosphate treatment has been evaluated for many applications including immobilization of contaminants in contaminated soils, stabilization of incineration and mine wastes, and as a waste form for heavy metal and mixed waste. As phosphates, certain contaminants have been shown to be thermodynamically stable and insoluble over most conditions encountered in the environment.[1-4] Contaminants amenable to phosphate treatment include lead, cadmium, zinc, copper, uranium, neptunium, plutonium, and europium. The solubility of these contaminants as phosphates is substantially less than in other forms. For example, the solubility of lead phosphate is more than 40 orders of magnitude lower than lead as a carbonate, oxide, or sulfate. Table 6.1.1 gives a list of metal phosphate solids along with their aqueous solubility.

In addition to decreased solubility, phosphate treatment decreases the bioavailability of contaminants.[7-11] Phosphate has been demonstrated to significantly reduce the bioavailability of cadmium, lead, copper, zinc, and arsenic.[12,13] Insoluble forms of lead, including phosphates, in lead mine waste may have resulted in unexpected low lead levels in the blood of children living in proximity.[9]

Dissolved, liquid forms of phosphate are used when fast kinetics are desired, while solid forms slowly release and provide a constant source of phosphate.[12,14] Dissolved forms include phosphoric acid and aqueous solutions of soluble phosphate salts; solid forms include natural phosphate rock and calcium apatite. The solubilization of the contaminant is often the rate-controlling step in phosphate treatment. Low pH favors faster kinetics because of the increased solubility of metal compounds.

Figure 6.1.1 illustrates different methods of implementing *in situ* treatment. Applying a solid or soluble phosphate compound to the surface or mixing the

---

TABLE 6.1.1
Aqueous Solubility of Metal Phosphate Compounds

Metal	Solid	Reaction	K_{sp} [a, b]
Pb	Hydroxypyromorphite	$Pb_{10}(PO_4)_6(OH)_2$	−76.8
	Chloropyromorphite	$Pb_{10}(PO_4)_6(Cl)_2$	−85.4
	Fluoropyromorphite	$Pb_{10}(PO_4)_6(F)_2$	−71.6
Ca	Hydroxyapatite	$Ca_{10}(PO_4)_6(OH)_2$	−55.9
	Fluorapatite	$Ca_{10}(PO_4)_6(F)_2$	−110.2
U	Autunite	$Ca(UO_2)_2(PO_4)_2 \cdot 10H_2O$	−50
	Chernikovite	$(H_3O)_2(UO_2)(PO_4)_2 \cdot 6H_2O$	−23
Sr	Strontium phosphate	$Sr_5(PO_4)_3(OH)$	−51.3
Cd	Cadmium phosphate	$Cd_3(PO_4)_2$	−32.6

[a] For a solid, $X_aY_bZ_c$, the solubility reaction is given by: $X_aY_bZ_c \leftrightarrow aX + bY + cZ$. The solubility product constant K_{sp} is defined as: $K_{sp} = [X]^a[Y]^b[Z]^c$.
[b] Data from References 1–7.

---

phosphate into the surface can treat the surface of a contaminated site. In a similar manner, the subsurface can also be treated by injection of phosphate into the contaminated soil. Groundwater contaminants can also be treated by either phosphate injection or by use of a permeable barrier. *In situ* stabilization has several advantages over conventional technologies, including 1) the ability to treat small and very large areas; 2) minimal exposure of workers to the contaminants; 3) low implementation and operational cost at depths of 5 to 50 m or more; 4) minimal required site excavation; and 5) no transportation or disposal cost for the contaminated soil. Two novel *in situ* stabilization techniques using soluble forms of phosphate have been proposed.[15,16]

The secondary minerals in the oxidized zones of lead ore deposits, as assemblages around ore bodies, in soil, sediments, and phosphate beds prove the stability of naturally occurring metal phosphates.[4,17] Studies indicate metal phosphate compounds can be extremely stable and remain unchanged for thousands to hundreds of thousands of years in different environmental settings.[2,3,18–22]

The most commonly investigated materials have been phosphoric acid, natural phosphate rock, bone char, and apatite. Treatment studies are also reported using organophosphorous, monocalcium phosphate, calcium oxyphosphate, and triple superphosphate. Phosphate fertilizers are readily available, inexpensive materials produced in very large quantities throughout the world. Phosphate rock is mined as a raw material for fertilizer production and as such is readily available and inexpensive. Bone charcoal is a mixture of activated carbon and apatite prepared by the calcination of animal bone in the absence of oxygen and is an inexpensive, widely available material used for the refinement of raw sugar.

Hydroxyapatite, $Ca_{10}(PO_4)_6(OH)_2$, is a calcium phosphate mineral that exhibits a high stability under reducing and oxidizing conditions and very low water solubility ($K_{sp} < 10^{-48}$) under alkaline conditions. Many substitutions can be made in the apatite

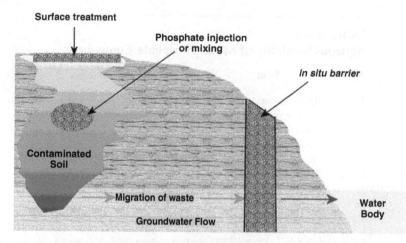

FIGURE 6.1.1 Phosphate treatment of contaminated soil using three different methods: surface treatment, subsurface treatment, and as an *in situ* permeable reactive or sorptive barrier.

structure, including Sr^{2+}, Na^+, and Mg^{2+} for Ca^{2+}; CO_3^{2-} and HPO_4^{2-} for PO_4^{3-}; and F^-, Cl^- and CO_3^{2-} for OH^-. High-purity hydroxyapatite can be produced synthetically by precipitation from oversaturated solutions of calcium and phosphate,[23] from high-temperature solid-state reactions,[24] or from calcination of animal bones.[25,26] Figure 6.1.2 presents transmission electron microscopy (TEM) images of synthetic hydroxyapatite and hydroxyapatite prepared by calcination of cattle bone at 500 to 1100°C. As calcination temperature increases from 500 to 1100°C, the average crystal size of the material increases from approximately 0.20 μm to 0.50 μm and the surface area decreases from 101 m²/g to 2.2 m²/g, indicating a large internal pore space in the material treated at 500°C. For the bone treated at 1100°C, small crystals of MgO can be seen on the surface (Figure 6.1.2f). Mg is a component of bone and is expelled from the apatite crystal at high temperatures and forms MgO.

A compilation of the available literature studies for heavy metal and radionuclide sorption, stabilization, and immobilization using phosphates is given in Table 6.1.2. The list includes laboratory studies, field trials, and reports on remediation of contaminated sites.

In conclusion, it has been well demonstrated through laboratory experiments, field trials, and remediation activities that phosphate treatment for stabilization and immobilization of heavy metal and radionuclide wastes is effective and economical. Phosphate treatment significantly reduces the solubility and bioavailability of contaminants in soil, groundwater, and wastes. Studies and applications of phosphate treatment have mainly centered on immobilization of contaminants in soil and removal of contaminants in groundwater. *In situ* treatment of contaminated soil is

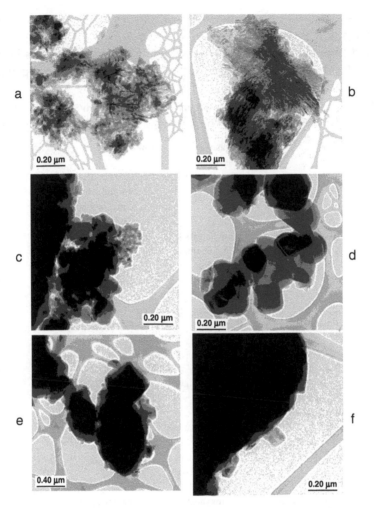

FIGURE 6.1.2 Transmission electron microscopy images: a. synthetic hydroxyapatite, and hydroxyapatites produced from cattle bone treated at b. 500°C, c. 700°C, d. 900°C, e. 1100°C, and f. small MgO crysals, hydroxyapatite from cattle bone treated at 1100°C.

a relatively new and very promising technology and economical alternative to conventional site excavation and disposal of contaminated soil in a waste repository. An examination of natural systems indicates phosphate treatment can be used to immobilize heavy metal and radionuclide contaminants for thousands to hundreds of thousands of years. There are inconsistencies in the reported mechanism of metal sorption and immobilization by phosphates. More work is needed in this area.

TABLE 6.1.2
Selected Treatment Studies for Heavy Metals and Radionuclides Using Phosphates; Includes Laboratory Studies, Field Trials, and Applications

Metal/ Metalloid Studied	Media/ Conditions	Phosphate Treatment	Mechanism/Solid Phases Formed	References
Actinides	Soil	Organo- phosphorous	Not determined	27
As, Cd, Pb, Zn	Soil	Phosphate rock	Replacement of As by P from soil binding sites increased mobility of As uptake by the plant	28a
As, Cd, Pb, Zn	Contaminated smelter soil	Phosphate rock, diammonium phosphate	Not determined	29
Cd	Soil	Monobasic calcium phosphate	Not determined	30
Cd	Aqueous solution	Bone char	Substitution of Cd for Ca and physical adsorption of Cd	31
Cd	Aqueous solution	Synthetic hydroxyapatite	Substitution of Cd for Ca through diffusion and ion exchange	32
Cd	Aqueous solution	Synthetic hydroxyapatite	Not determined	33
Cd	Aqueous solution	Synthetic hydroxyapatite prepared in the presence of Cd	Substitution of Cd for Ca	34
Cd	Aqueous solution; pH 3 and 9	Hydroxyapatite	Cd sorbed to apatite surface	35
Cd	Aqueous solution; pH 4.5 to 5	Synthetic hydroxyapatite	Substitution of Cd for Ca	36
Cd, Cu, Pb, Zn	Aqueous solution; pH 5 to 6	Synthetic hydroxyapatite	Substitution of Cd for Ca	32
Cd, Cu, Pb, Zn	Combustion ash from municipal solid waste incineration	Phosphoric acid	Formed calcium phosphates, tertiary metal phosphates, and apatite family minerals depending on the pH of the experiment	37
Cd, Cu, Pb, Zn	Municipal solid incineration waste	Not reported	Not determined	38

(continued)

TABLE 6.1.2 (CONTINUED)
Selected Treatment Studies for Heavy Metals and Radionuclides Using Phosphates; Includes Laboratory Studies, Field Trials, and Applications

Metal/ Metalloid Studied	Media/ Conditions	Phosphate Treatment	Mechanism/Solid Phases Formed	References
Cd, Cu, Pb, Zn	Municipal solid waste combustion dry scrubber residue; pH 12	Phosphoric acid	Formation of apatite, tertiary metal phosphates, and calcium phosphates with incorporated metals	15
Cd, Pb, U	Aqueous solution	Synthetic hydroxyapatite	Amorphous uranium phase formed; formation of hydroxypyromorphite and lead oxide; possible substitution of Pb for Ca; Cd reported substitution for Ca	39
Cd, Pb, Zn	Aqueous solution; pH range 1 to 12	North Carolina mineral apatite	Hydrocerussite $(Pb_3(CO_3)_2(OH)_2)$ and lead oxide fluoride (Pb_2OF_2) formed under alkaline conditions and pyromorphites under acidic condition; otavite $(CdCO_3)$ calcium hydroxide $(Cd(OH)_2)$, and zinc oxide (ZnO) in alkaline conditions	40
Cd, Pb, Zn	Soils and mine tailings; pH 5.5 to 8	Calcium oxyphosphate	Not determined	41
Cd, Se	Aqueous solution	Synthetic hydroxyapatite	Penetration in the intra-particle porosity generated by aggregation of Cd and Se crystals	42
Cd, Zn	Aqueous solution; pH 5.5 to 7.5	Synthetic hydroxyapatite	Sorption by metal complexation with surface functional groups and coprecipitation	43
Cr, Mn	Aqueous solution	Apatite from thermally treated fish bone	Not determined	44a, 44b

(continued)

TABLE 6.1.2 (CONTINUED)
Selected Treatment Studies for Heavy Metals and Radionuclides Using Phosphates; Includes Laboratory Studies, Field Trials, and Applications

Metal/ Metalloid Studied	Media/ Conditions	Phosphate Treatment	Mechanism/Solid Phases Formed	References
Cr, Cd, Hg, Ni, Pb, low-level mixed waste	Aqueous solution, solids	Phosphate ceramic waste form (magnesium potassium phosphate)	Macroencapsulation	45
Cu, Zn	Aqueous solution; pH 5	Bone char	Ion exchange and physical adsorption	31
Cu, Pb, Zn	Soil	Phosphoric acid, dicalcium phosphate, phosphate rock	Pyromorphite-like mineral, no phosphate mineral of Zn and Cu detected	28b
Eu, Ce	Aqueous solution	Synthetic hydroxyapatite	Substitution of Eu for Ca	46
Ni U	Aqueous solution	Synthetic hydroxyapatite	Not determined	47a
Ni	Soil	Synthetic hydroxyapatite	U and Ni associated with Al rich precipitates	48
Np	Aqueous solution	Synthetic hydroxyapatite	Not determined	16b
Pb	Aqueous solution	Meat and bone meal combustion residue	Surface complexation and precipitation	50
Pb	Aqueous solution	Synthetic hydroxyapatite	Initial formation of a solid solution of $Pb_{(10-x)}Ca_x(PO_4)_6(OH)_2$ converting to pure hydroxypyromorphite over time	51
Pb	Mine tailings	Phosphoric acid and lime	Tertiary metal phosphates, e.g., $(Cu, Ca_2)(PO_4)_2$, apatite, e.g., $Pb_5(PO_4)_3Cl$	52
Pb	Soil	Phosphoric acid	$Pb_5(PO_4)_3OH$, $Pb_5(PO_4)_3Cl$	10, 16, 53–61

(continued)

TABLE 6.1.2 (CONTINUED)
Selected Treatment Studies for Heavy Metals and Radionuclides Using
Phosphates; Includes Laboratory Studies, Field Trials, and Applications

Metal/ Metalloid Studied	Media/ Conditions	Phosphate Treatment	Mechanism/Solid Phases Formed	References
Pb	Soil, pH 2.7 to 7.1	Bonemeal	Formation of Pb and Zn phosphates	62
Pb	Soil	Triple super-phosphate, phosphate rock	Formation of pyromorphite-like minerals	63
Pb, Zn	Soil	Phosphoric acid	Not determined	64
Pb, Zn, Cd	Mine wastes	Apatite II™ — commercial apatite mineral product derived from fish bones	Not determined	65
Pu	Aqueous solution; pH 5 to 8	Synthetic hydroxyapatite	Theorized to be either ion exchange or physical adsorption	66
Sb	Aqueous solution; pH 6 to 11	Synthetic hydroxyapatite	Not determined	14
Se		Synthetic hydroxy-apatite, natural fluorapatite	Substitution of Se species for phosphate in hydroxyapatite	67
Se	0.1M KNO_3 solution, pH 5 to 11.5	Synthetic hydroxyapatite	Exchange of SeO_3^{2-} for phosphate	68
Sr	Aqueous solution	Synthetic hydroxyapatite	Incorporation of Sr into the apatite structure; exchange for Ca	69
Sr	Aqueous solution	Synthetic hydroxyapatite	Substitution of Sr for Ca	70
Th	Aqueous solution	Synthetic hydroxy-apatite, natural apatites and phosphate rock	Sr substitution for Ca; isolated cluster of UO_2 in the apatite structure	71
Tc, As	Aqueous solution	Synthetic hydroxyapatite with $SnCl_2$	Reduction and sorption	14

(continued)

TABLE 6.1.2 (CONTINUED)
Selected Treatment Studies for Heavy Metals and Radionuclides Using Phosphates; Includes Laboratory Studies, Field Trials, and Applications

Metal/ Metalloid Studied	Media/ Conditions	Phosphate Treatment	Mechanism/Solid Phases Formed	References
U, Sr	Synthetic acidic uranium mill effluent	Synthetic hydroxyapatite	Sorption and coprecipitation	72
U	0.1 M KNO$_3$ aqueous solution	Synthetic hydroxy-apatite, cattle bone hydroxyapatite prepared at 500, 700, 900, and 1100°C	Not determined	73
U	Aqueous solution	Natural fluorapatite	Not determined	74
U	Aqueous solution	Synthetic hydroxyapatite	At U sorbed < 4700 ppm formed as inner-sphere complex; at U sorbed > 7000 ppm formation of chernikovite	75
U	Aqueous solution, solids	Bone charcoal and bonemeal	Surface complex for U sorbed < 5500 ug U(VI)/g and at U sorbed > 5500 ug U(VI)/g chernikovite formation	76
U	Aqueous solution	Synthetic fluorapatite	Substitution of U(VI) for Ca	77
U, Ce	Soil	Synthetic hydroxyapatite	Not determined	47b
U, Pb	Aqueous solution; synthetic groundwater	Apatite II™ — commercial apatite mineral product derived from fish bones	For uranium formation of autunite; not determined for Ce	78
U, Pb, Pu	Aqueous solution	Apatite II — commercial apatite mineral product derived from fish bones	Formation of U and Pb phosphates	79

(continued)

TABLE 6.1.2 (CONTINUED)
Selected Treatment Studies for Heavy Metals and Radionuclides Using Phosphates; Includes Laboratory Studies, Field Trials, and Applications

Metal/ Metalloid Studied	Media/ Conditions	Phosphate Treatment	Mechanism/Solid Phases Formed	References
U, Pb, and Pu	Aqueous solution and soil	Apatite II™ — commercial apatite mineral product derived from fish bones	Formation of U, Pb, and Pu phosphates and other low-solubility phases	80
U, Pu, Am, As, Se, Tc	Aqueous solution	Natural and synthetic apatite, bone char	Not determined	81
Zn	Aqueous solution; pH 4.8 to 9.5	Natural fluorapatite	Substitution of Zn for Ca	82

REFERENCES

1. Nriagu, J.O. (1972). Lead orthophosphates. 1. Solubility and hydrolysis of secondary lead orthophosphate. *Inorg. Chem.* 11(10):2499–2503.
2. Nriagu, J.O. (1973). "Lead orthophosphates. 2. Stability of chloropyromorphite at 25 degrees C." *Geochim. Cosmochim. Acta* 37(3):367–377.
3a. Nriagu, J.O. (1974). "Lead orthophosphates. 4. Formation and stability environment." *Geochim. Cosmochim. Acta* 38(6):887–898.
3b. Nriagu, J.O. (1973). "Lead orthophosphates. 3. Stabilities of fluoropyromorphite and bromopyromorphite at 25 degrees C." *Geochim. Cosmochim. Acta* 37(7):1735–1743.
4. Nriagu, J.O. (1984). "Lead and lead-poisoning in antiquity — response." *Sci. Total Environ.* 37(2–3):267–268.
5. Grenthe, I., chairman (1992). *Chemical Thermodynamics of Uranium,* North-Holland, Amsterdam.
6. Link, W.F. and A. Seidell. (1979). *Solubilities of Inorganic and Organic Compounds,* Pergamon Press, New York.
7. Casteel, S.W., R.P. Cowart, C.P. Weis, G.M. Henningsen, E. Hoffman, W.J. Brattin, R.E. Guzman, M.F. Starost, J.T. Payne, S.L. Stockham, S.V. Becker, J.W. Drexler, and J.R. Turk (1997). "Bioavailability of lead to juvenile swine dosed with soil from the Smuggler Mountain NPL site of Aspen, Colorado." *Fundam. Appl. Toxicol.* 36(2):177–187.
8. Ruby, M.V., A. Davis, R. Schoof, S. Eberle, and C.M. Sellstone (1996). "Estimation of lead and arsenic bioavailability using a physiologically based extraction test." *Environ. Sci. Technol.* 30(2):422–430.
9. Davis, A., J.W. Drexler, M.V. Ruby, and A. Nicholson (1993). "Micromineralogy of mine wastes in relation to lead bioavailability, Butte, Montana." *Environ. Sci. Technol.* 27(7):1415–1425.

10. Yang, J., D.E. Mosby, S.W. Casteel, and R.W. Blanchar (2002). "In vitro lead bio-accessibility and phosphate leaching as effected by surface application of phosphoric acid in lead-contaminated soil." *Arch. Environ. Contam. Toxicol.* 43(4):399–405.

11. Bubb, J.M. and J.N. Lester (1991). "The impact of heavy-metals on lowland rivers and the implications for man and the environment." *Sci. Total Envrion.* 100:207–233.

12. Laperche, V., S.J. Traina, P. Gaddam, and T.J. Logan (1996). "Chemical and miner-alogical characterizations of Pb in a contaminated soil: reactions with synthetic apatite." *Environ. Sci. Technol.* 30(11):3321–3326.

13. Hettiarachchi, G.M., G.M. Pierzynski, and M.D. Ransom (2001). "In situ stabilization of lead using phosphorus." *J. Environ. Qual.* 30(4):1214–1221.

14. Zhang, P., J. A. Ryan, and J. Yang (1998). "In vitro soil Pb solubility in the presence of hydroxyapatite." *Environ. Sci. Technol.* 32:2763–2768.

15. Nash, K.L., M.P. Jensen, and M.A. Schmidt (1998). "Actinide immobilization in the subsurface environment by in-situ treatment with a hydrolytically unstable organo-phosphorus complexant: uranyl uptake by calcium phytate." *J. Alloys Compd.*

16a. Moore, R.C. (2003). "In situ formation of apatite for sequestering radionuclides and heavy metals." U.S. Patent 6,592,294 B1. Issued Jul. 15, 2003.

16b. Moore, R.C., K. Holt, H. Zhao, A. Hasan, N. Awwad, M. Gasser, and C. Sanchez (2003). "Sorption of Np(V) by synthetic hydroxyapatite." *Radiochim. Acta* 91:721–727.

16c. Moore, R.C., K.C. Holt, H. Zhao, A. Hasan, M. Hasan, R. Bontchev, F. Salas, and D. Lucero (2003). Anionic Sorbents for Arsenic and Technetium Species. SAND203–3360.

17. Eighmy, T.T., B.S. Crannell, L.G. Butler, F.K. Carteledge, E.F. Emery, D. Oblas, J.E. Krzanowski, J.D. Eudsen, E.L. Shaw, and C.A. Francis (1997). "Heavy metal stabi-lization in municipal solid waste combustion dry scrubber residue using soluble phosphate." *Environ. Sci. Technol.* 31:3330–3338.

18. Ruby, M. V., A. Davis, and A. Nicholson (1994). "In situ formation of lead phosphates in soils as a method to immobilize lead." *Environ. Sci. Technol.* 26:646–654.

19. Lippolt, H.J., M. Leitz, R.S. Wernicke, and B. Hagedorn (1994). "(Uranium plus thorium) helium dating of apatite — experience with samples from different geochem-ical environments." *Chem. Geol.* 112(1–2):179–191.

20. Scholten, L.C. and C.W.M. Timmermans (1996). "Natural radioactivity in phosphate fertilizers." *Fertilizer Res.* 43(1–3):103–107.

21. Panczer, G., M. Gaft, R. Reisfeld, S. Shoval, G. Boulon, and B. Champagnon (1998). "Luminescence of uranium in natural apatites." *J. Alloys Compd.* 277:269–272.

22. Jerden, J.L., A.K. Sinha, and L. Zelanzy (2003). "Natural immobilization of uranium by phosphate mineralization in an oxidizing saprolite-soil profile: chemical weather-ing of the Coles Hill uranium deposit, Virginia." *Chem. Geol.* 199(1–2):129–157.

23. Andronescu, E., E. Stefan, E. Dinu, and C. Ghitulica (2002). "Hydroxyapatite syn-thesis." *Key Eng. Mater.* 206–2:1595–1598.

24. Papargyris, A.D., A.I. Botis, and S.A. Papargyri (2002). "Synthetic routes for hydroxyapatite powder production." *Key Eng. Mater.* 206–2:83–86.

25. Kanno, T., Y. Motogami, M. Kobayashi, and T. Akazawa (1998). "Difference of carbonate ions incorporated into a cattle bone-originated and a chemically synthesized hydroxyapatites." *J. Ceram. Soc. Jpn.* 106(4) 432–434.

26. Wang, D.Z., J.H. Liu, and Q.W. Chen (2001). "Study on cuttlebone modified to hydroxyapatite and its nanoscale microstructure." *Rare Metal Mater. Eng.* 30:470–474.

27. Jensen, M.P., K.L. Nash, L.R. Morss, E.H. Appelman, and M.A. Schmidt (1996). "Immobilization of actinides in geomedia by phosphate precipitation." *ACS Symp. Ser.* 651:272–285.

28a. Cao, X.D., L.Q. Ma, and A. Shiralipour (2003). "Effects of compost and phosphate amendments on arsenic mobility in soils and arsenic uptake by the hyperaccumulator, Pteris vittata L." *Environ. Pollu.* 126(2):157–167.

28b. Cao, R.X., L.Q. Ma, M. Chen, S.P. Singh, and W.G. Harris (2003). "Phosphate-induced metal immobilization in a contaminated site." *Environ. Pollu.* 122:19–28.

29. Basta, N.T. and S.L. McGowen (2004). "Evaluation of chemical immobilization treatments for reducing heavy metal transport in a smelter-contaminated soil." *Environ. Pollu.* 127:73–82.

30. Theodoratos, P., N. Papassiopi, and Xenidis (2002). "Evaluation of monobasic calcium phosphate for the immobilization of heavy metals in contaminated soils from Lavrion." *J. Hazard. Mater.* B94:135–146.

31a. Cheung, C.W., C.K. Chan, J.F. Porter, and G. McKay (2001). "Combined diffusion model for the sorption of cadmium, copper, and zinc ions onto bone char." *Environ. Sci. Technol.* 35(7):1511–1522.

31b. Cheung, C.W., C.K. Chan, J.F. Porter, and G. McKay (2001). "Film-pore diffusion control for the batch sorption of cadmium ions from effluent onto bone char." *J. Colloid Interface Sci.* 234(2):328–336.

32. Mandjiny, S., K.A. Matis, A.I. Zouboulis, M. Fedoroff, J. Jeanjean, J.C. Rouchaud, N. Toulhoat, V. Potocek, P. Maireles-Torres, and D. Jones (1998). "Calcium hydroxyapatites: evaluation of sorption properties for cadmium ions in aqueous solution." *J. Mater. Sci.* 33:5433–5439.

33. Middelburg, J.J. and R.N.J. Comans (1991). "Sorption of cadmium on hydroxyapatite" *Chem. Geol.* 90(1–2):45–53.

34. Nounah, A., J. Szilagyi, and J.L. Lacout (1990). "Calcium-cadmium substitution in hydroxyapatites." *Annales de Chimie-Science Des Materiaux* 15(7–8):409–419.

35. Kozai, N, T. Ohnuki, S. Komareni, T. Kamiya, T. Sakai, M. Oikawa, and T. Satoh (2003). "Uptake of cadmium by synthetic mica and apatite: observation by micro-PIXE." *Nuclear Instruments & Methods in Physics Research Section B-Beam Interactions with Materials and Atoms* 210:513–518.

36. McGrellis, S., J.N. Serafini, J. JeanJean, J.L. Pastol, and M. Fedoroff (2001). "Influence of the sorption protocol on the uptake of cadmium ions in calcium hydroxyapatite." *Sep. Purif. Technol.* 24(1–2):129–138.

37. Crannell, B.S., T.T. Eighmy, J.E. Krzanowski, J.D. Eusden, E.L. Shaw, and C.A. Francis (2000). "Heavy metal stabilization in municipal solid waste combustion bottom ash using soluble phosphate." *Waste Manage.* 20(2–3):135–148.

38. Mizutani, S., H.A. van der Sloot, and S. Sakai (2000). "Evaluation of gas cleaning residues from MSWI with chemical agents." *Waste Manage.* 20(2–3):233–240.

39. Jeanjean, J., J.C. Rochaud, L. Tran, and M. Fedoroff (1995). "Sorption of uranium and other heavy metals on hydroxyapatite." *J. Radioanal. Nucl. Chem.* 201(6):529–539.

40. Chen, X.B., J.V. Wright, J.L. Conca, and L.M. Peurrung (1997). "Evaluation of heavy metal remediation using mineral apatite." *Water Air Soil Pollut.* 98(1–2):57–58.

41. Xenidis, A., C. Stouraiti, and I. Paspaliaris (1999). "Stabilisation of highly polluted soils and tailings using phosphates." Global Symposium on Recycling, Waste Treatment and Clean Technology, San Sebastian, Spain, Minerals, Metals & Materials Soc.

42. Badillo-Almarez, V.E., N. Toulhoat, P. Trocellier, and M. Jullien (2003). "Application of microanalytical techniques to the study of aqueous ion sorption phenomena on mineral surfaces." *Radiochim. Acta* 91(8):487–493.

43. Xu, Y., F. Schwartz, and S.J. Traina. (1994). "Sorption of Zn^{2+} and Cd^{2+} on hydroxyapatite surface." *Environ. Sci. Technol.* 28(8):1472–1480.

44a. Ozawa, M., K. Satake, and R. Suzuki (2003). "Removal of aqueous chromium by fish bone waste originated hydroxyapatite." *J. Mater. Sci. Lett.* 22(7):513–514.

44b. Ozawa, M., K. Satake, and S. Suzuki (2003). "Removal of aqueous manganese using fish bone hydroxyapatite." *J. Mater. Sci. Lett.* 22(19):1363–1364.

45. Singh, D., A.S. Wagh, M. Tlustochowicz, and S.Y. Jeong (1998). "Phosphate ceramic process for macroencapsulation and stabilization of low-level debris wastes." *Waste Manage.* 18:135–143.

46. Bidoglio, G., P.N. Gibson, E. Haltier, and N. Omenetto (1992). "Xanes and laser fluorescence spectroscopy for rare-earth speciation at mineral-water interfaces." *Radiochim. Acta* 58–59 pt. 1:191–197.

47a. Seaman, J.C, J.S. Arey, and P.M. Bertsch (2001). "Immobilization of nickel and other metals in contaminated sediments by hydroxyapatite addition." *J. Environ. Qual.* 30(2):460.

47b. Seaman, J.C., T. Meehan, and P.M. Bertsch (2001). "Immobilization of cesium-137 and uranium in contaminated sediments using soil amendments." *J. Environ. Qual.* 30(4):1206–1213.

48. Seaman, J.C., J.M. Hutchison, B.P. Jackson, and V.M. Vulva (2003). "In situ treatment of contaminated soils with phytate." *J. Environ. Qual.* 32(1):153–161.

49. Sugiyama, H., T. Watanabe, and T. Hiarayama (2001). "Nitration of pyrene in metallic oxides as soil components in the presence of indoor air, nitrogen, dioxide gas, nitrate ion, or nitrate ion under zenon irridation." *J. Health Sci.* 47(1):28–35.

50. Deydier, E., R. Guilet, and P. Sharrock (2003). "Beneficial use of meat and bone meal combustion residue: 'an efficient low cost material to remove lead from aqueous effluent'." *J. Haz. Mater.* 101(1):55–64.

51. Mavropoulos, E., A.M. Rossi, A.M. Costa, C.A.C. Perez, J.C. Moriera, and M. Saldanha (2002). "Studies on the mechanisms of lead immobilization by hydroxyapatite." *Environ. Sci. Technol.* 36:1625–1629.

52. Eusden, J.J. Jr., D., L. Gallagher, T.T. Eighmy, B.S. Crannell, J.R. Krzanowski, L.G. Butler, F.K. Carteledge, E.F. Emery, E.L. Shaw, and C.A. Francis (2002). "Petrographic and spectroscopic characterization of phosphate-stabilized mine tailings from Leadville, Colorado." *Waste Manage.* 22:117–135.

53. Ma, Q.Y., T.J. Logan, S.J. Traina, and J.A. Ryan (1994). "Effects of aqueous Al, Cd, Fe(II), Ni, and Zn on Pb immobilization by hydroxyapatite" *Environ. Sci. Technol.* 28(7):1219–1228.

54. Ma, Q.Y., T.J. Logan, and S.J. Traina (1995). "Lead immobilization from aqueous solutions and contaminated soils using phosphate rocks." *Environ. Sci. Technol.* 29(4):1118–1126.

55. Ma, L.Q., and G.N. Rao (1997). "Effects of phosphate rock on sequential chemical extraction of lead in contaminated soils." *J. Environ. Qual.* 26(3):788–794.

56. Rao, A.J., K.R. Pagilla, and A.S. Wagh (2000). "Stabilization and solidification of metal-laden wastes by compaction and magnesium phosphate-based binder." *J. Air Waste Manage. Assoc.* 50(9):1623–1631.

57. Boisson, J., M. Mench, J. Vangronsveld, A. Ruttens, P. Kopponen, and T. De Koe (1999). "Immobilization of trace metals and arsenic by different soil additives: evaluation by means of chemical extraction." *Commu. Soil Sci. Plant Anal.* **30**(3–4):365–387.

58. Melamed, R., X. Cao, M. Chen, and L.Q. Ma (2003). "Field assessment of lead immobilization in a contaminated soil after phosphate application." *Sci. Total Environ.* 305: 117–127.

59. Yang, J. et al. (2000). "Field treatment of phosphoric acid for in situ lead immobilization in soil." *Abstracts of Papers of the American Chemical Society*, March 26, 2000, v 219, pt 1, U732–U732.

60. Stanforth, R. and J. Qiu (2001). "Effect of phosphate treatment on the solubility of lead in contaminated soil." *Environ. Geol.* 41(1–2):1–10.

61. Yang, J. et al. (2001). "Lead immobilization using phosphoric acid in a smelter-contaminated urban soil." *Environ. Sci. Technol.* 35(17):3553–3559.

62. Hodson, M.E., E. Valsami-Jones, J.D. and Cotter-Howells (2000). "Bonemeal additions as a remediation treatment for metal contaminated soil." *Environ. Sci. Technol.* 34(16): 3501–3507.

63. Hettiarachchi, G.M., G.M. Pierzynski, and M.D. Ransom (2000). "In situ stabilization of soil lead using phosphorus and manganese oxide." *Environ. Sci. Technol.* 34(21): 4614–4619.

64. CotterHowells, J. and S. Caporn (1996). "Remediation of contaminated land by formation of heavy metal phosphates." *Appl. Geochem.* 11(1–2):335–342.

65. Williams, B.C., S.L. McGeehan, and N. Ceto (2000). "Metals-fixation demonstrations on the Coeur d'Alene River, Idaho — Final results." 7th International Conference on Tailings and Mine Waste.

66. Leyva-Ramos, R., L.A. Bernal-Jacome, R.M. Guerrero-Coronado, and L. Fuentes-Rubio (2001). "Competitive adsorption of Cd(II) and Zn(II) from aqueous solution onto activated carbon." *Sep. Sci. Technol.* 36(16):3673–3687.

67. Duc, M., G. Lefevre, J. Jeanjean, J.C. Rouchaud, F. Monteil-Rivera, J. Dumonceau, and S. Milonjic (2003). "Sorption of selenium anionic species on apatites and iron oxides from aqueous solutions." *J. Environ. Radioact.* 70(1–2):61–72.

68. Monteil-Rivera, F., M. Fedoroff, J. Jeanjean, L. Minel, M. Barthes, and J. Dumonceau (2000). "Sorption of selenite (SeO_3^{2-}) on hydroxyapatite: an exchange process." *J. Colloid Interface Sci.* 221(2):291–300.

69. Vukovic, Z., S. Lazic, I. Tutunovic, and S. Raicevic (1998). "On the mechanism of strontium incorporation into calcium phosphates." *J. Serbian Chem. Soc.* 63(5):387–393.

70. Lazic, S. and Z. Vukovic (1991). "Ion-exchange of strontium on synthetic hydroxyapatite." *J. Radioanal. Nucl. Chem.* 149(1):161–168.

71. Raicevic, S., I. Plecas, D.I. Lalovic, and V. Veljkovic (1999). "Optimization of immobilization of strontium and uranium by the solid matrix." *Mater. Res. Soc. Symp. Proc.* 556:135–142.

72. Landa, E.R., A.H. Le, R.L. Luck, and P.J. Yeich (1995). "Sorption and coprecipitation of trace concentrations of thorium with various minerals under conditions simulating an acid uranium mill effluent environment." *Inorg. Chim. Acta* 229:247–252.

73. Hasan, M.A., C.A. Sanchez, R.C. Moore, A. Hasan, K. Holt, T. Headley, H. Zhao, and F. Salas (2003). "Containment of uranium in the proposed Egyptian geologic repository for radioactive waste using hydroxyapatite." Sandia National Laboratories, SAND2003-3104C.

74. Ordonez-Regil, E., E.T.R. Guzman, and E.O. Regil (1999). "Surface modification in natural fluorapatite after uranyl solution treatment." *J. Radioanal. Nucl. Chem.* 240(2):541–545.

75. Fuller, C.C., J.R. Bargar, J.A. Davis, and M.J. Piana (2002). "Mechanisms of uranium interactions with hydroxyapatite: implications for groundwater remediation." *Environ. Sci. Technol.* 36:158–165.

76. Fuller, C.C., J.R. Bargar, and J.A. Davis (2003). "Molecular-scale characterization of uranium by bone apatite materials for a permeable reactive barrier demonstration." *Environ. Sci. Technol.* 37(20):4642–4649.

77. Rakovan, J., R.J. Reeder, E.J. Elzinga, D.J. Cherniak, C.D. Tait, and D.E. Morris (2002). "Structural characterization of U(VI) in apatite by X-ray absorption spectroscopy." *Environ. Sci. Technol.* 36(14):3114–3117.

78. Bostick, W.D., R.J. Stevenson, R.J. Jarabek, and J.L. Conca (1999). "Use of apatite and bone char for the removal of soluble radionuclides in authentic and simulated DOE groundwater (Reprinted from *Advances Environmental Research*, vol. 3, pg 488–498, 2000)." *Adv. Environ. Res.* 3(4):U9.

79. Lu, N.P., J.L. Conca, G. Parker, B. Moore, A. Adams, J. Wright, and P. Heller (2000). "PIMS remediation of metal contaminated waters and soils." *Applied Mineralogy,* Vols 1 and 2: Research, Economy, Technology, Ecology and Culture. Eds. D. Rammlmair, J. Mederer, T. Oberthur, R.B. Heimann, and H. Pentinghaus. Rotterdam, Netherlands p. 603–605.

80. Conca, J.L., N.P. Lu, G. Parker, B. Moore, A. Adams, J. Wright, and P. Heller (2000). "PIMS remediation of metal contaminated waters and soils." 2nd International Conference on Remediation of Chlorinated and Recalcitrant Compounds.

81. Thomson, B.M., C.L. Smith, R.D. Busch, M.D. Siegel, and C. Baldwin (2003). "Removal of metals and radionuclides using apatite and other natural sorbents." *J. Environ. Eng.* 129(6):492–499.

82. Brigatti, M.F., D. Malferrari, L. Medici, and L. Poppi (2002). "Reactions of Zn2+ aqueous solutions with fluor-hydroxyapatite: crystallographic evidence." *Neus Jahrbuch fur mineral ogie-monatshefte* (3):129–137.

6.2 CERAMICRETE: AN ALTERNATIVE RADIOACTIVE WASTE FORM

Arun Wagh

6.2.1 INTRODUCTION

Cement is a familiar material widely used in the construction industry and has been successful in stabilizing many wastes.[1] Conventional cement systems may only accommodate a low loading of the waste, resulting in large volume increases and high transportation and disposal costs. In addition, cement interacts with and is not always compatible with the waste, for example, acids. Ceramicrete is a rapid-setting phosphate grout system that comes close to producing ceramics by chemical bonding at room temperature and exhibits advantages of both vitrification and cement grout without their major drawbacks. Ceramicrete technology is based on a form of chemically bonded phosphate ceramic conceived and developed at Argonne National Laboratory (ANL) and holds promise as an alternative to vitrification and cement grouts. It can treat a wide range of radioactive waste streams at room temperature. Studies have demonstrated that Ceramicrete waste forms were comparable to vitrified waste forms in performance and can incorporate high loading of a wide variety of waste streams. This section provides an overview of the Ceramicrete technology, briefly discusses the solution chemistry behind chemical immobilization of radio-active waste streams, and explores the mechanism of physical encapsulation that isolates the contaminants from the environment.

6.2.2 THE CERAMICRETE PROCESS AND ITS CHEMISTRY

The Ceramicrete process is based on an acid–base reaction between calcined magnesium oxide (MgO) and a solution of potassium dihydrogen phosphate (KH_2PO_4). Details of the calcination procedure of MgO have been published.[2] The dissolution reactions of KH_2PO_4 and MgO are given by the following equations:[3]

$$KH_2PO_4 = 2H^+ + KPO_4^{2-} \qquad (1)$$

$$MgO + 2H^+ = Mg^{2+}(aq) + H_2O \qquad (2)$$

The reaction product is magnesium potassium phosphate ($MgKPO_4.6H_2O$) that is formed by dissolution of MgO in the solution of KH_2PO_4 and eventual reaction in the solution given by:

$$Mg^{2+}(aq) + KPO_4^{2-} + 6H_2O = MgKPO_4.6H_2O \qquad (3)$$

$MgKPO_4.6H_2O$ (referred to as MKP hereafter) acts as a binder that can be used as the matrix material to host any inorganic and to some extent even some organic waste material. This is an exothermic reaction and releases a large amount of heat.

6.2.3 MECHANISMS OF IMMOBILIZATION

Radioactive waste streams are mixed with the binder powders and water, and the reaction is allowed to occur by mixing the components for 30 min in a concrete mixer. The resulting slurry forms a smooth paste that can be poured into the storage containers and allowed to set. Once the mixing is done, exothermic reaction between the components starts heating the slurry. With this reaction the slurry thickens, and when the temperature reaches 55°C the entire slurry sets into a hard mass. Even in the hardened monolith, the exothermic reaction proceeds to produce additional heat. Typically a maximum temperature of 60°C in 2-l-size samples prescribed for storage of transuranic (TRU) waste streams, and 82°C in 55-gal-size monoliths used for waste streams where there are no concerns of criticality, have been noted, as shown in Figure 6.2.1. The samples attain sufficient structural integrity to be transported in approximately 2 hours, but actual curing continues for several weeks. Waste loadings of 40 to 80 wt% are typical, depending on the characteristics of a given waste stream. Liquids and sludge are easily incorporated because the process involves aqueous mixing.

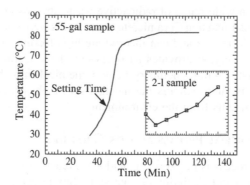

FIGURE 6.2.1 Variation of temperature during mixing of the Ceramicrete slurry at the center of the 55-gal drum. The inset shows the same in a 2-l-size container.

The binders and the amount of water needed for stabilization are poured into the container, and a motorized paddle is lowered into the container to mix the contents. Once the mixing is completed, the paddle is either removed or sacrificed with the container. Mixers at the 55-gal scale have been designed and built for this purpose (Figure 6.2.2). The slurry is then allowed to set or harden, and the container is closed for final disposition.

FIGURE 6.2.2 Photograph of a 55-gal scale stabilization of soil using Ceramicrete.

The following three mechanisms govern the immobilization of radioactive and hazardous contaminants in a stabilized Ceramicrete waste form: (1) chemical stabilization, (2) microencapsulation, and (3) macroencapsulation. The first two mechanisms work together to immobilize chemical constituents, while the third one is used to physically encapsulate large objects. Chemical stabilization results from conversion of contaminants in the waste to insoluble phosphate forms. This conversion depends on the dissolution kinetics of these components. In general, if these components are in a soluble or even in a sparsely soluble form, they will dissolve in the initially acidic Ceramicrete slurry and react with the phosphate anions in the same manner that MgO reacts. The resultant product will be an acid-phosphate that is insoluble in groundwater. On the other hand, if a certain radioactive component is not soluble in the acid slurry, it is already immobilized and will generally not dissolve in groundwater, because such components are less soluble at the higher pH of groundwater. Thus, the solubility of hazardous and radioactive components is key to chemical immobilization as phosphates.

MKP is a dense material that coats the individual particles of both the contaminants and the fillers in the waste and forms an impermeable cover to particles. This cover protects each particle from contact with groundwater and the surrounding media. In addition, MKP consolidates loose particles into a leach-resistant monolithic waste form.

6.2.4 S/S of Hazardous and Radioactive Wastes with Ceramicrete

Ceramicrete technology has been used to stabilize/solidify a range of waste streams (both simulated and actual) containing hazardous and radioactive contaminants.[2,4-8] Table 6.2.1 lists waste acceptance criteria and corresponding studies demonstrating compliance by Ceramicrete waste forms with those criteria. They are briefly described in the subsections that follow.

6.2.4.1 Stabilization of Hazardous Contaminants

The effectiveness of the Ceramicrete matrix to stabilize ash was demonstrated by simulating Pu-contaminated ash from the Rocky Flats site of DOE.[6] The simulated waste contained oxides of aluminum, calcium, iron, magnesium, potassium, and silicon, with silicon oxide comprising 41 wt% of the waste. The hazardous contaminants were Ba, Cr, Ni, and Pb, added as oxides at a concentration of several thousand parts per million. CeO_2 was added as a surrogate of PuO_2 and formed 11.2 wt% of the waste. In addition, carbon content was about 20 wt% of the waste.

The ash loading in the waste form was 54 wt%, dictated by a safeguard limit of 5 wt% loading of Pu (simulated by Ce in the test) in the final waste form. The density of the waste forms was 1.84 kg/l, open porosity was 4.4 vol%, and compressive strength was 60.0 MPa (8700 psi). The toxicity characteristic leaching procedure (TCLP) results in Table 6.2.2 list contaminant levels in the leachate well below the Universal Treatment Standard (UTS) limit, indicating effective Ceramicrete stabilization of the ash.

TABLE 6.2.1
Waste Acceptance Criteria and Examples of Ceramicrete Waste Forms that Meet the Criteria

Criterion	Example Selected in this Article
Leaching of contaminants	1. Hazardous metals: Wastewater and contaminated soil, simulated ash waste, Pb-lined gloves, Hg-contaminated bulbs
	2. Fission products:
	Tc: technetium partitioned from high-level waste tanks, debris from contaminated pipes from K-25 Gaseous Diffusion plant at Oak Ridge
	Cs: from salt supernate and sludge, silico titanates, and
	3. wastewater
	Radioactive components:
	Ra: Fernald silo waste
	Transuranics: simulated and actual Rocky Flats ash waste, wastewater
Leaching of salts	Simulated salt waste streams (both supernate and sludge)
Leaching of matrix components	Simulated salt waste streams (both supernate and sludge)
Physical properties	All of the above
Ignitability	Simulated salt waste
Radiolyses	U-Pu alloy, Pu-contaminated ash
Pyrophoricity	Oxidation of Ce_2O_3 to CeO_2

TABLE 6.2.2
Toxicity Characteristic Leaching Levels for Surrogate Waste Form (mg/L)

Element	Ba	Cr	Ni	Pb
Level in waste (ppm)	1077	5360	4890	8600
Leaching level (mg/L)	0.68	0.01	< 0.05	< 0.10
UTS limit (mg/L)	1.2	0.86	5.00	0.37

Another study stabilized soil contaminated with hazardous contaminants and low-level radioactive wastewater into a single waste form.[9] The wastewater provided the stoichiometric amount of water needed in the Ceramicrete slurry for the stabilization process. The total waste loading was 77 wt%, including the wastewater. The waste forms had an open porosity of 2.7 vol% and a density of 2.17 kg/l. Compressive strength was 33.9 MPa (4910 psi). Both TCLP and ANSI/ANS-16.1-2003 leaching test results on the waste forms indicate leach resistance well within regulatory requirements and a high leachability index (see Table 6.2.3). The Hg TCLP leachate concentration was 0.0015 mg/L, less than one tenth of the UTS limit of 0.025 mg/L.

TABLE 6.2.3

Levels of Contaminants in Soil and

Wastewater and Results of Leaching Tests

Contaminant	Cd	Cr	Pb	Hg
In soil (ppm)	1044	1310	2457	1002
TCLP result (mg/L)	0.18	0.13	< 0.2	0.0015
Leaching index	ND*	ND	16.4	16.6

* ND: Not detected, below detection limit.

For Ceramicrete, the waste is pretreated with sodium or potassium sulfide to stabilize Hg.[5]

6.2.4.2 Stabilization of Fission Products

The stabilization of some fission products such as ^{99}Tc and ^{137}Cs has been a challenge, since these contaminants are found mostly in salt waste streams and are readily soluble in groundwater. Several laboratory studies demonstrated that Ceramicrete technology can be used to effectively stabilize such fission products.

Technetium is highly mobile in its soluble Tc^{+7} oxidation state. Also, because it is volatile at moderately elevated temperatures, waste streams containing technetium must be stabilized at ambient temperatures. Ceramicrete stabilization of ^{99}Tc partitioned from high-level tank wastes was demonstrated by Singh et al.[10] The waste stream was a product of a complexation–elution process that separates ^{99}Tc from High-level waste (HLW) such as supernate from salt waste tanks at Hanford and Savannah River. Adding $SnCl_2$ to the binder mixture and then slurrying with water reduced technetium from its +7 to +5 valence state. The concentration of ^{99}Tc in the waste form varied from 40 to 900 ppm. The slurry was mixed for 20 min, set into a hard ceramic waste form, and cured three weeks before being leach tested. The ANSI/ANS-16.1-2003 leachability index on technetium has been consistently between 13.3 and 14.6, higher than the index for cesium and strontium from glass.[11] The waste forms had a compressive strength of about 30 MPa and proved durable in an aqueous environment. These results demonstrated the effectiveness of a reducing environment in the Ceramicrete matrix for contaminants that tend to oxidize and leach out.

In another study, debris waste, produced from scraping the internal surface of pipes from the K-25 Gaseous Diffusion plant at Oak Ridge destined for demolition, was stabilized. The main contaminants were technetium and uranium with technetium concentrations of 33,886 and 1,750 pCi/g (2020 and 104 ppm) in the two samples tested. Again $SnCl_2$ was used as the reductant to suitably stabilize technetium. The technetium leachability indices from the ANSI/ANS-16.1-2003 test were found to be 12 and 17.7, consistent with values obtained for the other waste streams described above. These examples illustrate the effectiveness of Ceramicrete stabilization of technetium by adding the reductant, $SnCl_2$.

Four studies have been reported on the stabilization of cesium using Ceramicrete technology. Cesium was added either as $CsNO_3$ or $CsCl$ in the first three tests, while the nature of cesium was not known for the fourth case but appeared to be a soluble form detected in the wastewater. The TCLP results indicate that cesium, though added as a nitrate or other soluble form, is immobilized. The leachability index is not as high as Bamba et al. observed for a glass waste form,[11] but is higher than was reported for a cement waste form.[12]

Spent crystalline silicotitanate (CST) resin encapsulated in a Ceramicrete matrix immobilized cesium well, but not as well as glass waste forms according to the PCT and MCC-1 tests.[13] These Ceramicrete waste forms for loaded CST may not have performed as well as glass because (1) PCT and MCC-1 tests are designed to test glass, and (2) not adding fly ash resulted in a porous Ceramicrete matrix (23 vol%). A Ceramicrete matrix that incorporates any form of ash, particularly fly ash, performs better.[5,14] The amorphous silica in fly ash forms silico-phosphate glass in the matrix, providing better strength and lower porosity.

6.2.4.3 Stabilization of Radioactive Elements

The oxides of actinides, even at high concentrations, are effectively encapsulated in the Ceramicrete matrix.[6] Often actinide oxides are found in a lower oxidation state such as oxides in the trivalent state (for example, U_2O_3 or Pu_2O_3). Such oxides have a higher solubility than their counterparts in a higher oxidized state. Once dissolved, they are likely to be oxidized in the aqueous phosphate slurry during formation of the stabilized matrix into insoluble oxidized compounds. X-ray diffraction patterns proved Ce was present in the less-soluble oxidation state (CeO_2), not the more-soluble lower oxidation state (Ce_2O_3).[6]

Radium-rich wastes from the Fernald Silos were stabilized at bench scale in a Ceramicrete matrix with a loading of 66 wt%.[15] The total specific activity of the waste was 3.85 µCi/g, with radium (^{226}Ra) at 0.477 µCi/g. Waste form samples were extracted per the TCLP test and the extract analyzed for specific activity. These data were used to estimate the level of encapsulation in the matrix. The alpha and beta activity of the leachate was 25 ± 2.5 and 9.81 ± 0.98 pCi/ml, respectively, compared to a total activity of 127 pCi/ml and radium activity of 16 pCi/ml that would be expected if all the activity leached during TCLP testing. Thus, most of the activity was not extracted, but a significant amount did leach from the Ceramicrete waste forms.

6.2.4.4 Salt Stabilization

The Ceramicrete slurry sets into a hard ceramic even in the presence of salts such as nitrates and chlorides. Studies produced monolithic Ceramicrete solids from concentrated sodium nitrate and sodium chloride solutions.[5,16]

6.2.5 EVALUATION OF RADIOLYTIC GAS GENERATION

The Ceramicrete matrix material, MKP, contains 6 moles of water for every mole of the magnesium potassium phosphate. Radiolytic decomposition of this water and

TABLE 6.2.4
Radiolysis Yield in Various Scenarios and in
Ceramicrete

Material Tested	G(H2) Moles H2/100ev
Pu^{239} in water	1.6
Tritiated water in concrete	0.6
FUETAP concrete	0.095
Hanford acid waste in FUETAP cement	0.43
TRU ash in Ceramicrete with 7.87 wt% Pu in ash	0.1
U–Pu alloy in Ceramicrete with 5.2 wt% Pu	0.13

of any organic compounds present in the waste may pressurize sealed containers during storage of the waste forms. For this reason, Pu-containing waste forms were subjected to gas generation tests by radiolysis. The results are summarized in Table 6.2.4 in terms of G values defined as the radiation chemical yield to the energy absorbed, expressed in terms of the number of molecules generated per 100 eV. Results on the first three examples given in Table 6.2.4 are obtained from the literature.[17-20] The last two provide the results for Ceramicrete waste forms. The U–Pu oxide mixture was a result of corrosion of a U–Pu alloy.[6] The TRU combustion residue was obtained originally from Rocky Flats. It was fully calcined for safe transport to Argonne; as a result, all organics and combustibles were completely incinerated and the plutonium concentrated. As Table 6.2.4 indicates, the G values for Ceramicrete waste forms are lower than for most grout systems and comparable to that for Formed Under Elevated Temperature and Pressure (FUETAP) concrete. These observations indicate that the gas yield is acceptable for the Ceramicrete waste forms.

6.2.6 SUMMARY

The various studies discussed in this section demonstrate that Ceramicrete is a versatile process for stabilization of hazardous and radioactive waste streams. The process not only chemically immobilizes the contaminants, but also microencapsulates them with enough leach resistance to pass standard regulatory leach tests and meet the waste acceptance criteria (WAC) at DOE disposal sites. Ceramicrete is also an excellent process for macroencapsulation of various contaminated objects. The waste form is a dense matrix, with good mechanical properties, does not degrade over time, is neutral to pH, converts flammable waste into nonflammable waste forms, generates tolerable amounts of gas from self-radiolysis, and can incorporate a variety of inorganic wastes, including solids, sludge, liquids, and salts. The technology does not target stabilizing organics at this time, though several tests have shown that it performs better than some methods. Organics are generally destroyed by combustion or other chemical means, not stabilized. The resulting ash can be immobilized in Ceramicrete.

REFERENCES

1. Connor, J. R., *Chemical Fixation and Solidification of Hazardous Wastes*, Van Nostrand Reinhold, New York (1990).
2. Singh, D., A. Wagh, L. Perry, and S. Y. Jeong, Pumpable/Injectable Phosphate Bonded Ceramics, U.S. Patent No. 6,204,214 issued March 20, 2001.
3. Wagh, A. S. and S. Y. Jeong, Chemically Bonded Phosphate Ceramics: Part I. A Dissolution Model of Formation. Vol. 18, No. 3, Sept. 2003, pp. 162–168, American Ceramic Society (2003).
4. Jeong, S. Y. and A. S. Wagh, Chemically Bonded Phosphate Ceramics — Cementing the Gap Between Ceramics and Cements, Vol. 18, No. 3, Sept. 2003, pp. 162–168, *Mater. Technol.* (2003).
5. Wagh, A.S., D. Singh, and S. Y. Jeong, Chemically Bonded Phosphate Ceramics, Chapter 7-3, in *Handbook of Mixed Waste Management Technology*, C. Oh, Ed., CRC Press, Boca Raton, FL (2001) pp. 6.3-1–6.3-18.
6. Wagh, A. S., R. Strain, S. Y. Jeong, D. Reed, T. Krause, and D. Singh, Stabilization of Rocky Flats Pu-Contaminated Ash within Chemically Bonded Phosphate Ceramics, *J. Nucl. Mater.* 265 (1999) 295–307.
7. Singh, D., S. Jeong, A. Wagh, J. Cunnane, and J. Mayberry, Chemically Bonded Phosphate Ceramics for Low-Level Mixed-Waste Stabilization, *J. Environ. Sci. Health,* A32(2) (1997) 527–541.
8. Wescott, J., R. Nelson, A. Wagh, and D. Singh, Low-Level and Mixed Radioactive Waste In-Drum Solidification, *Practice Periodical of Hazardous, Toxic, and Radioactive Waste Management,* Jan. (1998), pp. 4–7.
9. Wagh, A. S., S. Y. Jeong, D. Singh, R. Strain, H. No, and J. Wescott, Stabilization of Contaminated Soil and Wastewater with Chemically Bonded Phosphate Ceramics, Proc. WM 97, Tucson, AZ (1997); http://www.wmsym.org/wm97proceedings/sess29/29-0.6.htm.
10. Singh, D., A. S. Wagh, and S.-Y. Jeong, Method for Producing Chemically Bonded Phosphate Ceramics and for Stabilizing Contaminants Encapsulating Therein Utilizing Reducing Agents, U.S. Patent 6,133,498 (2000).
11. Bamba, T., H. Kamizono, S. Nakayama, H. Nakamura, and S. Tashiro, Studies of Glass Waste Form Performance at the Japan Atomic Energy Institute, in *Performance of High Level Waste Forms and Engineered Barriers Under Repository Conditions,* International Atomic Energy Agency Report IAEA-TECDOL-582 (1991) 165–190.
12. Wagh, A. S., D. Singh, S. Y. Jeong, D. Graczyk, and L. B. TenKate, Demonstration of Packaging of Fernald Silo I Waste in Chemically Bonded Phosphate Ceramic, Proc. WM '99: Conditioning of Operational and Decommissioning Waste, Tucson, AZ (Feb. 28–March 4, 1999); http://www.wmsym.org/wm97proceedings/sessC9%C9%.htm.
13. Langton, C. A., D. Singh, A. S. Wagh, M. Tlustochowicz, and K. Dwyer. Phosphate Ceramic Solidification and Stabilization of Cesium-Containing Crystalline Silicotitanate Resins, *Proc. 101st Ann. Mtg. of the American Ceramic Society,* Indianapolis (April 25–28, 1999).
14. Wagh, A. S., S. Y. Jeong, and D. Singh, High Strength Phosphate Cement Using Industrial Byproduct Ashes, *Proc. First Intl. Conf. on High Strength Concrete,* Kona, HI, A. Azizinamini, D. Darwin, and C. French, Eds., (1997) pp. 542–553.
15. Bibler, N. E. and E. G. Orebaugh, Radiolytic Gas Production from Tritiated Waste Forms — Gamma and Alpha Radiolysis Studies, Report No. DP-1459, Savannah River Laboratory (1977).

16. Wagh, A.S., M. D. Maloney, G. H. Thomson, and A. Antink, Investigations in Ceram-icrete Stabilization of Hanford Tank Wastes, *Proc. Waste Management '03*, Tucson, AZ (2003).

17. Dole, L. R., G. Rogers, M. Morgan, D. Stinton, J. Kessler, S. Robinson, and J. Moore, Cement-Based Radioactive Waste Hosts Formed Under Elevated Temperatures and Pressures (FUETAP Concrete) for Savannah River Plant High-Level Waste, Report No. ORNL/TM-8579, Oak Ridge National Laboratory (1983).

18. Dole, L. R. and H. A. Friedman, Radiolytic Gas Generation from Cement-Based Waste Hosts for DOE Low-Level Radioactive Wastes, Report No. CONF-860605-14, Oak Ridge National Laboratory (1986).

19. Siskind, B., Gas Generation from Low-Level Waste: Concerns for Disposal, Report No. BNL-NUREG-47144, Brookhaven National Laboratory (1992).

20. Bates, D. J., R. W. Goles, L. R. Greenwood, R. C. Lettau, G. F. Piepel, M. J. Schweiger, H. D. Smith, M. W. Urie, J. J. Wagner, and G. L. Smith, Vitrification and Product Testing of C-104 and AZ-102 Pretreated Sludge Mixed with Flowsheet Quantities of Secondary Wastes, PNNL-13452, Pacific Northwest National Labora-tory, Richland, WA (2001). Note that currently there is no accepted WAC for Yucca, but this report suggests a possible WAC for the disposal facility.

6.3 SULFUR POLYMER CEMENT (SPC)

Paul D. Kalb

Elemental sulfur is a thermoplastic material that can be melted at around 119°C, mixed with aggregates, and cooled to form a concrete. While its potential use as an alternative to conventional hydraulic cement probably dates back to ancient times, evidence of its use in Latin America as early as the 1600s to join metal and stone can still be seen today.[1-3] However, as the molten sulfur cools to a solid and then further cools to ambient temperature, it undergoes an allotropic solid phase trans-formation from the more loosely packed monoclinic crystalline form (S) above 96°C to the more densely packed orthorhombic crystal lattice form (S_α) below 96°C. Since the orthorhombic form has a slightly higher density, the solid undergoes a volume reduction of 6% following the phase transition, creating stress within the solid. This phenomenon leads to poor durability or mechanical failure as the stresses are relieved, especially under freeze/thaw cycling or if the solid is shocked.[4] These limitations hindered the more widespread use of sulfur until the relatively recent introduction of modified sulfur compounds, e.g., sulfur polymer.

Surpluses of elemental sulfur following World War I led researchers to begin to look for ways to modify the sulfur matrix and make it more durable and thus suitable for construction applications in place of conventional cement.[5] Although unsuccess-ful, this work paved the way for further development of modified sulfur binders. The advent of sulfur removal from petroleum and flue-gas desulfurization from coal-burning power plants that resulted from new environmental regulations in the early 1970s created even larger surpluses of by-product sulfur and encouraged new efforts to develop techniques for using sulfur for construction applications.

Work conducted at the U.S. Bureau of Mines (USBM)[2,6,7] and in Canada[8-10] and elsewhere[11] explored the use of organic modifiers to form polysulfides that plasticize the sulfur and suppress the solid phase transition from the monoclinic form (S) to the orthorhombic (S_α) form that occurs in elemental sulfur when cooled below 95°C. Various modifiers have been investigated including dicyclopentadiene (DCPD) and other organic oligomers, and silane coupling agents such as vinyltrimethoxysilane (VTMS) and vinyltriethoxysilane (VTES). The USBM focused on the addition of DCPD and got favorable results with mixtures of 95 wt% elemental sulfur, but found that at reaction temperatures (120 to 140°C) the organic reagent was unstable and depolymerized, resulting in a difficult-to-control exotherm and a viscous product. This was alleviated by substituting an additive mixture of 2.5 wt% DCPD and 2.5 wt% of other higher molecular weight oligomers (trimer through pentamer), yielding a stable, low-viscosity mixture that maintains the monoclinic crystalline form upon cooling. This SPC formulation (also referred to as modified sulfur cement) was patented and commercially licensed and is currently available from Martin Chemicals (Odessa, TX) and GRC Chempruf Concrete (Clarksville, TN) for approximately $0.12 per pound.[12]

Work conducted at Pacific Northwest National Laboratory (PNNL) revealed an effect of the thermal history and cooling rate on the crystalline structure, independent of the chemical modifiers that are added. They claim that the role of the polymer modification is to assist in the control of the microstructure on cooling, by facilitating the formation of plate-like microcrystals found in beta sulfur rather than the larger (millimeter- to centimeter-scale) crystals found in alpha sulfur. They found that very slow cooling (< 1.5°C/min) resulted in formation of alpha sulfur regardless of the polymer modification, whereas more rapid cooling (> 1.5°C/min) yielded the more desirable beta form. The literature does not contain other references to the impacts of cooling rate on the formation of stable sulfur. *The Sulphur Cement Concrete Design and Construction Manual* indicates that SPC concrete forms attain 80% of maximum strength within 1 day and most sulfur concrete data are reported following a 1-day cooling time.[13]

Concretes for construction applications formulated by the addition of aggregate materials to SPC have been developed and extensively tested. The addition of high-quality sand and quartz aggregate yields high-strength concretes that can be used in a number of applications in place of conventional hydraulic cement concretes, e.g., road paving, concrete blocks, walls and floors, support columns, pipes, and sewer systems. Compressive strengths of 48.2 to 68.9 MPa (7,000 to 10,000 psi) are typical for sulfur polymer concretes and are attained within hours compared with weeks for hydraulic portland cement concretes. Typical formulations and mechanical properties of SPC compared with portland cement concrete are summarized in Table 6.3.1.[14] Sulfur polymer concretes are extremely resistant to harsh chemical environments and can be used in applications where conventional concrete materials are subject to degradation. For example, SPC is resistant to corrosive electrolytic solutions, and mineral acids and salts that are found in many industrial applications (e.g., electroplating, metallurgical refining, and acid and battery production). In these cases, precast or custom-poured SPC tanks, slabs, and foundations can provide improved durability and performance. Figure 6.3.1 is a photograph of SPC and

TABLE 6.3.1
Properties of Typical Sulfur Concrete and Portland Cement Concrete[14]

	Sulfur Concrete[a]	Regular PCC[b]
Strength psi:		
Compressive	7,000–10,000	3,500–5,000
Splitting tensile	1,000–1,500	500
Flexural	1,350–2,000	535
Coefficient of thermal expansion (μin/in)/°C	14.0–14.7	12
Moisture absorption ... . pct	0.0–0.10	0.30–3.0
Air void content ... ... pct	3.0–6.0	4.0
Elastic modulus ... ... 10^6psi	4.0	4.0
Specific gravity	2.4–2.5	2.5
Linear shrinkage ... ...pct	0.08–0.12	0.06–0.10
Impact strength, ft lb:		
Compressive	100–119	81
Flexural	0.3–0.5	0.2
Mix proportions, wt. %:		
SPC	14–18	0
Water	0	6–9
Mineral filler	6–9	0
Portland cement	0	12–18
Sand	38–42	30
Coarse aggregate	33–37	45

[a] Properties obtained at age of 1 day.
[b] Properties obtained at age of 28 days.

FIGURE 6.3.1 Sulfur polymer cement concrete (on left) and conventional hydraulic portland cement concrete following a 2-week immersion in mineral acid.

conventional portland cement concrete following exposure to mineral acid. Performance of SPC under various chemical environments is summarized in Table 6.3.2.[3]

The mechanical and chemically resistant properties of SPC make it a good candidate matrix material for the microencapsulation and macroencapsulation of low-level radioactive, hazardous, and mixed wastes. In these solidification applications, a broad range of waste products is substituted for the aggregate used to form sulfur polymer concretes. For example, contaminated soil, sludges, evaporator concentrates, and incinerator ash have been successfully solidified in SPC. In a related process application, SPC can be made to react with the contaminants in the waste (e.g., toxic metals) to form low-solubility, leach-resistant compounds, which are then physically encapsulated within the waste form.

TABLE 6.3.2
USBM Industrial Testing Results of SPC Materials[3]

Environment	Performance
Sulfuric acid	Nonreactive
Copper sulfate–sulfuric acid	Nonreactive
Magnesium chloride	Nonreactive
Hydrochloric acid	Nonreactive
Nitric acid	Nonreactive
Zinc sulfate–sulfuric acid	Nonreactive
Slimes–electrolytic residue	Attacked by organics used to process residues
Nickel sulfate	Nonreactive
Vanadium sulfate–sulfuric acid	Nonreactive
Uranium sulfate–sulfuric acid	Nonreactive
Potash brines	Nonreactive
Manganese oxide–sulfuric acid	Nonreactive
Hydrochloric acid–nitric acid	Nonreactive
Mixed nitric–citric acid	Nonreactive
Ferric chloride–sodium chloride–hydrochloric acid	Nonreactive
Boric acid	Nonreactive
Sodium hydroxide	Nonreactive
Citric acid	Attacked by 10% NaOH
Acidic and biochemical	Nonreactive
Sodium chlorate–hypochlorite	Nonreactive
Ferric and chlorate ion	Attacked by solution at 50 to 60°C
Sewage	Nonreactive
Hydrofluoric acid	Nonreactive
Glyoxal–acetic acid–formaldehyde	Nonreactive
Chromic acid	Nonreactive
	Deteriorated at 80°C and 90% concentration, marginal at lower temperature and concentration

Research and development of SPC-based solidification technologies was initially investigated at Brookhaven National Laboratory (BNL),[15-21] which received a patent for SPC microencapsulation of radioactive, hazardous, and mixed wastes[22] and for Sulfur Polymer Stabilization/Solidification (SPSS).[23] Additional development of SPC for the treatment of waste has been conducted by the Commission of the European Communities,[24] Idaho National Environmental and Engineering Laboratory (INEEL),[25-27] Pacific Northwest National Laboratory,[13] Oak Ridge National Laboratory (ORNL),[28-31] and the University of Georgia.[32]

Polyethylene is a similar thermoplastic material that is heated to melt temperature, combined with waste to form a composite mixture, and then cooled to form a solid, monolithic final waste form. The melt viscosity of SPC is significantly lower (~25 mPa-s), however, making the processing of SPC easier to accomplish. For example, SPC can be processed in a variety of heated mixing vessels, whereas polyethylene requires processing in an extruder or thermokinetic mixer. A typical processing system consisting of a heat-jacketed vertical cone mixer with a conical screw that rotates on its axis and around the circumference of the vessel mixing is shown in Figure 6.3.2. The relatively slow orbital mixing action provides thorough mixing without entraining air in the liquid solution, which would result in a less-desirable, low-density product. Other processing options include low- and high-shear mixers and various types of heated pug mill mixers. Several modified heated pug mills, with hollow mixing shafts that allow circulation of hot oil or steam, were tested at INEEL.[26] Process temperatures range from 120 to 140°C to ensure proper mixing without overheating the SPC, which can lead to increased viscosity and breakdown of the SPC. Systems that allow processing under negative pressure are preferable to ensure containment of off-gas and to reduce the odors associated with molten sulfur.

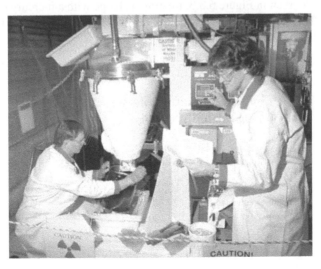

FIGURE 6.3.2 Pilot-scale process equipment for Sulfur Polymer Stabilization/Solidification (SPSS) treatment of mercury waste.

Capitalizing on the ability of sulfur to react with toxic metals and form stable sulfides led to the development of a related technology known as Sulfur Polymer Stabilization/Solidification. This technology first reacts the waste to chemically reduce the metals to metal sulfides and then physically encapsulates the stabilized waste in a microencapsulated final waste form. SPSS has been effectively applied for the treatment of elemental mercury and mercury-contaminated soil, sludge, and debris and has been licensed for the treatment of mercury generated as a result of mining operations.[23,33] While the U.S. Environmental Protection Agency (EPA) considered an Advanced Notice of Proposed Rulemaking that would allow SPSS treatment for disposal of elemental mercury, it has yet to approve any technology for this purpose.[34]

Sulfur polymer-microencapsulated final waste forms exceeded minimum recommended standards[37] for the Toxicity Characterization Leaching Procedure (TCLP) of the U.S. EPA and the short-term tests[35,36] recommended by the U.S. Nuclear Regulatory Commission (NRC) for low-level radioactive waste forms to demonstrate compliance with 10 CFR 61. Testing included compressive strength, water immersion, leaching, freeze/thaw durability, radiation stability, and biodegradation. For radioactive contaminants, leaching was conducted using the American Nuclear Society protocol, ANSI/ANS-16.1-2003, and the American Society of Testing Materials (ASTM) accelerated leaching test, ASTM D-1308. Sulfur polymer has a relatively low permeability and, consequently, as seen in Table 6.3.3, leachability indices of 10 to 14, compared to the NRC minimum leachability index criterion of 6.[38] SPC-microencapsulated mixed waste incinerator ash, containing as much as 7 wt% lead at waste loadings up to 43 wt% ash, passed TCLP.[19] For SPSS-encapsulated mercury and mercury-contaminated soil, sludge, and debris, mercury concentrations below the Universal Treatment Standard (UTS) level of 0.025 mg/L have been routinely achieved.[39,40] As seen in Figure 6.3.3, treating a sludge with a mercury concentration of 5000 mg/L resulted in leachates below this UTS limit for leachant pH ranging from 2 to 10.

TABLE 6.3.3
Leach Indices for Sulfur Polymer-Microencapsulated Waste Forms[38]

Waste Type	Waste Loading	Leachability Index	
		^{60}Co	^{137}Cs
Sodium sulfate	25	12.5	10.6
Evaporator concentrate	40	10.7	9.7
Incinerator ash	20	14.0	11.2
	40	14.6	11.1

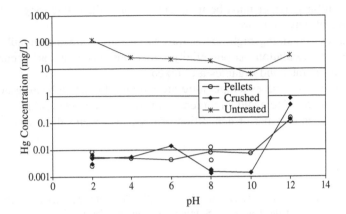

FIGURE 6.3.3 Leachability of mercury from sludge treated by SPSS under constant pH conditions. (Data from Reference 34.)

Advantages of sulfur polymer encapsulation of waste include:

- Its thermoplastic nature assures that waste forms produced using SPC binder result in a solid final waste form product on cooling to ambient temperature.
- The solidification process is independent of waste chemistry, unlike that of hydraulic cement, which is subject to potential interaction between waste and binder that can interfere with the hydration chemistry and result in solidification failures.
- Relatively high waste loadings have been achieved for SPC microencapsulation. For example, waste loadings of 40 dry wt% evaporator concentrate salts (e.g., sodium sulfate, boric acid), 43 dry wt% incinerator ash, and 80 dry wt% soil have been encapsulated in SPC, while still maintaining performance criteria required of treated radioactive and hazardous wastes.[17] From a processing perspective, waste loading efficiency is limited by the workability of the mix and the ability to form a homogeneous mixture of waste and binder. As with other processes, performance of the final waste form also constrains waste loading. Excessive waste loading or concentrations of contaminants may exceed the ability of the binder to effectively immobilize contaminants.

Limitations of the sulfur polymer encapsulation process include:

- Waste containing aqueous liquid must be pretreated to remove moisture, since the processing temperature for SPC is 120 to 140°C. Alternatively, the sulfur polymer must be maintained in a molten state long enough to drive off any moisture.

- Wastes containing high concentrations of nitrate are incompatible with SPC encapsulation, as potentially dangerous reactions can occur when sulfur, nitrate, and organics are mixed.
- Controls on process temperature are required to ensure heating does not exceed 150°C to avoid an increase in viscosity and potential for degradation of the polymer or generation of toxic hydrogen sulfide gas.

REFERENCES

1. Rybcsynsiki, W., A. Ortega, and W. Ali, Sulfur Concrete and Very Low Cost Housing, Proceedings of the Canadian Sulphur Symposium, University of Calgary, Calgary, Alberta, Canada, 1974.
2. McBee, W.C. and T.A. Sullivan, Development of Specialized Sulfur Concretes, RI 8346, United States Dept. of Interior, Bureau of Mines, 1979.
3. The Sulphur Institute, *Sulphur Polymer Cement Concrete Design and Construction Manual*, Washington, D.C., 1994.
4. Shrive, N.G., J.E. Gillott, I.J. Jordaan, and R.E. Loov, Freeze/Thaw Durability of Sulphur Concretes, University of Calgary, *Proceedings of Sulphur 81 Symposium*, 1981.
5. Bacon, R.F. and H.S. Davis, Recent Advances in the American Sulfur Industry, *Chemical and Metallurgical Engineering*, v.24, No. 2, 1921.
6. Sullivan, T.A. and W.C. McBee, Development and Testing of Superior Sulfur Concretes, RI-8160, United States Dept. of Interior, Bureau of Mines, 1976.
7. McBee, W.C., T.A. Sullivan, and B.W. Jong, Modified-Sulfur Cements for Use in Concretes, Flexible Pavings, Coatings, and Grouts, RI-8545, United States Dept. of Interior, Bureau of Mines, 1981.
8. Sulphur Innovations Ltd., Sulfurcrete, A Technical Brief, Calgary, Alberta, December 1980.
9. Egan, D.W., E. Nishimura, and R.T. Woodhams, Sulfurwood: a Fiber-Reinforced Syntactic Sulphur Foam, *Proceedings of Sulphur 81 Symposium*, 1981.
10. Vroom A.H., Sulphur Cements, Process for Making Same and Sulphur Concretes Therefrom, U.S. Patent, 4,058,500, 1977.
11. Anani, A.A., A. Halasa, A.A. Mobasher, and A. Nurita, New Approaches and Techniques to Sulphup Plasticization and Fire Retardation, *Proceedings of Sulphur 81 Symposium,* 1981.
12. McBee, W.C. and T.A. Sullivan, Concrete Formulations Comprising Polymeric Reaction Products of Sulfur/Cyclopentadiene, Oligomer/Dicyclopentadiene, U.S. Patent No. 4,348,313, Sept. 7, 1982.
13. P. Sliva, et al. Sulfur Polymer Cement as a Low-Level Waste Glass Matrix Encapsulant, PNNL-10947, Jan. 1996.
14. McBee, W.C., F.E. Ward, W.T. Dohner, and H. Weber, Utilization of Waste Sulfur in Construction Materials and as a Stabilization/Encapsulation Agent for Toxic, Hazardous, and Radioactive Waste, *Proceedings of the Utilization of Waste Material in Civil Engineering Construction*, September 13–17, 1992.

15. Colombo, P., P.D. Kalb, and M. Fuhrmann, Waste Form Development Program Annual Report, BNL-51756, Brookhaven National Laboratory, Upton, NY, 11973, September 1983.
16. Kalb, P.D. and P. Colombo, Polyethylene and Modified Sulfur Cement Solidification of LLW, *Proceedings of the Sixth Annual Participant's Information Meeting - DOE Low-Level Waste Management Program*, CONF-8409115, December 1984.
17. Kalb, P.D. and P. Colombo, Modified Sulfur Cement Solidification of Low-Level Wastes, Topical Report, BNL 51923, Brookhaven National Laboratory, Upton, NY, 11973, October 1985.
18. Kalb, P.D., J. Heiser, and P. Colombo, "Comparison of Modified Sulfur Cement and Hydraulic Cement for Encapsulation of Radioactive and Mixed Wastes," *Proceedings of the 12th Annual DOE Low-Level Waste Management Conference,* August 28–29, 1990, Chicago.
19. Kalb, P.D., J. Heiser, and P. Colombo, "Modified Sulfur Cement Encapsulation of Mixed Waste Contaminated Incinerator Fly Ash," *Waste Management*, Vol. 11, November 3, 1991, Pergammon Press, New York.
20. Kalb, P.D., J. Heiser, R. Pietrzak, and P. Colombo, "Durability of Incinerator Ash Waste Encapsulated in Modified Sulfur Cement," *Proceedings of the 1991 Incineration Conference*, Knoxville, TN, May 13–17, 1991.
21. Kalb, P.D. and J. Adams, "Mixed Waste Treatability Using Polyethylene and Sulfur Polymer Encapsulation Technologies," American Nuclear Society, *Proceedings of Spectrum '94*, Seattle, WA, August 14–18, 1994.
22. Colombo, P., P.D. Kalb, and J.H. Heiser, "Process for the Encapsulation and Stabilization of Radioactive Hazardous and Mixed Wastes," U.S. Patent No. 5,678,234, Oct. 14, 1997.
23. Kalb, P.D., D. Melamed, B. Patel, and M. Fuhrmann, "Treatment of Mercury Containing Waste," U.S. Patent No. 6,399,849, June 4, 2002.
24. Van Dalen, A. and J.E. Rijpkema, Modified Sulphur Cement: a Low Porosity Encapsulation Material for Low, Medium and Alpha Waste, Nuclear Science and Technology, EUR 12303, Commission of the European Communities, Brussels, 1989.
25. Darnell, G.R., Sulfur Polymer Cement, a New Stabilization Agent for Mixed and Low-Level Radioactive Waste, *Proceedings of the First International Symposium on Mixed Waste*, A.A. Mohhissi and G.A. Benda, Eds., University of Maryland, Baltimore, 1991.
26. Darnell, G.R., W.C. Aldrich, and J.A. Logan, Full-Scale Tests of Sulfur Polymer Cement and Non-Radioactive Waste in Heated and Unheated Prototypical Containers, EGG-WM-10109, Idaho National Engineering Laboratory, Feb. 1992.
27. Darnell, G.R., W.C. McBee, and E.B. McNew, Stabilization of Toxic Metal Oxides in Sulfur Polymer Cement, INEEL/EXT-98-00145, Idaho National Engineering and Environmental Laboratory, Feb. 1998.
28. Calhoun, C.L., Jr. and L.E. Nulf, "Sulfur Polymer Cement Encapsulation of RCRA Toxic Metals and Metal Oxides," *Proceedings of the American Chemical Society I&EC Special Symposium*, Atlanta, GA, September 1995.
29. Mattus, C.H. and A.J. Mattus, Evaluation of Sulfur Polymer Cement as a Waste Form for the Immobilization of Low-Level Radioactive or Mixed Waste, ORNL/TM12657, Oak Ridge National Laboratory, March 1994.
30. Calhoun, C.L., Jr., L.E. Nulf, and V.V. Federov, Sulfur Polymer Cement Encapsulation of Oily Matrix Mixed Low-Level Sludge, Y-DZ-2010, Oak Ridge Y-12 Plant, July 1996.

31. Mattus, C.H., Demonstration of Macroencapsulation of Mixed Waste Debris Using Sulfur Polymer Cement, ORNL/TM-13575, Oak Ridge National Laboratory, July 1998.
32. Lin, S.-L., J.S. Lai, and E.S.K. Chian, Modification of Sulfur Polymer Cement Stabilization and Solidification Process, *Waste Management*, Vol. 15 Nos. 5/6, pp. 441–447, Elsevier Science Ltd., Sept. 1995.
33. Fuhrmann, M., D. Melamed, P.D. Kalb, J.W. Adams, and L.W. Milian, "Sulfur Polymer Solidification/Stabilization of elemental mercury waste," *Waste Management Journal* 22 (2002) pp. 327–333.
34. U.S. Environmental Protection Agency, "Land Disposal Restrictions: Treatment Standards for Mercury-Bearing Hazardous Waste; Notice of Data Availability" 4481 Federal Register/Vol. 68, No. 19/Wednesday, January 29, 2003, pp. 4481–4489.
35. U.S. Nuclear Regulatory Commission, Licensing Requirements for Land Disposal of Radioactive Waste, in Title 10 of the Code of Federal Regulation, Part 61, USNRC, Washington, D.C., 1983.
36. U.S. Nuclear Regulatory Commission, Technical Position on Waste Form, Revision 1, in Final Waste Classification and Waste Form Technical Position Papers, USNRC, Washington, D.C., 1983.
37. Kalb, P.D., J. Heiser, R. Pietrzak, and P. Colombo, "Durability of Incinerator Ash Waste Encapsulated in Modified Sulfur Cement," Proceedings of the 1991 Incineration Conference, Knoxville, TN, May 13–17, 1991.
38. Kalb, P.D., "Sulfur Polymer Encapsulation," in *Hazardous and Radioactive Waste Treatment Technologies Handbook*, C.H. Oh, Ed., CRC Press, Boca Raton, FL, 2001.
39. Bowerman, B., J. Adams, P. Kalb, R.-Y. Wan, and M. LeVier, Using the Sulfur Polymer Stabilization/Solidification Process to Treat Residual Mercury Wastes from Gold Mining Operations, Society of Mining Engineers Conference, February 2003.
40. Adams, J.W., B.S. Bowerman, and P.D. Kalb, Sulfur Polymer Stabilization/Solidification (SPSS) Treatability of Simulated Mixed-Waste Mercury Contaminated Sludge, Waste Management 2002 Symposium, Tucson, AZ, 2002.

6.4 ENVIROSTONE™ GYPSUM CEMENT IN RADIOACTIVE WASTE STABILIZATION

Gerald W. Veazey and Earl W. McDaniel

6.4.1 INTRODUCTION

In the early 1980s Los Alamos National Laboratory (LANL) faced a number of challenges with respect to management of a variety of waste streams generated from the processing of plutonium for the Department of Energy's (DOE's) Nuclear Defense Program. At the end of the Cold War, these challenges became more acute as many process streams that were previously recycled were now designated as wastes for treatment and disposal. To meet these challenges, LANL staff reviewed a number of options available for addressing waste treatment and solidification. Most of these processes, however, addressed gamma radiation-based fission product waste or nuclear utilities waste. LANL, on the other hand, was involved with considerable quantities of transuranic* (TRU) waste, which possessed greater radioactive emissions from alpha-emitting sources. TRU waste is generated by plutonium-based DOE nuclear defense activities and is disposed of at the Waste Isolation Pilot Plant (WIPP) in Carlsbad, NM. Little work in the field of waste stabilization had been done and reported on the TRU waste.

In 1983 LANL chose to investigate a product manufactured by the United States Gypsum Company called Envirostone™** Gypsum Cement (hereafter referred to as Envirostone). The data available in 1983 was almost exclusively contained in a technical marketing document issued by U.S. Gypsum in 1982.[1] Envirostone is a finely ground, nonflammable powder composed of calcium sulfate hemihydrate, water-soluble melamine formaldehyde resin, and a small amount of ammonium chloride, which serves as a cross-linking agent to facilitate resin curing.[2] Envirostone is presented as a low-level radioactive waste (LLRW) solidification medium capable of solidifying a variety of waste streams associated with nuclear utilities operations for which portland cement has experienced difficulty in solidifying. The waste streams of primary interest were boric acid evaporator bottoms and ion exchange resins, as well as nonpolar organic liquids, such as waste oils, when used in conjunction with Envirostone Emulsifier. Brookhaven National Laboratory (BNL) independently confirmed the compatibility of Envirostone with emulsified oil.[3] The Envirostone product solidifies acidic waste best,[1] unlike portland cement that requires a basic pH for solidification. The U.S. Gypsum report preceded the issuance of 10 CFR Part 61.56, which detailed the NRC's stricter standards for waste form performance.[4] However, satisfactory performance was demonstrated in the 1980–1983 test period using the current versions of the following standards:[1]

- Leach resistance for ^{137}Cs, ^{144}Ce, ^{60}Co, and ^{85}Sr according to ANSI/ANS-16.1-2003[5]
- Compressive strength according to ASTM C472[6] and ASTM D560[7]
- Water impingement and 8-month water immersion testing

Favorable results were reported by U.S. Gypsum with actual wastes from the Kewaunee Nuclear Station, the Ginna Nuclear Station, and Lawrence Livermore National Laboratory (LLNL).[1] Envirostone™ solidified waste forms were accepted by the state agencies and operators of the three commercial LLRW disposal sites.[1]

6.4.2 Use of Envirostone™ at Los Alamos National Laboratory

LANL was interested in Envirostone™ as an alternative to portland cement for several TRU waste streams generated at the TA-55 Plutonium Facility. TRU waste form performance requirements are defined in the Waste Isolation Pilot Plant Waste Acceptance Criteria (WIPP-WAC).[8] Envirostone™ had not yet been evaluated

---

* TRU waste is any waste that contains alpha-emitting radionuclides of atomic number greater than 92 with a half-life exceeding 20 years and a specific activity of greater than 100 nCi/g.
** Envirostone™ is a registered trademark of United States Gypsum Company, Chicago.

against the WIPP-WAC, which contain a number of requirements specific to TRU waste. One restriction prevents accumulation of unsafe concentrations of radiolytically generated hydrogen gas inside the TRUPACT-II container. This restriction determines the maximum amount of special nuclear material (SNM) that can be present in a drum of cemented waste. Most requirements are similar to those for Class A LLRW forms. No minimum compressive strength is required for a TRU waste form, and free liquid is restricted to 1 vol%.

Although U.S. Gypsum did not formally market Envirostone as a TRU waste solidification medium, company representatives conducted limited small-scale tests at LANL that convinced the LANL technical staff that TRU waste solidification requirements are within the capability of Envirostone performance. Envirostone primarily solidified evaporator bottoms generated from nitric acid-based plutonium processing at TA-55. This waste liquid consisted of high concentrations of nitric acid (3 to 9 N) and a saturated nitrate salt solution,[9] including several cations later regulated by the EPA under the Resource Conservation and Recovery Act (RCRA)[10] as hazardous waste. portland Type I/II cement was used for this waste stream prior to the conversion to Envirostone. After converting to Envirostone, adjusting the liquid waste pH to about 3 prior to solidification was no longer required, resulting in no pretreatment dilution and higher loadings in the final solidified waste. LANL also began using Envirostone and Envirostone Emulsifier to solidify its non-polar organic liquid wastes from machining activities, which had previously been treated by absorption into an inert material. The disadvantage of the conversion to Envirostone was its relatively short shelf-life of 6 months[1] and a materials cost approximately 10 times that of portland Type I/II cement.[11]

6.4.2.1 Free-Liquid Generation from Envirostone™

Concerns with Envirostone performance arose in 1989 when waste forms produced with Envirostone generated free liquid several weeks after solidification.[12-14] The free liquid formed from 7 to 44 weeks after solidification and resulted in the generation of up to 15 l (4 gal) of liquid in a 55-gal waste drum.[15] The phenomenon was first observed when liquid seeped through a drum lid's venting filter, collected on the lid, and crystallized from evaporation (see Figure 6.4.1). The crystalline material consisted primarily of sodium salts, the most soluble cation in the original waste. The ^{239}Pu and ^{241}Am concentrations were ~10 ng per gram of crystalline material, consistent with their low solubility for the liquid pH of 5. This incident resulted in removal of WIPP certification for the TA-55 solidification process and a major effort to remove the free liquid from the drums.[15,16]

FIGURE 6.4.1 Drum of Envirostone product with free liquid residue on top.

A significant effort was initiated at LANL to find the cause of the phenomenon and prevent future occurrences.[12] The following two hypothetical mechanisms for producing free liquid caused by irradiation were investigated and eliminated: (1) a water-producing condensation reaction between previously unpolymerized melamine and formaldehyde components of the melamine-formaldehyde (MF) resin, and (2) reversal of the calcium sulfate hydration reaction ($CaSO_4.1/2H_2O + 3/2H_2O = CaSO_4.2H_2O$), an event that occurs at relatively low temperatures.[17] Gamma irradiation via a ^{60}Co source proved that Envirostone sample waste forms generated free liquid after absorbing a dose of 2.5 Mrads, but portland cement sample waste forms generated no free liquid after absorbing a dose of 81 Mrads.[12]

The generation of free liquid was ultimately shown to be reoxidation of hydrogen formed from radiolysis of interstitial water or organic polymer in the solidified waste form. Radiolysis of interstitial water has been investigated by several workers.[18-21] Hanford workers showed that free-liquid accumulation as a result of hydrogen or other gas generation due to radiolysis could be correlated to the waste form's compressive strength and permeability.[22] A computer model used these correlations to predict the occurrence of free liquid formed from reoxidation of hydrogen produced by radiolysis as a function of the waste form permeability, the gas generation rate, and internal pressurization.[23] Work at LANL with actual waste showed that increasing either the amount of Envirostone or mix time reduced the occurrence and volume of free liquid, presumably because of a reduction in pore size and permeability.[15]

6.4.2.2 Leaching Performance of Envirostone

LANL continued to use Envirostone as a solidifying medium at TA-55 until 1992, when the waste forms were determined to be hazardous for chromium toxicity by TCLP per RCRA standards, resulting in all TA-55 solidified waste forms being reclassified as mixed waste drums and complicating storage requirements.[24,25] Portland cement waste forms were not hazardous for chromium toxicity and had TCLP extract chromium concentrations an order of magnitude lower than was measured for Envirostone waste forms. The difference in chromium solubility at the high pH of cement and the low pH of the waste solidified in Envirostone or permeability differences in the waste forms may have caused this difference in TCLP performance. Figure 6.4.2 compares the TCLP performance of using Envirostone and portland cement waste forms. Solidification operations returned to using portland cement as the stabilization agent. No instances of free-liquid generation have been found in the portland cement solidified waste forms generated since that time.

6.4.3 EXPERIENCE WITH ENVIROSTONE AT ROCKY FLATS

Envirostone was used at Rocky Flats Environmental Technology Site (RFETS) for the immobilization of TRU-contaminated oils and halogenated solvents.[25] The process used in this operation was called the Organic and Sludge Immobilization System (OASIS). In this operation Envirostone, water, emulsifier, and oil were mixed in a glovebox with an in-drum mixer to stabilize the organic wastes and produce an

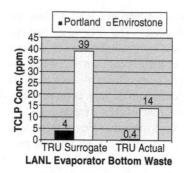

FIGURE 6.4.2 Comparison of chromium TCLP performance for portland cement and Envirostone.

immobilized, solid waste form. Several hundred drums of solidified TRU organic waste were generated by OASIS to meet the WIPP-WAC before being discontinued.[25] No cases of free-liquid generation were reported from this waste form. However, testing at Argonne National Laboratory (ANL)-West, RFETS, and Idaho National Environmental and Engineering Laboratory (INEEL) showed the OASIS waste form to exceed the permissible WIPP Waste Acceptance and transportation (TRUPACT-II) limits for hydrogen and volatile organic gases.[25] These characteristics would later become a safety problem and posed a logistical problem that impacted the decommissioning schedule of RFETS.[25]

6.4.4 ADDITIONAL INSTANCES OF ENVIROSTONE USE

Envirostone was either investigated for or used in a few other TRU and LLRW applications. In 1996 Envirostone was examined for its ability to retard the migration of ^{237}Np and ^{99}Tc from the WIPP repository.[26] Envirostone was found to have inadequate absorption potential for the task. Concern was also expressed that Envirostone contained sufficient nutrients to support microbial growth and activity that could impact radionuclide migration, corrosion, water chemistry, and hydrology.[26] Presently, Envirostone remains listed in the DOE/WIPP TRUCON Codes as being a solidification material/absorbent for some TRU wastes. It is listed at LLNL for use with small amounts of TRU solvents and oil-based liquids and at ANL-East for TRU organic wastes and aqueous and homogeneous inorganic wastes.[27,28]

Three cases were found of Envirostone being used for the stabilization of low-level Class B or C radioactive waste. The NRC approved its use for 1 year ending March 3, 1989, for use with a single unidentified waste stream.[29] The other cases took place at the Richland, WA, and Beatty, NV, commercial low-level radioactive disposal sites, where Envirostone was used for a few years for Class A, B, and C LLRW streams.[29,30] The NRC did not officially grant its approval for the use of Envirostone in these cases, but deferred to approvals granted by each state. By 1991 Envirostone use for Class B and C LLRW had been discontinued due to NRC prohibition.[15]

At the current time, Envirostone use seems to be restricted to Class A LLRW. In 1997, it was on the approved solidification list for Class A LLRW at four commercial facilities.[31] It is also currently being used as a stabilization agent at the Containerized Waste Facility in Utah by Envirocare of Utah, Inc. for Class A containerized LLRW.[31] No cases have been reported of free-liquid generation from LLRW forms produced with Envirostone. Apparently, the lower rate at which LLRW produces radiolytic H_2 is within the ability of Envirostone's pore structure to contain the interstitial liquid.

6.4.5 CONCLUSION

Envirostone entered the market at a time that saw myriad different stabilization technologies surface to meet increasing regulatory requirements. Its formulation incorporated gypsum and polymer products in an attempt to gain the advantages of both materials. Due to the different chemistry of the gypsum reaction, Envirostone was compatible with several nuclear power plant waste streams with which portland cement was not, such as boric acid evaporator bottoms. It was only formally marketed for Class A LLRW. However, attempts were made to expand into stabilization of Class B and C LLRW and TRU-level radioactive waste with some limited success. It was ultimately discontinued for TRU use due to issues that allowed excessive leaching and free-liquid generation from radiolytic H_2. A limited shelf-life[1] and a high cost compared to portland cement[11] also lessened the attractiveness of Envirostone in relation to portland cement.

REFERENCES

1. Envirostone™ Gypsum Cement: Solidification Medium for Low Level Radioactive Wastes, T.L. Rosenstiel, M.D. Joss, and R.G. Lange, United States Gypsum Company, Chicago, Jan. 1982.
2. U.S. Patent, Rosenstiel et al., Jan. 3, 1984. 2 U.S. Patent 4,424,148 issued Jan. 3, 1984 to U.S. Gypsum Co. for Envirostone™.
3. D.E. Clark, P. Colombo, and R.M. Neilson, Jr., "Solidification of Oils and Organic Liquids," BNL-51612, Brookhaven National Laboratory, Upton, NY, July 1982.
4. U.S. Nuclear Regulatory Commission, *10 Code of Federal Regulations Part 61.55,* Waste Classification-Low-Level Waste Licensing Branch Technical Position on Radioactive Waste Classification, Rev. 0, May 1983.
5. *"Measurement of Leachability of Solidified Low-Level Radioactive Wastes by a Short-term Test Procedure,"* ANSI/ANS-16.1-2003, American Nuclear Society, LaGrange, IL, 2003.
6. American Society for Testing and Materials, "Standard Test Methods for Physical Testing of Gypsum, Gypsum Plasters and Gypsum Concrete," ASTM C 472-73, ASTM International, Philadelphia, 1973.
7. American Society for Testing and Materials, Standard Test Methods for Freezing and Thawing Compacted Soil-Cement Mixtures," ASTM D 560-82, ASTM International, Philadelphia, 1982.
8. U.S. Department of Energy, "Contact-Handled Transuranic Waste Acceptance Criteria for the Waste Isolation Pilot Plant," WIPP/DOE-02-3122, Revision 0.1 July 25, 2002.

9. Veazey, G.W., "Real-Waste TCLP Comparison for Cements," NMT-2-PROC-93-053, Los Alamos National Laboratory, Los Alamos, NM, April 23, 1993.
10. U.S. Environmental Protection Agency, 40 CFR Land Disposal Restrictions, Washington, D.C., May 1992.
11. Guidance for Low-Level Radioactive Waste (LLRW) and Mixed Waste (MW) Treatment and Handling, EM-1110-1-4002, U.S. Army Corps of Engineers, June 30, 1997.
12. Waste-Form Development for Conversion to Portland Cement at Los Alamos National Laboratory Technical Area 55, LA-13125, Oct. 1996.
13. Foxx, C.L., "Status Report on Cement Drum 52105," NMT-17:89-078, Los Alamos National Laboratory, Los Alamos, NM, Feb. 20, 1990.
14. Veazey, G.W., "Development of Free Liquid in Cement Drums," NMT-7:90-67, Los Alamos National Laboratory, Los Alamos, NM, Feb. 20, 1990.
15. Veazey, G.W., "The Cement Solidification Systems at LANL," *Proceedings of Workshop on Radioactive, Hazardous, and/or Mixed Waste Sludge Management*, Knoxville, TN, Dec. 4–6, 1990. Lomenick, T.F., Ed., CONF-901264.
16. Veazey, G.W., "Strategic Plan for Cement Free-Liquid Problem," NMT-7:91-448, Los Alamos National Laboratory, Los Alamos, NM, Dec. 2, 1991.
17. Shalek, P., "Documentation Re Possible Free-Liquid Mechanisms," NMT-2-FY96-262, LANL, Los Alamos, NM, June 27, 1996.
18. Offermann, P., "Calculation of the Radiolytic Gas Production in Cemented Waste," *Scientific Basis for Nuclear Waste Management XII*, Eds., W. Lutze and R.C. Ewing, Vol. 127, Material Research Society, Pittsburgh, PA, 1989, pp. 461–468.
19. Bibler, N.E., "Radiolytic Gas Production from Concrete Containing Savannah River Waste," DP-1464, Savannah River Laboratory, Aiken, SC, Jan. 1978.
20. Mockel, H.J. and Köster, R.H., "Gas Formation during the Gamma Irradiation of Cemented Low- and Intermediate-Level Waste Products," *Nuclear Technology*, 59, pp. 494–497, Dec. 1982.
21. Powell, W.J., "Gas Generation from Tank 102-AP Simulated Waste, and Grout Preparation," WHC-SD-WM-RPT-083, Westinghouse Hanford Company, Richland, WA, Feb. 4, 1994.
22. Powell, W.J. and H.L. Benny, "Liquid Return from Gas Pressurization of Grouted Waste," *Spectrum '94 Proceedings,* Atlanta, GA, American Nuclear Society, Inc., La Grange Park, IL, Aug. 14–18, 1994.
23. Roblyer, S.P., "Grout Disposal Facility Gas Concentrations," WHC-WM-ER-RPT-151, Westinghouse Hanford Company, Richland, WA, Feb. 24, 1993.
24. Punjak, W., "Cementation Chromium Leach Study," NMT-2-C-92-088, Los Alamos National Laboratory, Los Alamos, NM, July 30, 1992.
25. Request for Statement of Qualifications for Orphan Waste at RFETS, 8/26/2002, Federal Business Opportunities, FBO Daily Issue of Aug. 28, 2002, FBO #0269, SN00149869-W 20020828/020826213220.
26. Waste Package Supports and Sorptive Inverts/Materials Selection, Mined Geologic Disposal System (MGDS) Waste Package Development Department, BBA000000-01717-0200-00055 REV 00, Civilian Radioactive Waste Management System, Aug. 1997.
27. U.S. Department of Energy, "CH-TRU Waste Content Codes (CH-TRUCON)," DOE/WIPP 01-3202, Aug. 1998, Rev. 0.
28. U.S. Department of Energy, "WIPP RCRA Part B Application," DOE/WIPP 91-005, Dec. 1999.

29. U.S. Nuclear Regulatory Commission, "Limitations on the Use of Waste Forms and High Integrity Containers for the Disposal of Low-Level Radioactive Wastes," NRC Information Notice No. 89-27, March 8, 1989.

30. Use of Solidification and Sorbent Media at U.S. Ecology Disposal Facilities, Appendix E, U.S. Ecology Inc., 1989.

31. Kerr, T.A., "A Comparison and Cross Reference of Commercial Low Level Radioactive Waste Acceptance Criteria," DOE/LLW-239, April 1997.

32. Envirocare of Utah Containerized Waste Facility Waste Acceptance Criteria, Rev. 4, Sept. 19, 2003.

6.5 HYDROCERAMIC CONCRETES

Darryl D. Siemer

6.5.1 INTRODUCTION

The U.S. federal government selected vitrification over 20 years ago as the baseline technology to treat reprocessing waste for disposal. Vitrification has proven expensive within the DOE Complex[1] such that the latest request for proposal from DOE-ID encourages contractors to consider undefined alternative technologies.[2] Great Britain applied C-S-H-based cements to their reprocessing waste from the Cold War that was not too radioactive (< 300 watts/m^3) to damage concrete,[3] and this same approach could be used in the U.S.

The U.S. erected institutional barriers to this approach, including QC protocols that tacitly assume contaminants are released when the waste form matrix dissolves, similar to glass. Consequently, current U.S. waste acceptance criteria (WAC) are generally keyed to the corrosion (leach) rates of a candidate material's bulk constituents (usually sodium), not to those of its toxic or radioactive components. Concretes made with C-S-H, calcium aluminate, or magnesium phosphate-based cements will not pass such tests because: (1) such waste is largely comprised of water-soluble sodium salts with which conventional cements do not form insoluble minerals; (2) with the exception of the calcined waste stored at INEEL, DOE has not used waste pretreatment technologies (e.g., calcination) that make these high-sodium salty wastes more compatible with conventional grouts; and (3) the leach tests are preceded by size-reducing operations (grinding, crushing, and so on) that deliberately destroy the specimen's physical integrity and compromise physical encapsulation of soluble constituents. This approach undercuts a chief advantage of any cementitious technology: the ability to make large monolithic waste forms relatively simple and cheap.

The hydroceramic (HC) process[4,5] addresses the problem of leachable salts by trapping sodium salt molecules within an aluminosilicate "cage mineral" (zeolite, sodalite, cancrinite, and so on) assemblage formed during the curing step. HC uses the same approach developed for the Formed Under Elevated Temperature and Pressure (FUETAP) cement waste form,[6] but substituting metakaolin and sodium hydroxide for portland cement. The HC process cooks the waste with an aqueous solution containing high concentrations of silicates, aluminates, and hydroxides, converting the reactive ions into poorly soluble minerals. The addition of a small amount of sulfide or powdered vermiculite to the formulation further enhances chemical fixation of certain key contaminants (e.g., cadmium, chromium, technetium, and cesium).

6.5.2 HYDROCERAMICS VERSUS GEOPOLYMERS

HCs represent a subset of the "geopolymeric" (GP) concretes mentioned in Chapter 4, in that both are made by activating a powdered aluminosilicate substrate (pozzolan) with a basic solution. However, both the formulations and finished products are different, as shown in Table 6.5.1.

Table 6.5.2 characterizes several aluminosilicate concretes, the raw materials from which they were made, and a pelletized, commercial, sodium-form "Zeolite A." Two of the concrete specimens contained 50 wt% of a calcine produced by "steam reforming" a slurry of an INEEL liquid waste simulant with kaolin[7] at 720°C (the gross composition approximates nepheline). The data in Table 6.5.2 illustrate the following points:

1. All aluminosilicate concretes exhibit a great deal more cation exchange capacity (zeolitic behavior) than do the materials from which they are made.
2. Sodium-based activators generate more ion exchange (IX) capacity than do potassium-based activators.
3. Concretes made with sodium-based activators do a better job of retaining their alkali when leached with water.
4. Metakaolin produces products with higher IX capacity and greater leach resistance than does Class F fly ash.
5. There is little correlation between the concrete's physical strength and its leach resistance.

The physical characteristics (strength, appearance, and porosity) of HC concretes resemble those of portland cement-based radwaste concretes (grouts). The leach behavior of constituents (most fission product elements, uranium, TRU, the alkaline earths, fuel cladding metals, plus RCRA-characteristic metals) is also similar to other waste forms that generate an alkaline buffer solution under leach test conditions.[8] Again, HC concrete's outstanding characteristic is that it surpasses the performance of DOE's benchmark glass using the current standard leach tests.

6.5.3 MANUFACTURE OF HYDROCERAMIC CONCRETE

6.5.3.1 Materials

Raw HC grout consists of a mixture of waste (typically 30 wt% dry-basis), metakaolin (calcined kaolin), ~5% powdered vermiculite (to enhance ^{137}Cs fixation), ~0.5 wt% sodium sulfide (redox buffer and RCRA metal precipitant), plus ~10% sodium

TABLE 6.5.1
Comparison of Hydroceramic and Geopolymeric Concretes

	Hydroceramic	Geopolymer
Pozzolanic substrate	Metakaolin only: no other single aluminosilicate pozzolan or combination of individual alumina/silica sources works as well	GPs have been made with virtually every conceivable combination of alumina-containing pozzolans: raw clay, calcined clay, fly ash, powdered feldspar, etc. commercial GP cements also often contain a substantial amount of a C-S-H forming component, usually granulated blast furnace slag (GBFS)
Activating solution	Concentrated (typically 8–20 M) NaOH solution only: choose a NaOH concentration that simultaneously provides enough alkali (see Formulation guidelines) and satisfies the mixture's water demand	Strong (typically 8 M) solution of sodium or potassium hydroxide solution; usually contains a good deal (1–4 M) of soluble silica, too
Formulation guidelines* (approximate)	1. The mix should contain roughly equal amounts (atom-wise) of sodium, aluminum, and silicon 2. Emulate cancrinite; i.e., less than 25% of the sodium should be present in forms (salts) other than oxide, hydroxide, aluminate, or silicate 3. At least 20% of the sodium should initially be present as NaOH	1. The $SiO_2:M_2O$ ratio of the activating solution should be between 4:1 and 6.6:1 2. The pozzolanic substrate must provide some base-soluble alumina 3. The mixture's overall $Al_2O_3:SiO_2$ ratio should be between 1:5.5 and 1:6.5
Curing conditions	Various hydrothermal: e.g., 1 month at 95°C; 10 hours in a culinary-type (~118°C) autoclave; or 1 hour in a 200°C autoclave	Not hydrothermal; typically a few weeks at room temperature or a few hours at 50–80°C
Chemical characteristics	Behaves like cancrinite: i.e., exhibits substantial cation exchange capacity but doesn't readily release alkali to pure water	Exhibits a good deal of cation exchange capacity but does not resist water leaching of its alkali
Physical characteristics	Generally chalk-like, unconfined compressive strength typically 5–15 MPa	Hard, brittle, often glassy
Primary virtues	Passes DOE's leach tests, low overall water solubility, high cation exchange capacity, compatibility with silicate rock (i.e., repository environment)	Superior to ordinary portland cement for both repository construction and backfill/overpack applications

* The GP "formulation rules" are taken from Davidovit's U.S. patent (# 4 349 386, 1982) – other workers have since demonstrated that the numeric ratios quoted above are overly restrictive.

TABLE 6.5.2
Characteristics of Aluminosilicate Concretes

Specimen[a] (Type, Pozzolan, Activator, Cure)	UCS[b] MPa	Meq/gram[c] Alkali (M+)	Fraction Water-Leachable[d] M+,%	Ion Exchange Capacity[e] meq/g	IX cap/M+ %
HC, MK,18.7 M NaOH, 45 min @ 200°C	14.5	5.9	6.1	1.12	19.1
GP, MK+calcine, 7.9 M Na + 4.1 M Si, 3 hr @ 70°C	9.0	4.36	17.8	1.11	25.3
GP, FA+calcine, 7.9 M Na + 4.1 M Si, 3 hr @ 70°C	11.0	4.4	33	0.25	5.6
GP, FA, 8 M KOH, 3 hr @ 70°C	10.6	1.1	87.5	0.047	4.3
GP, FA, 8 M K + 2 M Si, 3 hr @ 70°C	25	1.06	71.8	0.106	9
GP, MK, 8 M NaOH, 3 hr @ 70°C	6.6	2.45	38.8	0.62	25.1
GP, MK, 8 M Na + 2 M Si, 2 hr @ 200C	3.4	2.54	12.2	0.81	31.5
Commercial Type 4A-8 Mole Sieve	NA	4.5	6.0	2.63	58.3
MK only	NA	< 0.1	NA	0.033	NA
FA only	NA	< 0.1	100	0.001	0.01
Raw "steam reformer" calcine	NA	5.97	3.9	0.013	0.2

[a] HC = hydroceramic; GP = geopolymer; MK = metakaolin (42% Al_2O_3); FA = Handy™ Class F fly ash (15% Al_2O_3); "calcine" = "bed" fraction of the product made by fluidized bed steam-reformation of an INEEL SBW stimulant slurried with kaolin; activators were made by boiling silica fume in concentrated Na/K hydroxide solutions and then diluting to the indicated M+/Si concentrations.
[b] Unconfined compressive strength determined with a homemade tester and tiny specimen (for comparative purposes only).
[c] Total alkali in the formulation.
[d] Determined by passing 10 bed volumes of 1 M NH_4Cl through the same powder, rinsing with water, displacing the ammonium ion with ten bed volumes of 1 M $NaNO_3$, and then determining it in the eluate.
[e] Determined by measuring the amount of alkali in 25 bed volumes of ~90°C water that had been passed through the powdered (50- to 212-micron) sample.

hydroxide dissolved in enough water to produce a stiff paste. The metakaolin does not have to be particularly pure nor ground fine. Suitable material can be cheaply produced (~$70/ton) by processing crude kaolin in a "cool" (~700°C) cement kiln, then grinding it, just as clinker is ground into portland cement. The caustic "activator" need not be pure either. The "waste" NaOH existing within the DOE Complex can be profitably employed to make HC waste forms. For example, the INEEL recently reacted 150,000 gallons of liquid metal reactor coolant (sodium) with water, producing enough NaOH to convert its entire inventory of calcined reprocessing waste into HC waste forms, a scenario that treats both wastes for off-site disposal.

6.5.3.2 Mixing

Because HC grouts tend to set rapidly, mixing should be done quickly with equipment that is capable of forcefully ejecting the product, e.g., a mixer-extruder. The mold (canister) should be situated on a vibrating table to float out air bubbles before the grout sets. The fresh grout will set, or harden, within a few hours at room temperature.

6.5.3.3 Curing

The curing rate is governed by the size of the metakaolin particles, the amount of hydroxide in the formulation, and temperature. If the metakaolin particles are too large, salt molecules will not diffuse into them before their surfaces are effectively sealed with reaction products. If they are too small, the grout is apt to exhibit excessive water demand, set too rapidly, and produce a physically weak and overly porous product.

Figures 6.5.1 and 6.5.2 illustrate the effects of different curing regimens on the fractions of sodium and nitrate leached from a typical HC by the Product Consistency Test (PCT). HC waste forms could be fully cured within a few hours at a temperature (140°C) corresponding to a steam pressure of 0.23 MPa (35 psi). Domestic water heater tanks are built to withstand pressures up to ~1 MPa (150 psig,) and represent a cheap source of grout canisters. An alternative is to store canisters at atmospheric pressure for several months in an insulated storage shed at a temperature approximating that of boiling water. Although ~100°C may produce a different mineral structure than higher temperatures and pressures, the lower temperatures still produce waste forms capable of passing current standard leach tests.

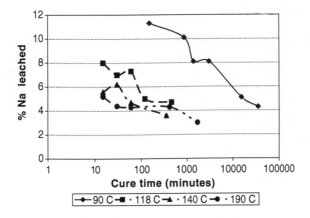

FIGURE 6.5.1 Curing Effect on PCT Sodium Leachability.

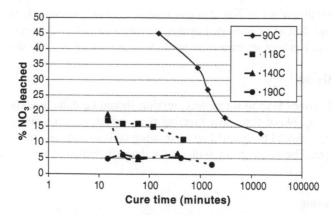

FIGURE 6.5.2 Curing Effect on PCT Nitrate Leachability.

6.5.3.4 Waste Pretreatment

The primary load-limiting characteristic of the HC process is that cancrinite (or sodalite) represents maximal waste loading; i.e., no more than 25% of the formulation's sodium can be in forms other than hydroxide, silicate, or aluminate. Since sodium speciation of most DOE salt wastes is inconsistent with this guideline, efficient implementation of the process would require pretreatment. The most straightforward way to render a high-nitrate/nitrite salt waste into something compatible with the process would be to slurry it with a water-soluble carbohydrate (sugar or starch, about 35 grams per mole of nitrate) plus enough kaolin to suppress CO_2 retention (about one third of the total needed to make the ultimate product) and then calcine it. A rotary kiln would probably prove to be more reliable than a fluidized bed reactor for this purpose.

6.5.4 LEACH BEHAVIOR

6.5.4.1 Intrinsic Solubility

$NaNO_3$-cancrinite ($Na_8Al_6Si_6O_{24} \bullet 2NO_3 \bullet xH_2O$) is the prototypical hydroceramic "cage mineral."[9,10] Heating condensed silica fume in 8 M sodium hydroxide containing excess aluminum nitrate for 16 hours at 90°C in air-tight containers resulted in ~1-μ particles with a chemical composition consistent with cancrinite having eight waters of hydration. Table 6.5.3 lists the equilibrium concentrations of sodium, aluminum, and nitrate in 25-ml aliquots of various solutions heated with 0.11 gram-aliquots of the powder.

These data support the following conclusions:

1. Cancrinite (HC) water solubility is lowest at near-neutral pH.
2. The predominant cations in normal groundwater (sodium, potassium, calcium, and magnesium) suppress cancrinite (HC) solubility.

TABLE 6.5.3
Effect of Various Solutes on the Water Solubility of Pure Sodium Nitrate-Cancrinite

Leachant	pH/cond (µS/cm)	ppm Na	ppm Al	ppm Si	ppm NO$_3^-$
Pure water	9.3/426	100	52	64	118
Same powder, fresh water	9.2/366	86	65	77	107
DI water + 0.25 g dry metakaolin	9.0/218	55	34	38	13
Silica-saturated water (initially 38 ppm SiO$_2$)	9.0/459	111	46	96	133
0.1 M NaCl	8.5/dnd	dnd	6.8	3.5	15
0.05 M MgCl$_2$	8.1/dnd	77	1.5	3.2	36
0.05 M CaCl$_2$	8.1/dnd	133	4.8	0.7	18
0.002 M Ca(HCO$_3$)$_2$	7.1/440	63	0.5	6.3	15
0.1 M LiOH	dnd	633	66	120	421
0.015 M Al(NO$_3$)$_3$	4.3/dnd	810	dnd	280	dnd
0.13 M KNO$_3$	8.8/dnd	271	13.8	4.1	dnd
INEEL tap water	8.5/477	72	1.9	5.9	20

* dnd = did not determine

3. Dissolved silica alone (i.e., silica present as H$_4$SiO$_4$ — not as an anion and thereby accompanied by an equivalent number of cations) does not affect cancrinite (HC) solubility.

This, in turn, suggests that dissolution (leaching) of cancrinite (HCs) involves both ion exchange and hydrolysis. Like most silicate soil minerals, cancrinite particles behave rather like sodium-form "weak acid" cation exchange resin; i.e., exhibiting substantial cation exchange capacity and having a high affinity for the hydronium (H$^+$) ion. Consequently, when cancrinite is exposed to pure water, hydrolysis ensues to provide H$^+$, which displaces an equivalent amount of sodium ion into the solution (100 ppm Na 0.004 M). The majority of the hydroxide ion simultaneously produced attacks the mineral's backbone to displace an equivalent amount of anions (silicate, aluminate, and nitrate) — the remainder serves to slightly boost the solution's pH (to ~9.3, [OH$^-$] 0.00002 M). When the water contains salts, the mass action effect of the cations suppresses hydrolysis (and, consequently, dissolution of the mineral) by occupying its cation exchange sites. The most important conclusion to be drawn from this is that the solubility of HC waste forms would be considerably lower in typical desert groundwater (like that under Yucca Mountain) than in the pure water leachants specified by most leach protocols.

6.5.4.2 Standard Leach Protocols

6.5.4.2.1 *Product Consistency Test (PCT)*
Most current WACs specify use of the PCT.[11] For "high" waste forms, the Yucca Mountain WAC decrees that a candidate material must lose a lesser fraction of its

TABLE 6.5.4
7-Day PCT Results

Material	Normalized Sodium Leach Rate (g/m²/day)
EA glass	0.96
PUREX glass	0.73
SRL-131 glass	0.67
HC#1 NaAlO$_2$/NaOH/ Crude metakaolin	0.40
HC#2 NaOH, NaNO$_3$ (25 wt% of the Na), Crude metakaolin	0.39 (2.6 wt% of the NO$_3^-$ leached)
HC#3 38 wt% alumina-type INEEL calcine/NaOH/DEA/Crude metakaolin	0.40
HC#4 46 wt% INEEL zirconia calcine/NaOH/Crude metakaolin	0.43
HC#6 NaOH/Englehard Metamax™ metakaolin, 9-hr cure @ 200°C	0.13 (ANSI/ANS-16.1-2003 LI$_{Na}$ = 11.6)

bulk constituents (for a borosilicate glass that means sodium, boron, lithium, and silicon) than does a benchmark "Environmental Assessment" (EA) glass. Because the PCT involves 1-week exposure of a relatively large specific surface area (SSA) to a small volume (V) of 90°C water (i.e., 1 g of 75- to 150-µm sample particles per 10 cc water — SSA/V ≈ 30 cm^{-1}), it generates an estimate of the material's solubility under conditions that promote back reaction or saturation. Table 6.5.4 compares several HCs with "representative" radwaste-type glasses[12] with respect to the dissolution of sodium (their most readily solubilized common component) under PCT conditions. The HCs outperformed these glasses even more for constituents such as ^{90}Sr and TRU.

To make the PCT more relevant, more conservative, and more accurate for these sorts of applications, the following modifications were used for the HC samples tested:

1. No minimum particle size (< 150 µ).
2. No prewashing of the sample powder.
3. For the surface area, use the geometric surface area for 150-micron spheres. (This underestimates an HC sample's "true" [BET] surface area by about three orders of magnitude.)

The reasons for these changes are as follows: a) calcines and calcine-containing concretes are usually inhomogeneous at the relevant size scale (75 to 150 µm), which means that the deliberate screening out of a specific size subsample is apt to generate a nonrepresentative result; b) a brief "prewash" meant to remove surface dust from 75- to 150-µm glass particles is apt to seriously bias results generated from a

TABLE 6.5.5
TCLP of HC Concrete Made with Sugar-Calcined
INEEL SBW

Metal	μg/g in the Leachate	UTS Limit (μg/g)	Concentration in Calcine (μg/g)
As	< 0.002	5	10.8
Ba	0.35	7.6	48
Cd	0.13	0.14	1372
Cr	0.023	0.86	950
Pb	< 0.1	0.37	1500
Se	< 0.002	0.16	6.9
Ag	< 0.1	0.3	1510

chemically inhomogeneous, intrinsically porous material (most calcines and calcine-containing concretes); and c) since the nominal purpose of performing this test is to compare one material to another (e.g., concrete vs. EA glass), assuming vastly different surface areas will produce misleading (grossly biased) results.

6.5.4.2.2 Toxicity Characteristic Leaching Procedure (TCLP)

Most INEEL reprocessing wastes, calcined or raw, fail the TCLP,[13] typically for cadmium, mercury, chromium, and lead. Table 6.5.5 lists the results of a TCLP test of an HC made with a sugar-calcined INEEL "sodium bearing waste" simulant, which had been doped with unrealistically high levels of several RCRA metals.

6.5.4.2.3 MCC-1 Leach Test

FUETAP concrete performed poorly at retaining ^{137}Cs as measured by the MCC-1 leach protocol.[14] Consequently, improving ^{137}Cs leach performance measured by that particular test was a key goal during the HC development. The MCC-1 exposes a small monolith of known composition and geometric surface area to a relatively large volume of 90°C distilled water for 28 days. Normalized leach rates (gram/m^2/day) for each constituent are calculated from the concentrations measured in the leachates.

Table 6.5.6 compares 28-day MCC-1 leach performance of two HC-type concretes with those of four radwaste glasses and a recently developed "glass bonded sodalite" ceramic.[15] The first HC specimen (HC#1) contained 42 wt% of a representative INEEL zirconia-type calcine, crude metakaolin (a cement-extender pozzolan produced at a local cement plant), vermiculite, a small amount of sodium sulfide, plus sodium hydroxide. The second (HC#2) was similar except that the waste simulant was a mix of all three major types of INEEL calcines (alumina, zirconia, and fluorinel-blend). The HCs surpassed all of the vitrified materials with respect to Cs leach performance.

The relative ranking of glasses and HCs with respect to sodium dissolution is often different with the PCT and MCC-1 tests. The reason for this is that a chunk of HC presents several orders of magnitude more surface area to the leach water

TABLE 6.5.6
28-day MCC-1 Leach Rates (g/m²/day) of Two HCs, Four Glasses, and a Hot-Isostatically Pressed Glass Ceramic*

Material	Na	Cs	Sr	Al	Si
EA glass*	1.4	1.6	0.20	0.57	1.1
SRL-131*	1.1	0.89	0.14	0.50	0.86
WV-39-2*	4.7	5	2.0	3.85	3.8
JSSA*	0.39	0.1	0.27	0.18	0.31
ANLW glass ceramic*	0.39	0.39	0.14	0.57	0.14
HC#1	13	0.096	0.012	1.6	0.54
HC#2	15	0.07	0.024	0.72	0.75

* Glass and glass ceramic leach data taken from Reference 15.

than does an equal-sized piece of glass. This means that a glass may out-perform the HC in short-duration leach tests even if the former is intrinsically much more water soluble. At the end of a 7-day PCT, the composition of the leachant exposed to powdered HC represents steady-state (saturation). In contrast, the leach rate of EA glass at that time is rapidly increasing (EA glass dissolution is autocatalytic under PCT conditions). If EA glass were to be ground to a size that presents as much BET surface to a leachant as does < 150-μm HC powder, it would totally decompose (dissolve) within 1 hour under PCT leach conditions.

6.5.4.2.4 ANSI/ANS-16.1-2003

Table 6.5.7 gives the results of an ANSI/ANS-16.1-2003 leach test[16] performed on one of the HC specimens (HC#1) described above. The goal of the ANSI/ANS-16.1-2003 is to determine the mobility of individual constituents *within* the sample. A small monolithic specimen of known composition and surface area is repeatedly leached with a relatively large volume of fresh deionized water. The fractions of key sample constituents in the resulting leachates are combined with the specimen's physical size and the leach times to generate effective diffusion coefficients (D, in

TABLE 6.5.7
ANSI/ANS-16.1-2003 HC Leachability Indices

Time (hrs)	Na	Cs	Zr	Sr	NO₃⁻
Leachability Index	9.9	13.7	> 15.4	14.3	10.4
Total wt% leached	8.26	< 0.0099	< 0.0025	0.015	10.5

</> values are based upon the detection capabilities of the analytical instrumentation: ICPAES for all metals except Cs, graphite furnace AAS for Cs, and ion chromatography for nitrate.

units of cm^2/s) for each component during each interval. The test results are generally reported in terms of a leachability index (LI), where $LI = -(\Sigma log_{10} D)/n$ for each constituent. Note that two of the most readily solubilized bulk constituents of U.S. reprocessing waste (sodium and nitrate) evinced diffusivities four orders of magnitude lower (better) than the usual U.S. radwaste grout WAC standard (10^{-6} cm^2/sec or a LI of 6.0). Note also that the concrete's polyvalent cationic components (e.g., Zr and Sr) evinced diffusivities several orders of magnitude even lower. A substantial drop in nitrate mobility (D) was observed between the first and subsequent leach intervals indicating that the curing process had trapped ~90% of the nitrate in "cage molecules." The 10% left in the pore water leached ("diffused") at a rate ($D \sim 10^{-8}$ cm^2/sec) typical of "good" conventional grout.

6.5.4.2.5 Vapor Hydration Test (VHT)

Currently, the most realistic glass durability test is the Vapor Hydration Test[17] because of the reasonable assumption used about how water would interact with a "naked" (breached canister) waste form in a vadose-zone repository such as Yucca Mountain and the quantitative transformation of glass to "alteration products." The VHT is performed by suspending a wafer of sample above the bottom of a stainless steel vessel with a platinum wire. After addition of sufficient water to saturate the gas phase at the desired temperature (typically 200°C), the vessel is hermetically sealed and put into a preheated oven for a time ranging from several days to several weeks. The thin film of liquid water that forms on the surface of the specimen is maintained by the condensation of steam. In practice, the corrosion rate of the glass remains low until the concentrations of ions leaching into that film reach a level that initiates the formation/precipitation of stable minerals under such conditions, primarily felspathoids (analcime, sodalite, cancrinites, and so on), zeolites, and silicates. Because the hydroxide ion tends to remain (accumulate) in the water, the overall glass corrosion (alteration) rate gradually increases until it becomes limited by the availability of the rate-limiting constituent of the product-forming reactions (usually aluminate) in the remaining glass. After the test is complete, the corrosion/alteration rate is determined by measuring the thickness of the remaining (unaltered) glass layer with a microscope.

Since VHT conditions are essentially identical to those used to cure HC concretes, a properly made specimen will not change (corrode/alter) when tested. Furthermore, since the application of successively more rigorous hydrothermal curing conditions tends to improve the leach performance of HCs[5], any slight changes that do occur are apt to be for the better. Since HC-type concretes are cured under conditions generally considered to represent a "worst case" repository scenario (hydrothermal), it is reasonable to conclude that they would prove to be more rugged than glasses if those conditions were actually realized. A fundamental difference between HC and high-level glass formulations is that the latter do not contain sufficient aluminum to form an intrinsically insoluble mineral assemblage with 100% of the alkali present. This means that a natural "alteration" process cannot convert these glasses into low solubility minerals.

6.5.5 APPLICATION

The HC process was specifically developed to deal with INEEL reprocessing waste. INEEL waste is uniquely suited for cementitious solidification because overwhelming amounts of salt-forming caustic were never added to it and ~95% of it was calcined. Like most of DOE's high-level waste, its total radionuclide content (~30 watts/m^3) is well under that of some of the "intermediate level" wastes routinely converted to 0.5-m^3 concrete waste forms at Great Britain's THORP facility. A portion of the reprocessing waste generated at other DOE fuel reprocessing facilities could be directly processed into HC-type concrete (e.g., Scientific Review Panel's strongly caustic "sludge washing" solutions[18] or the sludges now destined for its glass melter), but the majority (the "supernates" and "salt cakes") is unfit for making anything but "saltstone" (or glass) unless first pretreated in a way that re-speciates the sodium. "Steam reforming" is the only calcination technology still viable in the DOE Complex that could conceivably be employed to accomplish this.

REFERENCES

1. Bell, J., *Nuclear Technology*, 130, 89, 2000.
2. "Idaho Clean-up Project Statement of Work," DE-RP07-03ID14516, February 2004.
3. Palmer, J. D. and Fairhall, G. A., *Cement and Concrete Research*, 22, 325, 1992.
4. Siemer, D. D., Olanrewaju, J., Scheetz, B. E., Krishnamurthy, N., and Grutzeck, M. W., Development of Hydroceramic Waste Forms, *Ceram. Trans.*, 119, 383–390, 2001.
5. Siemer, D. D., Olanrewaju, J., Scheetz, B. E., and Gruzeck, M. W., Development of Hydroceramic Waste Forms for INEEL Calcined Waste, *Ceram. Trans.*, 119, 391–398, 2001.
6. Dole, R. L. et al., Cement-Based Radioactive Hosts Formed Under Elevated Temperatures and Pressures (FUETAP Concretes) for Savannah River Plant High-Level Defense Waste, ORNL/TM-8579, March 1983.
7. Soelberg, N., R., Marshall, D. W., Bates, S. O., and Taylor, D. D., Phase 2 THOR Steam Reforming Tests for Sodium-Bearing Waste Treatment, INEEL/EXT-04-01493, January 30, 2004.
8. Conner, J. R., *Chemical Fixation and Solidification of Hazardous Wastes*, Van Nostrand Reinhold, New York, p. 13, 1990.
9. Barney, G. S., Fixation of Radioactive Wastes by Hydrothermal Reactions with Clays, *Advances in Chemistry Series* 153, American Chemical Society, Washington, D.C., pp. 108–125, 1976.
10. Brownell, L. E., Kindle, C. H., and Theis, T. L., Review of Literature Pertinent to the Aqueous Conversion of Radionuclides to Insoluble Silicates with Selected References and Bibliography (revised), ARH-2731 Rev., 1973.
11. Jantzen, C. M. et al., Characterization of the Defense Waste Processing Facility (DWPF) Environmental Assessment (EA) Glass Standard Reference Material, WSRC-TR-92-346, Rev 1, June 1, 1993.
12. Shi-Ben, Xing and Pegg, I. L., Effects of Container Materials on PCT Leach Test Results for High-Level Nuclear Waste Glasses, *Mat. Res. Soc. Symp., Proc., Vol. 333*, pp. 557–564 (Scientific Basis for Nuclear Waste Management XVII), 1994.
13. U.S. EPA, SW-846, *Test Methods for Characterization of Solid Waste: Physical/Chemical Methods, 3rd Edition*, Method 1311, 1996.

14. Leach Testing using MCC-1-1P and MCC-3S Test Methods, PNL Technical Procedure PSL-417-LCH, 1989.

15. Benedict, R. W. and McFarlane, H. F., *Radwaste Magazine*, pp. 23–28, July 1998.

16. American Nuclear Society, American National Standard Measurement of the Leachability of Solidified Low-Level Radioactive Wastes by a Short-Term Test Procedure, ANSI/ANS-16.1-1986.

17. PNNL Procedure No. GDL-VHT, Rev. 1, is a "proceduralized" version of the VHT.

18. Bao, Y., Kwan, S., Siemer, D. D., and Grutzeck, M. W., Binders for Radioactive Waste Forms Made From Pretreated Calcined Sodium Bearing Waste, *J. Mater. Sci.*, 39, 481–488, 2004.

7 Interactions between Wastes and Binders

Julia A. Stegemann

CONTENTS

7.1 INTRODUCTION

Wastes treated by stabilization/solidification (S/S) may originate from a great variety of industrial processes (Chapter 3). Inorganic wastes tend to be more compatible with cementitious binders; most wastes treated by S/S with cement contain metal contaminants in an inorganic matrix composed mainly of calcium, aluminum, and silicon, such as dusts from air pollution control systems, sludges, and soil. USEPA has acknowledged that organic compounds interfere with cement-based S/S, particularly when the organic concentration exceeds 1% total organic carbon by mass.[1] However, there are numerous instances in which organic wastes have been solidified with cement,[2] and many primarily inorganic wastes contain some organic contamination. In fact, although some wastes can be successfully blended or treated with cement in higher proportions than others, there are virtually no limits on their physical and chemical characteristics, other than that they must be liquids or finely divided solids at ambient temperature. Wastes may contain some of the chemicals, e.g., $CaCl_2$ or gypsum, that are used as additions and admixtures in conventional cement and concrete applications (see also Chapter 8), but also contain many different compounds that would not otherwise be blended with cement, including harmful contaminants. In the context of this chapter, all components of wastes are considered "impurities," i.e., potentially reactive materials that would not ordinarily be present in a commercial cement.

In S/S, interaction of wastes with binders is of interest from two main standpoints:

1. Interferences of impurities with cement hydration, including setting and strength development, and matrix durability
2. Immobilization of contaminants

These aspects are linked, in the sense that development of a strong, durable matrix of low permeability is usually important for contaminant immobilization, and is therefore a goal of S/S treatment. The focus of this chapter is on both beneficial and deleterious interactions that can occur between waste components and hydraulic binders ("cement").

7.2 INTERFERENCES WITH BINDER HYDRATION

7.2.1 INTRODUCTION

The detailed process of cement hydration still remains a subject of intensive study and differs for different types of cement, as discussed in Chapter 4. In general, it proceeds by dissolution of the anhydrous cement phases from the surface of the cement particles in the mixing water, followed by precipitation of hydration products to build a strong cohesive matrix. Impurities may alter normal cement hydration at different stages; therefore, different compounds have different, and sometimes multiple, mechanisms of interaction with the binder. By changing the chemistry of the evolving pore solution with respect to composition, pH, and ionic strength, impurities can change:

- The solubility of the anhydrous phases
- The dissolution kinetics of the anhydrous phases
- The nucleation rate of the hydration products
- The growth rate of the hydration products
- The composition of the hydration products
- The morphology of the hydration products

These effects are often concentration dependent and can vary according to the curing conditions. When impurities with different interference mechanisms as pure compounds are combined, as is likely to occur in both waste utilization and solidification applications, the overall effect can be difficult to predict.[3]

Setting and hardening are physical manifestations of the chemical process of hydration of cements. Taylor[4] defines setting as "stiffening without significant development of compressive strength," hardening as "significant development of compressive strength." Thus, interferences with cement hydration may result in the following bulk effects:

- Acceleration/activation of setting or hardening (including flash setting, in which the matrix loses its plasticity immediately upon mixing)
- False setting (in which matrix plasticity is lost quickly upon mixing, but can be recovered by additional mixing)
- Altered water demand
- Retardation of setting or hardening (including complete inhibition of hydration)
- Altered strength development (including matrix disruption)
- Altered pore solution composition

From a use standpoint, acceleration of setting and false setting may cause handling difficulties, leading to equipment failure and poor workability. Increased water demand, which may be caused by acceleration of setting or false setting, or simply by altered rheological characteristics of the mix, can result in a more porous structure and poor durability, as can retardation. Altered strength development is often thought of as taking the form of increased or decreased early strength, but later strength may also be affected, and even a product that appears to set and gain strength normally at first can deteriorate rapidly later.

Numerous examples of interactions of impurities with cement, in either S/S or cement and concrete products, may be found in the literature, and have also been discussed in a number of review articles.[2,3,5-10] Findings from the literature are summarized in the following sections, for inorganic compounds, organic compounds, and other impurities. These results have been generated by experiments using pure compounds; numerous case studies also exist in the literature for treatment of real wastes. Because of the heterogeneity and complex structure and composition of real wastes, it is more difficult to generalize about these, but an attempt to address this issue is made in Section 7.2.5.

It should be noted that many of the observations in the literature are based on studies of the physical manifestations of setting and strength development, or sometimes on measurements of heat evolved by hydration reactions, but it is stressed that the effects discussed above do not simply change the rate of setting or hardening, but often alter the hydration products. At the simplest level, retardation decreases the quantities of the hydration products at a given time, whereas acceleration increases them. However, impurities in cement pastes may also alter the proportions of the hydration products, change the Ca/Si ratio of the C-S-H, create solid solutions with the hydration products, or create entirely new hydration products. Several recent studies have shown effects of wastes and contaminants on leachate pH and S/S product acid neutralization capacity, which are indicative of changes to the hydration products.[11-14] Such changes can have important consequences for immobilization of contaminants by all of the mechanisms discussed in the following Section 7.3, but this is an aspect of S/S of which there have been few investigations.

Since it is clear from the foregoing that cements respond, often dramatically, to small additions of impurities, it is possible to use additives and admixtures to adjust the hydration characteristics of an S/S product to mitigate undesirable effects of waste components. This is discussed further in Chapter 8.

7.2.2 EFFECTS OF INORGANIC COMPOUNDS ON CEMENT HYDRATION

Table 7.1 and Table 7.2 summarize the effects observed for a number of inorganic and heavy metal compounds, respectively, on cement hydration, as described in the literature.[2-6,8,9,12,15-24] The terms "acceleration" and "retardation" have been applied to describe effects on setting, hardening, or both, as this is how they are used in the literature. It should be noted that this is a somewhat misleading practice, as an impurity may cause, for example, initial acceleration, followed by delayed hardening, or vice versa. Conflicting reports about the effects of different impurities may be at least partly attributable to this practice.

The information in Table 7.1 and Table 7.2 reflects the preponderance of investigations concerning portland cements in the literature, but may sometimes also be true for other cement types, or blended cement systems. For example, it has been shown that metal cations such as Zn^{2+}, Pb^{2+}, Cd^{2+}, and Cr^{6+} also decrease the strength of alkali-activated slag cement.[25] Some of the effects are discussed in more detail in the following notes, which refer to the superscripts in the tables:

A. As the most reactive phases in portland cement, and those most important for setting and initial strength development, C_3A and C_3S are accelerators, and also tend to be the focus of action by other accelerators and retarders. Other reactive calcium aluminates or calcium aluminate cements are also accelerators for portland cement, and vice versa. These species act as activators or accelerators for pozzolans. Cement kiln dust or lime may also be used as activators for pozzolans.

B. Alkali carbonates illustrate the unpredictable nature of some compounds that interfere with the setting of cement. Low proportions of alkali carbonates ($< 0.1\%$) have been found to retard the setting of portland cement; an increased amount results in flash setting, and further increased amounts may have no effect on setting, whereas flash setting occurs also at very high proportions. The interference mechanism is not well understood, but it has been suggested that the effect of carbonates is due in part to the production of thaumasite $[Ca_6(Al,Si)_2(SO_4)_2(CO)_3.(OH,O)_{12}.24H_2O]$, rather than ettringite $[Ca_6Al_2(SO_4)_3(OH)_{12}.26H_2O]$.

C. In general, sodium and potassium salts are thought to raise the pH and cause precipitation of amorphous CH that interferes with C_3A hydration, and yet many alkali salts act as accelerators of portland cement. They may be added as activators to cements containing pozzolanic materials (Chapter 4), where they increase the solubility of the unhydrated phases. Some salts, such as NaCl, accelerate at low concentrations, but retard at very high concentrations.

D. Sodium silicate is a popular additive in waste solidification. Providing both silicate and a high pH, it may be used as an accelerator or activator for pozzolanic materials, or to consume excess water. Particularly in the latter instance, or in the event that excessive amounts are added, formation of silica gel may lead to a physically unstable matrix, with swelling and shrinkage caused by changes in the humidity of the surrounding environment.

E. Sulfates also have several possible effects on portland cement hydration. Possibly leading to acceleration or retardation by reaction with C_3A and C_4AF, they may also lead to false or flash setting, by forming gypsum instead of ettringite, or matrix destruction through delayed formation of ettringite, which has a high volume because of its waters of hydration. Thiosulfate is also reported to be an accelerator. Chloride salts can also form voluminous chloroaluminates that are destructive to the matrix if their formation is delayed. Both chlorides and sulfates, as well as carbonates and other anions, can also result in matrix destruction if the solubility of one of their salts in the pore solution is exceeded and crystallization results.[6,26,27]

F. More than a couple of percent of MgO can destroy the cement matrix by gradually hydrating to more voluminous $Mg(OH)_2$; $MgSO_4$ also reacts to form more voluminous products, gypsum and $Mg(OH)_2$, as well as degrading C-S-H.

G. Matrix destruction may also be caused by evolution of gas; for instance, from reaction of aluminum metal or ammonia at high pH.

H. Whereas soluble chromium salts accelerate hydration of portland cement, chromium (III) oxide has little effect on setting. The review conducted by Mattus and Mattus[3] indicated that chromium substitutes for silicon in C-S-H (see also Section 7.3.4), but that the final strength of the matrix is decreased. Chromate, CrO_4^{2-} is thought to act similarly to sulfate and form chromoaluminate crystals ($C_6A[CrO_4]_3H_y$ or $C_4A[CrO_4]H_y$, see also Section 0) that coat C_3A grains. Like sulfate, chromate can also act as an accelerator.

I. The effect of $ZnSO_4$ illustrates the importance of considering the combined effect of both the anion and cation. Whereas Zn in most compounds is generally thought of as a retarder of portland cement, $ZnSO_4$ has been found to be an accelerator at concentrations less than 2.5% and a retarder at concentrations between 2.5 and 5.5%; it completely inhibits cement hydration at higher concentrations.

J. Boric acid is an example of a compound that has a strong accelerating effect on setting of portland cement, but retards hardening. Hydroxides, carbonates, silicates, and aluminates accelerate setting, but retard hardening or leave it unaffected; halides and nitrates accelerate both setting and hardening.

In order to generalize regarding the results of different investigations, separate out the effects of cations and anions, and look for interaction effects, data from 12 literature studies of pure compound additions to portland cement paste were

TABLE 7.1
Summary of Effects of Inorganic Compounds on Cement Hydration

-\+	Sodium or Potassium	Calcium	Magnesium	Other
Silicate	$Na_2O \cdot xSiO_2$ A[3,8],D	$3CaO \cdot SiO_2$[A] A[3,8] $2CaO \cdot SiO_2$[A]		Most A[19,J]
Aluminate	$NaAlO_2$ A[3,8]	$CaAl_2O_4$[A] A $3CaO \cdot Al_2O_3$[A] A[3,8] $4CaO \cdot Al_2O_3 \cdot Fe_2O_3$ R[15] Other A[3,8]		Most A,AA[3,8], A[19,J] Metal[G] D[3,8]
Oxide		CaO A,D[15]	MgO[F] A,D[3,8]	Fe,Al A[18]
Hydroxide	NaOH A[3,8] KOH	$Ca(OH)_2$[A] A[18]		Most A[3,8,19,J]
Carbonate	Na_2CO_3[B] R,A,0,AA[18] K_2CO_3[B] $NaHCO_3$ R,A[3,8] $KHCO_3$	$CaCO_3$ A,R[15]		Most R(C₃A)[3,8], A[19,J]
Sulfate	Na_2SO_4 A[3,8] K_2SO_4 R[20]	$CaSO_4 \cdot 2H_2O$ R,D[3,8] $CaSO_4 \cdot \frac{1}{2}H_2O$ A,F,D[3,8] $CaSO_4$ A,F,D[3,8] $CaK_2(SO_4)_2 \cdot H_2O$ A[15]	$MgSO_4$[F] D[3,8]	Most[E] A,F[3,8] R(C₃A)[3,8]
Chloride	NaCl A[3,8], R[4,20]	$CaCl_2$ R(<1%)[18] A(~2%)[3,8] AA(>3%)[18]	$MgCl_2$ A,D[3,8]	$AlCl_3$ A[3,8] NH_4Cl[G] R(<2%),A(>2%),D[3,8] Most R(<1%)[3,8] A(>1%)[3,8] AA(3–5%)[3,8] D A[19,J]

Fluoride	NaF	$A^{3.8}$	CaF_2	Most	$0^{3.8}$	$A^{3.8}$	Most	$R(C_3S)^{3.8}$ $A(C_3A)^{3.8}$ $0^{3.8}, A^{19}$
Other	Most[c]	$A(C_3S)^{3.8}$	Most	$A^{3.8}$	$A^{3.8}, R^5$			
	Most	A						
	Iodate	$R(C_3A)^{3.8}$						
	$NaNO_2$	R^9, A^5						
	$Na_4P_2O_7$	R^{19}						

Species	Effect
Fe^{3+}	$D^{3.8}$
Li,Cs	$A^{3.8}$
S	R^9
Elemental metals[G]	D^9
Acetate	$A^{3.8}$
Phosphate	$R, I^{3.8,19}$
Borate	$A, R^{19}, R, I^{2,3.8}$
Br^-	$A^{3.8,19}$
Thiosulfate	$A^{3.8}$
NO_3^-	$0, A, R^{3.8}, A^{19}$
$Na_2C_2O_4$	$R^{3.8}$

0 = no effect, A = accelerator, AA = flash set, F = false set, R = retarder, I = inhibitor, D = matrix destroyer.

NB: Superscript numbers refer to references; superscript letters refer to notes in text.

TABLE 7.2
Summary of Effects of Heavy Metal Compounds on Cement Hydration

−\+	Chromium(III)	Cadmium	Mercury	Lead	Zinc	Other Metals
Oxide	Cr_2O_3[H] R[3,8,16]	CdO R[16]		PbO R[16,19]	ZnO R[3,8,19,24]	
Sulfate					$ZnSO_4$[I] A,R,I[3,8]; Most	$BaSO_4$ R[3,8]; Most A[3,8]
Chloride	$CrCl_3$[H] A[16]	$CdCl_2$ 0; R; A[16]	$HgCl_2$ A[3,8]		$ZnCl_2$ R-set[12]; A-UCS[12]	$NiCl_2$ A(1–3%)[22]; $CuCl_2$ A[20]
Nitrate	$Cr(NO_3)_3$[H] A[12,16,20,53]	$Cd(NO_3)_2$ 0[3,8]; R[6]	$Hg(NO_3)_2$ R[3,8]	$Pb(NO_3)_2$ R(<2%)[3,8,53]; A(5–8%)[3,8]; R(>10%)[3,8]; 0[12]	$Zn(NO_3)_2$ R(>2%)[3,8]; R-set[20,53]; A-UCS[20]	Most A[3,8]; Mn R[53]
Other	Most A(C_3A)[3,8]	$Cd(OH)_2$ 0,A[12]; Most A[2]; R[19]		Most R[2,3,8,19]	Most Metal[g]; D[3,8]	Ba A[3,8],R(CO_3)[23],O[12]; Co A[3,8]; Cu R[3,8,12]; La A[3,8,17]; Ni A[3,8],O[12]; Sb R[3,8]; Sn R[19]; Ti,Mn O[19]; Fe,Co,Ni O[19]; Chromate[h] A,R(C_3A)[3,8]; Arsenate R[3,8,21]; Vanadate R[21]

NB: Superscript numbers refer to references; superscript letters refer to notes in text.

collected and used to construct neural network models of unconfined compressive strength (UCS) as a function of mix composition.[28] The following ranking was determined for UCS values predicted for addition of the contaminants, on an equimolar basis: at 7 days, Cl ~ Cr(III) > NO_3^- ~ Cd > control > Zn ≥ Ni > Pb > Cu >>Ba; at 28 days, Cl > Cr(III) > NO_3^-~ control ≥ Zn ≥ Cd > Ni > Pb > Cu >> Ba. Application of the best neural network to other data suggested that Cs is a retarder and Cr(VI) has no effect. No trends could be discerned for Hg, K, Mn, Na, and SO_4^{2-} with the available data.

The effects observed in the literature for calcium aluminate cements,[17,29-45] which are less commonly used in waste management due to the superior properties of calcium silicate-based cements, are presented separately in Table 7.3. Unfilled areas in the tables indicate an absence of available information. Some of the effects are discussed in more detail in the following notes, which refer to the superscripts in Table 7.3:

K. There is agreement in the literature that Li salts, with the single exception of Li_2BO_2, are the strongest accelerators for calcium aluminate cements of all types and are capable of causing flash setting. Li salts are thought to act by creating nucleation substrates for the hydration products.

L. However, for the other anions and cations in Table 7.3, varied effects on calcium aluminate cement have been reported. Several authors have tried to rank the effects of various cations and anions, but there is no agreement among them.[34,38] Other authors have tried to come up with general statements to describe the behavior, but these tend to be subject to exceptions. For instance, Parker[29] stated that alkalis are accelerators and acids are retarders of calcium aluminate cement, yet sulfuric acid is known to be an accelerator; Sharp et al.[46] suggested that compounds that increase the C/A ratio act as accelerators, yet $CaCl_2$ is known to be a retarder.

M. The possible diversity of the effects is illustrated by observations for calcium sulfate by Bayoux and coworkers,[43] who found gypsum to have no effect on setting time of calcium aluminate cement, whereas hemihydrate was an accelerator and anhydrite a retarder. The authors suggested that the relations between the dissolution kinetics of these different forms of calcium sulfate controls both the kinetics and the products of hydration.

N. It has been suggested that CH acts as an accelerator for calcium aluminates by increasing the C/A ratio of the solution, and driving precipitation of the calcium-rich phases C_2AH_8 and C_4AH_{13}.[46] portland cement is thought to be an accelerator because hydration of C_2S and C_3S generates CH.[7]

Again, neural network analysis was used to construct models of setting time as a function of mix composition using existing data for pure compound additions to calcium aluminate cements from 17 references.[47] The following ranking of induction times was determined, on an equimolar basis: Ba > Ca ≥ Sr ≥ Mg ≥ K ~ Na > control >> Li and citrate ≥ Br⁻ >> SO_4^{2-} > Cl⁻ > control ≥ CO_3^{2-} ≥ OH⁻ > PO_4^{3-} ~ NO_3^- > AlO_2^-. It was also shown that the reproducibility of setting time measurement is poor, which may account for some of the differences in the findings in the literature.

TABLE 7.3
Summary of Effects of Inorganic Compounds on Calcium Aluminate Cements

$-/+$	Li^K	Na^L	K^t	Mg^L	Ca^L	Ba^L	Other
Br^-	A^{39}	A^{39},R^{34}	A^{39}				
SO_4^{2-L}	$A^{35,39}$	$A^{18,37,39}$ 0^{35} $0,A^{29}$	$A(>30°C)^{37}$ $R(15°C)^{37}$ $0^{29,35},a^{39}$	A^{37} A,R^{29} $0^{35,34}$	A^{30},a^{34} $A,0,R^{43,M}$ 0^{29}	0^{35}	$A(0.5–1\%)^{30}$ $R(<0.25\%)^{30}, R(>0.5\%)^{33}$ $H,Fe-A^{18}$
Cl^{-L}	$A^{29,32,34}$	a^{37} $R^{18,32},r^{29}$ a,R^h $R(<28°C)^{41}$ $0(>28°C)^{41}$	a^{37},A^{36} $R^{18,32}$ $A,0,R^{34}$ A,R^{29}	A^{37},a^{32} R^{29} $a,0,R^{34}$ $R(<x\%)^{38}$ $A(>x\%)^{38}$	$r^{37},R^{29,32,34}$ a^{36} A,R^{34}	$0(15°C)^{37},A^{36}$ $R(>30°C)^{37}$ $R^{18,29}$ $R(<x\%)^{38}$ $A(>x\%,28°C)^{38}$	most-A^{36} Cs-A^{32} H-$R^{18,30}$ $Zn>Cu>Hg>Mn>Sr$ $>Ni>NH_4>Fe>Co$- R^{32}
CO_3^{2-L}	$A^{32,34,35,39,44,45}$	$A^{18,30,39},0^{35,34,45}$	$0^{35},A^{30,39}$	0^{35}			
OH^{-L}	$A^{32,34,35,44,45}$	$a,r^{34,45},A^{18,29,30,35}$	a^{34},A^{30}	$0^{35},R^i$	$A^{f,18,29,30,34,46,N}$	$R^{29,30},A^{35}$	Sr-A^{35}
NO_3^{-L}	$A^{35,39},AA^{34}$	$A^{39},0^{35},R^{18}$	A^{39},a,r^{34}	$0^{35},R^i$		$r,A^{29},0^{35}$	
Other	all-A,$AA^{18,29,32,35,40}$ BO_2-0^{35}	a^{40} F-a^{35} Silicate-A^{30} $(PO_3)_6{}^{6-}$ - $A^{31,46}$ $R(12°C)^{46}$ $0(20°C)^{46}$ $A(>28°C)^{46}$	$a^{40},0^{35},R^{34}$ Silicate-A^{30}				Rb,Cs-A^{40} Sr-R^{34} Portland cement-$A^{30,N}$ NH_4-R^{34} Seawater-R^{42} Acids-R^{29}

AA = flash set, A = accelerator, 0 = no effect, a/r = weak accelerator/retarder, R = retarder.

NB: Superscript numbers refer to references; superscript letters refer to notes in text.

7.2.3 Effects of Organic Compounds on Cement Hydration

Many organic compounds are added as deliberate admixtures to cement and concrete. A selection of these and their effects, taken from Massazza and Testolin[5], are summarized in Table 7.4. Other organic compounds typically found in waste include oil and grease, chlorinated hydrocarbons (e.g., trichlorobenzene), chelating agents (e.g., ethylene diaminetetraacetate [EDTA]), phenols, glycols, alcohols, and carbonyls. The effects of organic compounds on portland cement hydration are variable and highly concentration dependent. For instance, triethanolamine and sugars are normally retarders for C_3S hydration, but can act as both retarders and accelerators

TABLE 7.4
Summary of Effects of Organic Additives on Cement Hydration

Organic Compound	Purpose/Effect on Cement Hydration
Amines:	Grinding aids
Ethanolamines:	Increased water demand, retarded set, may increase early strength
Monoethanolamine	Accelerated set and hardening
Diethanolamine, triethanolamine	Set accelerator, retarder
Melamine derivatives	Set accelerators, retarders, plasticizers
Alcohols	Grinding aids, retarders
Lignosulfonates and their derivatives	Grinding aids, plasticizers; Accelerated or retarded set Gain in final strength for ordinary portland cement
Polycyclic sulfonates	Superplasticizers; retarded set
Lignin compounds	Grinding aids, accelerated set
Fatty acids and their salts (acetic, ascorbic, citric, formic, butanoic, decanoic, hexanoic, oleic, oxalic, propanoic)	Grinding aids, set accelerators (particularly calcium formate), retarders, plasticizers
Alternative treatment:	
Sugars:	Set accelerator, retarder
Lactose	
Raffinose	
Saccharose	
Sucrose	
Phenolic resins	Accelerators
Asphaltenes	Grinding aids, accelerators
Diethylene glycol	Grinding aid; Unchanged or accelerated set
Ethylene glycol (surfactant)	Grinding aid; retarded set
Hydroxycarboxylic acids	Accelerators, retarders, plasticizers
Formaldehyde	Accelerator
Cellulose derivatives	Retarders

TABLE 7.5
Summary of Effects of Organic Contaminants on Cement Hydration

	Effect on	
Organic Compound	Setting	Hardening
Phenol		Strong retarder[49]
Oil		Strong retarder[49]
Grease		Strong retarder[49]
Hexachlorobenzene		Minor retarder[49]
Trichloroethylene		Minor retarder[49]
Toluene	Minor retarder[48]	
Cyclooctane	Moderate retarder[48]	
Hexanol	Strong retarder[48]	

for C_3A hydration, and will cause a flash setting at high concentrations.[3,9] Nestle et al.[48] reported that polar solvents delay hydration to a much greater degree than nonpolar solvents; findings for specific contaminants studied by them and others[49] are summarized in Table 7.5. Many organic compounds also result in progressive deterioration of cement products over time.[6,8]

Most organic compounds are retarders of calcium aluminate cements.[8] Citrate salts, in particular, are commonly used retarders. However, Bier et al.[44] present a way of using Li and citrate together to optimize the setting characteristics of calcium aluminate cements such that retardation does not necessarily result, whereas Baker and Banfill[42] use trilithium citrate as an accelerator, and trilithium citrate had almost no effect in experiments conducted by Damidot et al.[45]

7.2.4 OTHER EFFECTS

Although pozzolans are accelerators of C_3S hydration,[50] they themselves hydrate slowly, and pozzolanic cements tend to gain strength more slowly than portland cement, unless an activator, such as CH or an alkali,[51,52] is used. Pozzolans may mitigate the effects of other impurities,[e.g.,53-55] and are known to reduce the hydraulic conductivity and improve the strength and durability of the final product in the longer term. When added to calcium aluminate cements, pozzolans help to form C_2ASH_8, which resists conversion to less voluminous C_3AH_6.[56-58]

Finally, hydration of a cement is also affected by the physical properties of the cement and any impurities, i.e., crystallinity, crystal defects, and the particle size and surface area of the different mineral phases. Reactivity of binders increases with fineness, but fine impurities, such as clay or colloidal matter, can retard setting.[9]

7.2.5 EFFECTS OF WASTES

The literature contains a large number and variety of case studies concerning S/S treatment of a wide variety of industrial wastes. In fact, the MONOLITH database of cement-based product properties[59] (available from http://www.civeng.ucl.ac.uk/?ID=210) includes 3333 products containing real industrial wastes (as of March 2001). Aside from portland cement, which was the most common binder, and water, these products contain 20 other binders or additives, and 230 different waste types. The diversity of formulations makes it difficult to observe trends in the effects of waste on cement, aside from the general fact that industrial wastes usually have a deleterious effect on cement. However, refinement of this data set into data subsets with common elements was attempted as part of a European project.[60] It was found that the most common waste types were municipal solid waste incinerator (MSWI) ash (European Waste Catalogue Codes 19 01 00, 19 01 01, and 19 01 03), plating sludges (EWC Codes 11 01 03 and 19 02 01), and steel industry dusts (EWC Codes 10 02 03 and 10 09 04). Overall, there were 540 products from 74 references composed of portland cement and these most common waste types. Elimination of products with missing data resulted in 253 products from 18 references, which were used to construct neural network models of UCS as a function of mix composition.[61] As expected, substantial decreases in UCS were caused by all wastes (Figure 7.1).[62] The effect was nonlinear to varying degrees, depending on the waste type, with the greatest decrease caused initially by approximately 12% plating sludge, 40% foundry dust, 58% other ash, and 72% MSWI fly ash by mass of dry product. It seems likely that the maximum waste additions used in modelling are the practical limits of waste addition to portland cement, i.e., 50% plating sludge or electric arc furnace dust, 64% foundry dust, 92% other ash, and 85% MSWI fly ash by mass of dry product.

As has been discussed in Chapter 4, blended cements may respond differently to waste addition than portland cement. For instance, it has been shown that increasing electric arc furnace dust concentrations retarded the setting of alkali-activated slag cement more severely than the setting of portland cement.[63]

7.3 CONTAMINANT IMMOBILIZATION

7.3.1 INTRODUCTION

A cement-based S/S product provides a dense physical matrix of low permeability, which constitutes a physical barrier to leaching. In addition, the matrix is composed of highly alkaline crystalline and amorphous phases (Chapter 4) that maintain a high pH in the pore solution (Chapter 10). Contaminants, particularly metals, may be immobilized as precipitates under these alkaline conditions, or can also be taken up into cement hydration products, or sorbed onto their surfaces, or onto admixtures added for this purpose.

Thus, contaminant immobilization will be discussed in the following sections under three headings:

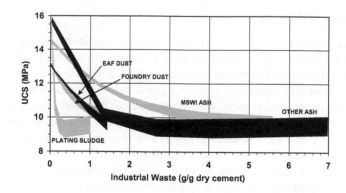

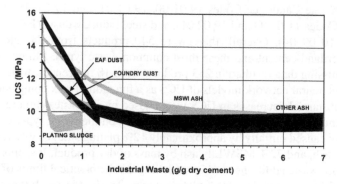

FIGURE 7.1   Response graphs showing predicted UCS as a function of industrial waste content for five different waste types.[62]

- Physical encapsulation
- Precipitation as a separate phase
- Uptake by the cement hydration products

These mechanisms can be considered to represent a continuum; physical encapsulation can take place at scales from centimeters to nanometers, until it becomes indistinguishable from sorption; increasing saturation of sorption sites leads to uptake in the cement hydration products by substitution, coprecipitation, or precipitation. Usually, several, or even all, of these mechanisms apply simultaneously.

7.3.2   PHYSICAL ENCAPSULATION

At the most basic level, contaminants can be physically encapsulated in the S/S product matrix, in the form of waste particles, such as dusts from thermal processes, or residues from pretreatment, including spent sorbents, spent ion exchange resin, sludge flocs from pretreatment of metals in solution by precipitation, or particles of contaminated metal or soil.[64]

In most cases, inorganic encapsulated particles will react with the cement matrix to some degree. Dusts from thermal processes, such as MSWI fly ash or electric arc

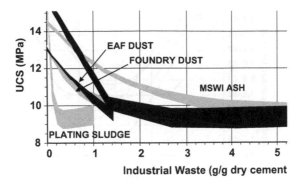

FIGURE 7.2   Encapsulated sludge particle.[65]

furnace dust, may contain a mixture of soluble species, such as chlorides and sulfates, and less soluble species, such as oxides. Depending also on reaction kinetics, the former may dissolve during or soon after mixing, releasing contaminants for other interaction with the cement hydration products, whereas the latter may dissolve more gradually and participate in secondary reactions during cement hardening and curing. The degree of reaction of precipitates from pretreatment of liquid wastes will also depend on their solubility; sulfides can be expected to have low solubility unless subjected to oxidizing influences, and many phosphates, carbonates, silicates, and hydroxides also have low solubility. Figure 7.2 shows an example of encapsulation of a hydroxide sludge floc in a cementitious matrix,[65] where little reaction with the cement appears to have taken place. However, energy-dispersive x-ray analyses of this sample did show the presence of contamination within the surrounding cement-based matrix, suggesting that some of the sludge dissolved during mixing, or that migration of contaminants from the sludge floc took place. Encapsulated sorbents and ion exchange beads will also release contaminants into the pore solution, possibly leading to other interactions with the cement hydration products. Encapsulated metals, such as shredded reactor fuel cladding, can hydrolyze over time to create metal hydroxides and acid or hydrogen gas. Some soil components, such as clay, may exhibit pozzolanic reactivity with cement over time, as discussed in Chapter 4.

Encapsulation of droplets of organic liquids that are not miscible with water has also been observed.[48,66]

Immobilization of encapsulated contaminants can be compromised by physical deterioration of the matrix, as either cracking due to physical stresses or advanced matrix dissolution (Chapter 10) can lead to exposure of encapsulated materials. Formulations containing high concentrations of soluble salts or high-ettringite cements[67] are particularly vulnerable to the latter.

7.3.3 PRECIPITATION

Whereas metal contaminants may have been precipitated as a pretreatment prior to S/S with cement, a dissolved metal ion may also be precipitated during treatment with cement, if its concentration in the pore solution of the resulting cement-based product exceeds the saturation concentration with respect to a metal salt. Thus,

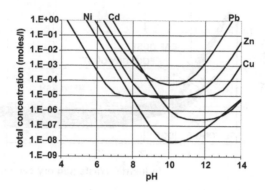

FIGURE 7.3 Cd, Cu, Ni, Pb, and Zn hydroxide solubility at 25°C in dilute solution, as a function of pH (based on data in the MINTEQ database[70]).

precipitation of contaminants in cement-based matrices is dependent on the pore solution concentration of the metal cation (i.e., not the total metal concentration) and the concentration of candidate anions with which it could form a precipitate. Due to the alkaline nature of cement hydration products, the most likely anion to form a metal precipitate is the hydroxide ion, OH^-.

Figure 7.3 shows the pH-dependent solubility of Cd, Cu, Ni, Pb, and Zn hydroxides. The initial pore solution pH of portland and blended cement-based matrices can range from about 11.9 (for pH control by C-S-H[68]), to higher than 13 (for high-alkali products); hydrated calcium aluminate cement also has a pore solution pH greater than 12.[69] The theoretical concentration ranges of these metal contaminants over a pH range from 12 to 13.5, calculated on the basis of equilibrium with hydroxide[70] (without adjustment for ionic strength), are shown in columns 2 and 3 of Table 7.6 (in mg/L). It is notable that Zn and Pb are several orders of magnitude more soluble than Ni and Cd at the pH values prevailing in the pore solution of cement-based products, while the solubility of Cu is intermediate. Given a typical saturated S/S product porosity of 57% vol,[65] the corresponding quantities of metals that can be dissolved in the pore solution, on the basis of the dry mass of the S/S product (i.e., in mg/kg dry product), are estimated in columns 4 and 5 of Table 7.6. In fact, because hydrated cement also has a significant capacity to sorb metals, it may be possible to add higher quantities of metals to a solidified product without precipitating hydroxides, but lower quantities than these values will be dissolved in the pore solution, or immobilized by another mechanism (Section 7.3.4).

It has been demonstrated that the concentrations of Cd, Cu, Ni, Pb, and Zn in near-equilibrium laboratory leachates from actual cement-based products generally do not exceed the solubility limits seen in Figure 7.3. However, unexpectedly high metal solubility can sometimes be observed if complexing agents, such as might be present in plating wastes, or dissolved organic matter, are present. For instance, the presence of organic cement admixtures has been shown to enhance the solubility of radionuclides.[71] Also, metals may not precipitate as their single-metal hydroxides, but may form other phases such as calcium zincate $(CaZn_2(OH)_6)$,[72] or other mixed hydroxides, e.g., [73–75] leading to different solubility curves. Hsiao et al.[76] observed

TABLE 7.6
Estimated Contaminant Solubility in Cement Pore Solutions

Metal	Pore Solution Concentration (mg/L)*		Soluble Amount (mg/kg dry product)**	
	Minimum (pH 12)	Maximum (pH 13.5)	Minimum (pH 12)	Maximum (pH 13.5)
Ni	0.005	0.1	0.07	0.1
Cd	0.02	0.09	0.03	0.1
Cu	0.6	9	0.8	12
Zn	10	1400	13	1900
Pb	100	66000	***	***

* Based on equilibrium with hydroxide in dilute solution (Figure 7.3).
** Assuming 57% saturated porosity.
*** Actual Pb solubility has been shown to be lower than predicted by hydroxide solubility.[77]

respeciation of soluble $CuCl_2$ to precipitated $Cu(OH)_2$, and also oxidation of Cu^+ to Cu^{2+} in MSWI fly ash solidified with portland cement.

Contaminants that form oxyanions, such as As, Cr(VI), Mo, Sb, and Se, can form precipitates with calcium. However, Johnson[77] reports that while AsO_4^{3-} and MoO_4^{2-} are known to form calcium salts of low solubility at high pH, the solubility of calcium salts of AsO_3^{3-}, CrO_4^{2-}, SeO_3^{2-}, and SeO_4^{2-} is higher, and that of Sb and V salts is unknown. Both U(VI) and Sn(IV) have been shown to precipitate as calcium salts, but the solubility of the former is two orders of magnitude greater than the latter.[78] Phosphates may also precipitate with calcium, to form apatite $[Ca_5(PO_4,CO_3)_3F]$.[79]

7.3.4 UPTAKE BY CEMENT HYDRATION PRODUCTS

In examining the measured pore solution or leachate concentrations of metal contaminants with respect to the solubility of their hydroxides (Figure 7.3), these can often be observed to lie below their maximum solubilities.[77] Such observations indicate that, in these cases, the total contaminant concentration in the product is lower than the capacity of the matrix for uptake of the contaminant, such that the leachate concentration is controlled by partitioning between the solid and liquid, rather than metal hydroxide solubility. Thus, the pore solution or leachate concentration does not reach saturation with respect to formation of the metal hydroxide. Leachate concentrations below the limit of solubility in equilibrium with hydroxide are particularly noticeable for Pb, whose solubility is an order of magnitude lower than the solubility in equilibrium with hydroxide at pH 10 to 13.5, and several orders of magnitude lower at more acidic pH.[77] However, Pb remains by far the most soluble of these metals in the alkaline environment provided by a cement matrix.

Based on leaching data alone, it is difficult to know the mechanism by which the matrix takes up contaminants. It is assumed that surface sorption is responsible at lower contaminant concentrations, and uptake into the hydration products through solid solution occurs as contaminant concentrations increase and surface sorption sites are saturated.

As described in Chapter 4, the main hydration products of portland cement are C-S-H (approximately 50% on a dry mass basis), $Ca(OH)_2$ (approximately 20% on a dry mass basis), as well as hydrated calciumalumino-ferrites and calciumsulfo-aluminates, including ettringite and monosulfate. The main hydration product in a blended (pozzolanic) cement is C-S-H. Contaminants may be taken up by sorption onto the surface of the C-S-H, or by substitution for other ions in these hydration products. Substitution of contaminants for other ions in a mineral structure is known as isomorphic substitution, which is a form of solid solution. The extent to which this can occur depends on the similarity between the charge, size, and geometry of the contaminants and the substituted ions.

C-S-H is nearly amorphous, and is often referred to as C-S-H gel, but there are some indications that it has some short-range order. Several structural models for C-S-H have been proposed.[4,80-83] Generally speaking, it is postulated that the short-range order consists of chains and sheets of SiO_4 tetrahedra, interlayered with CaO. SiO_4 groups may be missing or substituted (e.g., by AlO_4), which can result in local charge imbalances, i.e., potential binding sites. The extent of silicate polymerization increases with age. Experiments with tobermorite, a crystalline calcium silicate hydrate mineral, show it to have a high capacity for exchange of Ca with Co, Ni,[84] Cu, and Mg.[85] On this basis, it is thought that cation exchange can occur in the CaO interlayer of C-S-H; replacement of Ca by Mg has been reported,[50] which suggests that the potential for replacement by other alkaline earth elements, such as Sr, exists. Particularly, since the amount of Ca in C-S-H is known to vary, low Ca C-S-H may be able to take up other cations in place of Ca (Chapter 4). However, recent work found that Zn taken up by C-S-H was bound to the silicate layers, rather than exchanging for Ca.[86,87] The C/S ratio was not affected, indicating that the uptake mechanism is one of incorporation rather than substitution. Johnson[77] also reports work by Moulin and Pointeau that suggest uptake of Pb in C-S-H by the same mechanism, and work by Rigo that observed uptake of Cu. She also mentions that C-S-H may be able to take up anions into its structure, because of the charge imbalances resulting from its complex structure. Indeed, a review by Gougar et al.[88] reports significant uptake of I^- and Cl^-. Recent work by Wieland and coworkers has shown significant uptake of stannate and uranate [Sn(IV) and U(VI)] by C-S-H and a structural model of the binding mechanism has been proposed on the basis of X-ray absorption spectroscopy studies.[78]

Hydrated calcium sulfoaluminates, which form in calcium silicate-based cements, and can also form in calcium aluminate cements containing waste, have the general formulas $C_6(A,F)X_3.H_y$ (also known as AFt, which includes ettringite) and $C_4(A,F)X.H_y$ (also known as AFm, which includes monosulfate) where X is one formula unit of a doubly charged anion, or two formula units of a singly charged anion.[4] AFt has a columnar structure, with $C_6(A,F)$ columns, and $X_3.H_y$ channels,[88] resulting in needle-like crystals, whereas the structure of AFm is derived from the hexagonal platelet structure of $Ca(OH)_2$, and consists of layers of $C_4(A,F)$, with interlayers of $X.H_y$ (see also Chapter 9). It appears from the general formulae that AFt and AFm minerals may take up a variety of anion impurities without changing their structures. Exploitation of this characteristic for immobilization of anions, such as Cl^-, NO_3^-, and SO_4^{2-}, and metalloid oxyanions, such as AsO_3^{3-}, AsO_4^{3-}, CrO_4^{2-}, MoO_4^{2-}, SeO_3^{2-}, SeO_4^{2-}, TeO_4^{2-}, VO_4^{3-}, and WO_4^{2-}, in ettringite has been proposed[89-91] and increasingly followed up with research. CrO_4^{2-}—AFt has been shown to form readily,[90,92-94] and both CrO_4^{2-}—AFt and CrO_4^{2-}—AFm have been observed in cement-based products containing real wastes.[95] Uptake of SeO_4^{2-} [96] and AsO_4^{3-} [97,98] by ettringite has also been demonstrated, and Zhang and Reardon[99] demonstrated high uptake of four oxyanions by ettringite, with the order of preference $B(OH)_4^- > SeO_4^{2-} > CrO_4^{2-} > MoO_4^{2-}$; uptake of these oxyanions by hydrocalumite (a form of AFm) was even higher, but did not follow the same order. Johnson[77] reports that synthesis of AFt containing SO_3^{2-}, NO_3^-, SeO_4^{2-}, and CO_3^{2-} is only possible using a laboratory method that does not resemble conditions in cements, whereas MoO_4^{2-}—AFt and Cl—AFt could not be formed at all. However, MoO_4^{2-}—AFm has been synthesized[100] and Cl—AFm, Friedel's salt, is commonly found in hardened cement pastes and concretes. However, solubility of chlorides remains high in cement-based products as elsewhere.

AFt in particular is also known to take up other trivalent cations, such as Cr^{3+}, Ni^{3+}, and Co^{3+} in place of Al or Fe.[88] C_3A may also react to produce hydrogarnet, $C_3AS_xH_{6-2x}$,[101] which is known to take up Fe, and has been shown to take up Cr^{3+}.[102] Gougar et al.[88] also review reports that divalent cations, such as Ba^{2+}, Cd^{2+}, Co^{2+}, Hg^{2+}, Ni^{2+}, Pb^{2+}, Sr^{2+}, and Zn^{2+}, can replace Ca^{2+} in AFt minerals, but the evidence for this is not strong. Johnson[77] reports several studies that indicate surface sorption is also an important mechanism for uptake of oxyanions by AFt and AFm. It should be noted that ettringite has been shown to have poor acid neutralization capacity and resistance to acid attack,[11,67] which limits its long-term durability in the environment, and the usefulness of immobilization mechanisms based on AFt. Studies of iodine uptake in calcium aluminate cement blended with $Ca(OH)_2$ and $CaSO_4$ showed transformation of $4CaO.Al_2O_3.13H_2O$ to $3CaO.Al_2O_3.CaI_2.12H_2O$.[103,104]

Examination of Cs retention by the hydrated $CaO-SiO_2-Al_2O_3$ system indicated that uptake was greatest in a matrix low in Al_2O_3 and nearly equimolar in SiO_2 and CaO content, which was attributed to the formation of zeolitic phases.[105] Experiments with stabilization of an alkaline low-level radioactive waste simulant using cement, coal fly ash, and attapulgite clay also showed formation of zeolites and concluded that contaminants were incorporated in these phases.[79] However, other work indicates that alumina-substituted calcium silicate hydroxy hydrate selectively takes up Cs.[106]

7.4 SUMMARY

The components of wastes have been demonstrated to have significant effects on properties related to the handling, durability, and leaching of cement-based products, which must be taken into account in formulation development. Contaminant leachability is controlled by both physical and chemical mechanisms. An improved understanding of chemical immobilization mechanisms is gradually developing, but many of the details remain unknown, particularly for less common contaminants. However, it is clear that the chemical environment provided in a cement-based product, which dictates contaminant solubility, can be adjusted to a limited degree by formulation optimization.

ACKNOWLEDGMENTS

I would like to acknowledge helpful comments received from the editors and the support of the Centre for Ecology and Hydrology Wallingford in writing this chapter.

REFERENCES

1. USEPA (1990), 40 CFR, June 1, page 22568.
2. Trussell, S. and Spence, R.D. (1994), A review of solidification/stabilization interferences, *Waste Management*, 14, 6, pp. 507–519.
3. Mattus, C.H. and Mattus, A.J. (1996), Literature review of the interaction of select inorganic species on the set and properties of cement and methods of abatement through waste pretreatment, in *Stabilization and Solidification of Hazardous, Radioactive and Mixed Wastes: 3rd Volume*, ASTM STP 1240, T. Michael Gilliam and Carlton C. Wiles, Eds., American Society for Testing and Materials.
4. Taylor, H.F.W. (1990), *Cement Chemistry*, Academic Press, London.
5. Massazza, F. and Testolin, M. (1980), Latest developments in the use of admixtures for cement and concrete, *Il Cemento*, 2, pp. 73–146.
6. Jones, L.W. (1990), Interference Mechanisms in Waste Stabilization/Solidification Processes — Literature Review, EPA/600/2-89/067, US Army Corps of Engineers Waterways Experiment Station, Vicksburg, MS.
7. Cox, J.D. and Sharp, J.H. (1994), The use of admixtures with calcium aluminate cements, *Proceedings of the ConChem International Exhibition and Conference*, Karlsruhe, Germany, pp. 381–398.
8. Mattus, C.H. and Gilliam, T.M. (1994), A Literature Review of Mixed Waste Components: Sensitivities and Effects upon Solidification/Stabilization in Cement-based Matrices, ORNL/TM-12656, Oak Ridge National Laboratory, Oak Ridge, TN.
9. Conner, J.R. (1997), *Guide to Improving the Effectiveness of Cement-based Stabilization/Solidification*, C.M. Wilk, Ed., Portland Cement Association, Skokie, IL.
10. Hills, C.D. and Pollard, S.J.T. (1997), The influence of interference effects on the mechanical, microstructural and fixation characteristics of cement-solidified hazardous waste forms, *Journal of Hazardous Materials*, 52, pp. 171–191.
11. Stegemann, J.A., Caldwell, R.J., and Shi, C. (1994), Response of Various Solidification Systems to Acid Addition, presented at WASCON '94, Maastricht, The Netherlands.
12. Stegemann, J.A., Perera, A.S., Cheeseman, C., and Buenfeld, N.R. (2000), 1/8 fractional factorial design investigation of metal effects on cement acid neutralisation capacity, *Journal of Environmental Engineering*, 126, 10.
13. Stegemann, J.A. and Buenfeld, N.R. (2002), Prediction of leachate pH for cement paste containing pure metal compounds, *Journal of Hazardous Materials*, B90, pp. 169–188.
14. Polettini, A., Pomi, R., and Sirini, P. (2002), Fractional factorial design to investigate the influence of heavy metals and anions on acid neutralization behavior of cement-based products, *Environmental Science and Technology*, 36, 7, pp. 1584–1591.

15. Fletcher, P., Coveney, P.V., Hughes, T.L., and Methven, C.M. (1994), Predicting the quality and performance of oilfield cements using artificial neural networks and FTIR spectroscopy, Society of Petroleum Engineers, SPE 28824.

16. Bhatty, J.I. and West, P.B. (1996), Stabilization of heavy metals in Portland cement matrix: effects on paste properties, in *Stabilization and Solidification of Hazardous, Radioactive and Mixed Wastes: 3rd Volume*, ASTM STP 1240, T. Michael Gilliam and Carlton C. Wiles, Eds., American Society for Testing and Materials.

17. Barret, P., Bertrandie, D., Casabonnemasonnave, J.M., and Damidot, D. (1992), Short-term processes of radionuclide immobilization in cement — a chemical approach, *applied geochemistry supplement 1*, pp. 109–124.

18. Lea, F.M. (1971), *The Chemistry of Cement and Concrete*, Chemical Publishing Company, Inc., New York.

19. Lieber, W. (1973), Wirkung anorganischer Zusätze auf das Erstarren und Erhärten von Portland zement, Zement-Kalk-Gips, 2, pp. 75–79.

20. Polettini, A. and Pomi, R. (2003), Modelling heavy metal and anion effects on physical and mechanical properties of Portland cement by means of factorial experiments, *Environmental Technology*, 24, 2, pp. 231–239.

21. Cartledge, F.K. (1993), *Solidification/Stabilization Interferences and Difficult Waste Types*, Gulf Coast Hazardous Substance Research Center, Beaumont, TX.

22. Zamorani, E., Sheikh, I., and Serrini, G. (1989), A study of the influence of nickel chloride on the physical characteristics and leachability of Portland cement, *Cement and Concrete Research*, 19, pp. 259–266.

23. Dumitru, G., Vazquez, T., Puertas, F., and Blanco-Varela, M.T. (2000), Influence of BaCO3 on hydration of Portland cement, *Materials and Construction*, 49, 254, pp. 43–48.

24. Fernandez Olmo, I., Chacon, E., and Irabien, A. (2001), Influence of lead, zinc, iron (III) and chromium (III) oxides on the setting time and strength development of Portland cement, *Cement and Concrete Research*, 31, 8, pp. 1213–1219.

25. Malolepszy, J. and Deja, J., Effect of Heavy Metals Immobilization on Properties of Alkali-Activated Slag Mortars, in *Proceedings of the 5th International Conference on the Use of Fly Ash, Silica Fume, Slag & Natural Pozzolans in Concrete*, SP-153, 2, pp.1087–1102, Milwaukee, WI, 1995.

26. Wastewater Technology Centre (1992), *Engineering Properties Testing of Solidified Residues*, Report for Ontario Waste Management Corporation, Toronto, Ontario, Canada.

27. Malone, P.G., Poole, T.S., Wakeley, L.D., and Burkes, J.P. (1997), Salt-related expansion reactions in Portland-cement-based wasteforms, *Journal of Hazardous Materials*, Volume 52, 2–3, pp. 237–246.

28. Stegemann, J.A. and Buenfeld, N.R. (2002), Prediction of unconfined compressive strength of cement paste with pure metal compound additions, *Cement and Concrete Research*, 32, 6, pp. 903–913, 2002.

29. Parker, T.W. (1954), The constitution of aluminous cement, 3rd International Symposium on the Chemistry of Cement, London.

30. Robson, T.D. (1967), *High Alumina Cements and Concretes*, John Wiley & Sons, New York.

31. Barret, P., Benes, C., Bertrandie, D., and Moisset, J. (1980), Comportement de divers phosphates avec des constituants des ciments, in *Proceedings of the 7th International Congress on the Chemistry of Cement*, Paris, 3, pp. 175–180.

32. Rodger, S.A. and Double, D.D. (1984), The chemistry of hydration of high alumina cement in the presence of accelerating and retarding admixtures, *Cement and Concrete Research*, 14, pp. 73–82.

33. Banfill, P.F. (1986), The effect of sulphate on the hydration of high alumina cement, *Cement and Concrete Research*, 16, pp. 602–604.

34. Currell, B.R., Grezeskowlak, R., Midgley, H.G., and Parsonage, J.R., (1987), The acceleration and retardation of set high alumina cement by additives, *Cement and Concrete Research*, 7, pp. 420–432.

35. Novinson, T. and Crahan, J. (1988), Lithium salts as set accelerators for refractory concretes: correlation of chemical properties with setting times, *ACI Materials Journal*, January/February, pp. 12–17.

36. Murat, M. and Sadok, E.H. (1990), Role of foreign cations in solution in the hydration kinetics of high alumina cement, in *Calcium Aluminate Cements*, R.J. Mangabhai, Ed., E. & F.N. Spon, Chapman & Hall, London, pp. 155–166.

37. Griffiths, D.L., Al-Qasar, A.N., and Mangabhai, R.J. (1990), Calorimetric studies on high alumina cement in the presence of chloride, sulphate and seawater solutions, in *Calcium Aluminate Cements*, R.J. Mangabhai, Ed., E. & F.N. Spon, Chapman & Hall, London, pp. 167–177.

38. Nilforoushan, M.R. and Sharp, J.H. (1995), The effect of additions of alkaline earth metal chlorides on the setting behavior of a refractory calcium aluminate cement, *Cement and Concrete Research*, 25, pp. 1523–1534.

39. Matusinovic, T. and Vrbos, N. (1993), Alkali metal salts as set accelerators for high alumina cement, *Cement and Concrete Research*, 23, pp. 177–186.

40. Matusinovic, T. and Curlin, D. (1993), Lithium salts as set accelerators for high alumina cement, *Cement and Concrete Research*, 23, pp. 885–895.

41. Nilforoushan, M.R. and Sharp, J.H. (1999), The setting behaviour of a refractory calcium aluminate cement in the presence of alkali chloride admixtures, unpublished data, Department of Engineering Materials, University of Sheffield, England.

42. Baker, N.C. and Banfill, P.F. (1990), Properties of fresh mortars made with high alumina cement and admixtures for the marine environment, in *Calcium Aluminate Cements*, R.J. Mangabhai, Ed., E. & F.N. Spon, Chapman & Hall, London, pp. 142–151.

43. Bayoux, J.P., Bonin, A., Marcdargent, S., and Verschaeve, M. (1990a), Study of the hydration properties of aluminous cement and calcium sulphate mixes, in *Calcium Aluminate Cements*, R.J. Mangabhai, Ed., E. & F.N. Spon, Chapman & Hall, London, pp. 320–334.

44. Bier, T.A., Mathieu, A., Espinosa, B., and Marcelon, C. (1995), Admixtures and their interactions with high range calcium aluminate cement, presented at the UNITECR Congress, Japan, Technical Paper F/95, Lafarge Aluminates, Paris, 6p.

45. Damidot, D., Rettel, A., and Capmas, A. (1996), Action of admixtures on fondu cement: Part 1. Lithium and sodium salts compared, *Advances in Cement Research*, 8, 31, pp. 111–119.

46. Sharp, J.H., Bushnell-Watson, S.M., Payne, D.R., and Ward, P.A. (1990), The effect of admixtures on the hydration of refractory calcium aluminate cements, in *Calcium Aluminate Cements*, R.J. Mangabhai, Ed., E. & F.N. Spon, Chapman & Hall, London, pp. 127–141.

47. Stegemann, J.A. and Buenfeld, N.R. (2001), Neural network modelling of the effects of inorganic impurities on calcium aluminate cement setting, *Advances in Cement Research*, 13, 3, pp. 101–114.

48. Nestle, N., Zimmermann, C., Dakkouri, M., and Niessner, R. (2001), Action and distribution of organic solvent contaminations in hydrating cement: time-resolved insights into solidification of organic waste, *Environmental Science and Technology*, 35, 24, pp. 4953–4956.

49. Minocha, A.K., Jain, N., and Verma, C.L. (2003), Effect of organic materials on the solidification of heavy metal sludge, *Construction and Building Materials*, 17, 2, pp. 77–81.

50. Massazza, F. (1998), Pozzolana and pozzolanic cement, *Lea's Chemistry of Cement and Concrete, 4th ed.*, P.C. Hewlett, Ed., Arnold, London, pp. 471–631.

51. Shi, C. and Day, R.L. (1995), Microstructure and Reactivity of Natural Pozzolans, Fly Ash and Blast Furnace Slag, in *Proceedings of 17th International Cement Microscopy Conference*, Calgary, Canada, April 23–27, pp. 150–161.

52. Shi, C. and Day, R.L. (2000), Pozzolanic reactions in the presence of chemical Activators – Part I: Reaction kinetics, *Cement and Concrete Research*, 30, 1, pp. 51–58, 2000.

53. Gervais, C. and Ouki, S.K. (2002), Performance study of cementitious systems containing zeolite and silica fume: effects of four metal nitrates on the setting time, strength and leaching characteristics, *Journal of Hazardous Materials*, 93, 2, pp. 187–200.

54. Jun, K.S., Hwang, B.G., Shin, H.S., and Won, Y.S. (2001), Chemical characteristics and leachability of organically contaminated heavy metal sludge solidified by silica fume and cement, *Water Science and Technology*, 44, 2–3, pp. 399–407.

55. Wang, S.Y. and Vipulanandan, C. (1996), Leachability of lead from solidified cement-fly ash binders, *Cement and Concrete Research*, 26, pp. 895–905.

56. Majumdar, A.J., Edmonds, R.N., and Singh, B. (1990a), Hydration of calcium aluminates in presence of granulated blast furnace slag, in *Calcium Aluminate Cements*, R.J. Mangabhai, Ed., E. & F.N. Spon, Chapman & Hall, London, pp. 259–271.

57. Majumdar, A.J., Singh, B., and Edmonds, R.N. (1990b), Hydration of mixtures of cement fondu aluminous cements and granulated blast furnace slag, *Cement and Concrete Research*, pp. 197–208.

58. Majumdar, A.J. and Singh, B. (1992), Properties of some blended high-alumina cements, *Cement and Concrete Research*, 22, pp. 1101–1114.

59. Stegemann, J.A. (2001), MONOLITH — a database of cement/waste properties, Environmentally Preferred Materials, Volume II, in *Proceedings of ICMAT 2001*, Singapore, July 1–6, 2001, T.J. White and J.A. Stegemann, Eds., Materials Research Society, Singapore.

60. Stegemann, J.A., Butcher, E.J., Irabien, A., Johnston, P., de Miguel, R., Ouki, S.K., Polettini, A., and Sassaroli, G. (Eds.) (2001), *Neural Network Analysis for Prediction of Interactions in Cement/Waste Systems — Final Report*, Contract No. BRPR-CT97–0570, Commission of the European Community, Brussels, Belgium.

61. Stegemann, J.A. and Buenfeld, N.R. (2003), Prediction of unconfined compressive strength of cement pastes containing industrial wastes, *Waste Management*, 33, 4, pp. 322–333.

62. Stegemann, J.A. and Buenfeld, N.R. (2004), Mining of existing data for cement-solidified wastes using neural networks, *Journal of Environmental Engineering*, 30, 5.

63. Shi, C., Stegemann, J.A., and Caldwell, R.J. (1997), An examination of interference in waste solidification through measurement of heat signature, *Waste Management*, 17, 4, pp. 249–255.

64. Roy, A., Eaton, H.C., Cartledge, F.K., and Tittlebaum, M.E. (1992), Solidification/stabilization of hazardous waste: evidence of physical encapsulation, *Environmental Science and Technology*, 26, 7, pp. 1349–1353.
65. Stegemann, J.A. and Coté, P.L. (1991), *Investigation of Test Methods for Solidified Waste Evaluation — A Cooperative Program*, Environment Canada Report EPS 3/HA/8, ISBN 0-662-18280-4, Ottawa, Ontario, Canada.
66. Butler, L.G., Owens, J.W., Cartledge, F.K., Kurtz, R.L., Byerly, G.R., Wales, A.J., Bryant, P.L., Emery, E.F., Dowd, B., and Xie, X. (2000), *Environmental Science and Technology*, 34, 15, pp. 3269–3275.
67. Stegemann, J.A. and Shi, C. (1997), Acid resistance of different monolithic binders and solidified wastes, *Waste Materials in Construction: Putting Theory in Practice, Studies in Environmental Science 71*, J. Goumans, J. Senden, and H. van der Sloot, Eds., Elsevier Science B.V., Amsterdam, pp. 551–562.
68. Greenberg, S.A. and Chang, T.N. (1965), Investigation of the colloidal hydrated calcium silicates. II. Solubility relationships in the calcium oxide-silica-water system at 25°C, *Journal of Physical Chemistry*, 69, 1, pp. 182–188.
69. Scrivener, K.L., Cabiron, J.L., and Letourneux, R. (1999), High-performance concretes from calcium aluminate cements, *Cement and Concrete Research*, 29, 8, pp. 1215–1223.
70. Allison, J.D., Brown, D.S., and Novo-Gradac, K.J. (1990). MINTEQA2/ PRODEFA2 — A geochemical assessment model for environmental systems. Version 3.0 user's manual. Environmental Research Laboratory, Office of Research and Development, U.S. Environmental Protection Agency, Athens, GA.
71. Greenfield, B.F., Ilett, D.J., Ito, M., McCrohon, R., Heath, T.G., Tweed, C.J., Williams, S.J., and Yui, M. (1998), The effect of cement additives on radionuclide solubilities, *Radiochimica Acta*, 82, pp. 27–32.
72. Cocke, D.L. and Mollah, M.Y. (1993), The chemistry and leaching mechanisms of hazardous substances in cementitious solidification/stabilization systems, in *Chemistry and Microstructure of Solidified Waste Forms*, R.D. Spence, Ed., Lewis Publishers, Boca Raton, FL, pp. 187–242.
73. Roy, A., Eaton, H.C., Cartledge, F.K., and Tittlebaum, M.E. (1991), Solidification stabilization of a heavy-metal sludge by a Portland-cement fly-ash binding mixture, *Hazardous Waste & Hazardous Materials*, 8, 1, pp. 33–41.
74. Scheidegger, A.M., Wieland, E., Scheinost, A.C., Dahn, R., and Spieler, P. (2000), Spectroscopic evidence for the formation of layered Ni-Al double hydroxides in cement, *Environmental Science and Technology*, 34, 21, pp. 4545–4548.
75. Johnson, C.A. and Glasser, F.P. (2002), Hydrotalcite-like minerals (M2Al(OH)(6) (CO3)(0.5).XH2O, where M = Mg, Zn, Co, Ni) in the environment: synthesis, characterization and thermodynamic stability, *Clays and Clay Minerals*, 51, 1, p. 108.
76. Hsiao, M.C., Wang, H.P., and Yang, Y.W. (2001), EXAFS and XANES studies of copper in a solidified fly ash, *Environmental Science and Technology*, 35, 12, pp. 2532–2535.
77. Johnson, C.A. (2002), *Metal Binding in the Cement Matrix: an Overview of our Current Knowledge*, Department of Water Resources and Drinking Water, Water-Rock Interaction Group, EAWAG, for Cemsuisse, Switzerland.
78. Wieland, E., Bonhoure, I., Fujita, T., Tits, J., and Scheidegger, A.M. (2003), Combined wet chemistry and EXAFS studies on the radionuclide immobilisation by cement and calcium silicate hydrates, *Geochimica et Cosmochimica Acta*, 67, 18, pp. A532–A532.

79. Olson, R.A., Tennis, P.D., Bonen, D., Jennings, H.M., Mason, T.O., Christensen, B.J., Brough, A.R., Sun, G.K., and Young, J.F. (1997), Early containment of high-alkaline solution simulating low-level radioactive waste in blended cement, *Journal of Hazardous Materials*, 52, 2–3, pp. 223–236.

80. Grutzeck, M., Benesi, A., and Fanning, B. (1989), Silicon-29 magic angle spinning nuclear magnetic resonance study of calcium silicate hydrates, *Journal of the American Ceramic Society*, 72, 4, pp. 665–668.

81. Richardson, I.G. and Groves, G.W. (1992), Models for the composition and structure of calcium silicate hydrate (C-S-H) gel in hardened tricalcium silicate pastes, *Cement and Concrete Research*, 22, pp. 1001–1010.

82. Kwan, S., LaRosa-Thompston, J., and Grutzeck, M.W. (1996), Structures and phase relations of aluminum-substituted calcium silicate hydrate, *Journal of the American Ceramic Society*, 79, 4, pp. 967–971.

83. Odler, I. (1998), Hydration, setting and hardening of Portland cement, in *Lea's Chemistry of Cement and Concrete, 4th ed.*, P.C. Hewlett, Ed., Arnold, London, pp. 241–297.

84. Komarneni, S., Roy, R., and Roy, D.M. (1986), Pseudomorphism in xonotlite and tobermorite with Co^{2+} and Ni^{2+} exchange for Ca^{2+} at 25°C, *Cement and Concrete Research*, 16, pp. 47–58.

85. Shrivastava, O.P. and Glasser, F.P. (1986), Ion-exchange of 11-Å tobermorite, *Reactivity Solids*, 2, pp. 261–268.

86. Ziegler, F., Scheidegger, A.M., Johnson, C.A., Dahn, R., and Wieland, E. (2001), Sorption mechanisms of zinc to calcium silicate hydrate: X-ray absorption fine structure (XAFS) investigation, *Environmental Science and Technology*, 35, 7, pp. 1550–1555.

87. Tommaseo, C.E. and Kersten, M. (2002), Aqueous solubility diagrams for cementitious waste, *Environmental Science and Technology*, 26, 13, pp. 2919–2925.

88. Gougar, M.L., Scheetz, B.E., and Roy, D.M. (1996), Ettringite and C-S-H Portland cement phases for waste ion immobilization: a review, *Waste Management*, 16, 4, pp. 295–303.

89. Hassett, D.J., Pflughoeff-Hassett, D.F., Kumarathasan, P., and McCarthy, G.J. (1989), Ettringite as an agent for the fixation of hazardous oxyanions, in *Proceedings of the Twelfth Annual Madison Waste Conference on Municipal and Industrial Waste*, Madison, WI, September 20–21.

90. Kumarathasan, P., McCarthy, G.J., Hassett, D.J. and Pflughoeft-Hassett, D.F. (1990), Oxyanion substituted ettringites: synthesis and characterization; and their potential role in immobilization of SAS, B, Cr, Se, and V, Fly Ash and Coal Conversion By-Products: Characterization, Utilization, and Disposal II, in *Materials Research Society Symposium Proceedings*, 178, pp. 83–104.

91. McCarthy, G.J., Hassett, D.J., and Bender, J.A. (1992), Synthesis, crystal chemistry and stability of ettringite, a material with potential applications in hazardous waste immobilisation, in *Materials Research Society Symposium Proceedings*, 245, pp. 129–140.

92. Poellmann, H., Auer, S., Kuzel, H.J., and Wenda, R. (1993), Solid-solution of ettringites 2. Incorporation of $B(OH)^{4-}$ and CrO_4^{2-} in $3CaO.Al_2O_3.3CaSO_4.32H_2O$, *Cement and Concrete Research*, 23, 2, pp. 422–430.

93. Myneni, S., Traina, S.J., and Waychunas, G. (1996), Sorption and coprecipitation of CrO_4 in ettringite $(Ca_6A_{12}(SO_4)_3(OH)1_2,26H_2O)$, *Abstracts of Papers of the American Chemical Society*, 211: 89-GEOC.

94. Perkins, R.B. and Palmer, C.D. (2000), Solubility of $Ca_6Al(OH)_62(CrO_4)_3.26H_2O$, the chromate analog of ettringite; 5–75 degrees C, *Applied Geochemistry*, 15, 8, pp. 1203–1218.

95. Palmer, C.D. (2000), Precipitates in a Cr(VI)-contaminated concrete, *Environmental Science and Technology*, 34, 19, pp. 4185–4192.

96. Hassett, D.J., McCarthy, G.J., Kumarathasan, P., and Pflughoefthassett, D. (1990), Synthesis and characterization of selenate and sulfate-selenate ettringite structure phases, *Materials Research Bulletin*, 25, 11, pp. 1347–1354.

97. Myneni, S.C., Traina, S.J., Logan, T.J., and Waychunas, G.A. (1997), Oxyanion behaviour in alkaline environments: sorption and desorption of arsenate in ettringite, *Environmental Science and Technology*, 31, pp. 1761–1768.

98. Myneni, S.C., Traina, S.J., Waychunas, G.A., and Logan, T.J. (1998), Vibrational spectroscopy of functional group chemistry and arsenate coordination in ettringite, *Geochimicha et Cosmochimica Acta*, 62, 21–22, pp. 3499–3514.

99. Zhang, M. and Reardon, E.J. (2003), Removal of B, Cr, Mo, and Se from wastewater by incorporation into hydrocalumite and ettringite, *Environmental Science & Technology*, 37, 13, pp. 2947–2952.

100. Kindness, A., Lachowski, E.E., Minocha, A.K., and Glasser, F.P. (1994a), Immobilization and fixation of molybdenum(VI) by Portland cement, *Waste Management*, 14, 2, pp. 97–102.

101. Damidot, D. and Glasser, F.P. (1995), Investigation of the CaO-Al2O3-SiO2-H2O system at 25°C by thermodynamic calculations, *Cement and Concrete Research*, 25, 1, pp. 22–28.

102. Kindness, A., Macias, A., and Glasser, F.P. (1994b), Immobilization of chromium in cement matrices, *Waste Management*, 14, 1, pp. 3–11.

103. Toyohara, M., Kaneko, M., Ueda, H., Mitsutsuka, N., Fujihara, H., Murase, T., and Saito, N. (2000), Iodine sorption onto mixed solid alumina cement and calcium compounds, *Journal of Nuclear Science and Technology*, 37, 11, pp. 970–978.

104. Toyohara, M., Kaneko, M., Mitsutsuka, N., Fujihara, H., Saito, N., and Murase, T. (2002), Contribution to understanding iodine sorption mechanism onto mixed solid alumina cement and calcium compounds, *Journal of Nuclear Science and Technology*, 39, 9, pp. 950–956.

105. Bagosi, S. and Csetenyi, L.J. (1998), Caesium immobilisation in hydrated calcium-silicate-aluminate systems, *Cement and Concrete Research*, 28, 12, pp. 1753–1759.

106. Shrivastava, O.P. and Shrivastava, R. (2000), Cation exchange applications of synthetic tobermorite for the immobilization and solidification of cesium and strontium in cement matrix, *Bulletin of Materials Science*, 23, 6, pp. 515–520.

8 Stabilization/ Solidification Additives

Steve Hoeffner, Jesse Conner, and Roger Spence

CONTENTS

8.1 INTRODUCTION

The commonly used binder system for stabilization and solidification (S/S) of wastes is described in Chapters 4 through 6. As discussed in Chapter 7, many ingredients in waste may interfere with the setting and hardening in different ways. In order to improve the efficacy of S/S of concerned contaminants in the waste, many additives have been tried out in S/S processes, and a number of them are used commercially, especially at treatment, storage, and disposal facilities (TSDFs). Additives can be grouped into three basic categories: (1) metal stabilizers, (2) organic stabilizers, and (3) processing and anti-inhibition aids.

Reaction products that form can be dependent on competing reactions and on the order of addition of reactants. The rate of addition, degree of mixing, and temperature may also affect the nature of the reaction products. Thus, the nature, dosage, and timing for addition of an additive may have a significant effect on the performance of the S/S waste forms.

8.2 METAL STABILIZATION

In the management of hazardous waste, more is known regarding the treatment of metals than about the fixation, destruction, or immobilization of any other hazardous constituent group. Metals cannot be destroyed and so must be converted to their least soluble or reactive form to prevent reentry into the environment. The mechanisms for metal stabilization consist of pH control and buffering, speciation/precipitation/re-speciation, oxidation/reduction, and sorption/ion exchange. Usually a combination of mechanisms is active and the metals are often not present as simple compounds.

8.2.1 PH CONTROL AND BUFFERING

Amphoteric metals form compounds that are soluble in both acidic and basic solutions.[1] Most metals are amphoteric and exhibit minimal solubility somewhere in the range of pH 8 to 11,[2] as shown in Figure 7.3. It should be understood, however, that these curves were determined for the pure metal hydroxides in water. The actual solubility minima for individual metals will vary somewhat in complex systems. For precipitation to occur with optimum effect, the binding reagent should be added in such proportions as to ensure that optimum pH conditions are maintained for the main contaminants present.[3] If there are several metal contaminants, the final desired pH must be a compromise because no two metals have minima at the same pH. If this is not adequate, speciation or other treatment alternatives may be needed.

In order to adjust the pH of the system within a range where heavy metals have the lowest solubility, additives can be added to control and buffer pH, including acids, alkalis, lime, caustic soda, soda ash, and ferrous sulfate. Acids can be used to solubilize metals so that treatment methods can be effective. If required, the most common acids used in S/S are sulfuric and hydrochloric.

In addition to pH adjustment, lime supplies additional calcium for reaction and can react with certain interfering organics (see Section 8.4). Flyash and kiln dust can also be used, among other things, for pH adjustment. Buffers, such as calcium carbonate, sodium carbonate, soda ash, and magnesium oxide, are used to keep the pH within the desired range.

Many metal contaminants of concern form highly insoluble hydroxides in a basic cement matrix. This is by far the most common method used to reduce solubility of metals. The alkalinity of this matrix can be supplemented with sodium hydroxide and other alkali hydroxides, if necessary.

8.2.2 Speciation/Precipitation/Re-Speciation

For metals where pH adjustment and hydroxide precipitation is insufficient, some other additives are used to preferentially speciate metals into less soluble forms. Additives used to speciate, precipitate, and re-speciate include sulfides, sulfur and organo-sulfur compounds; soluble silicates and rice hull ash; carbonates; phosphates (especially important for lead) and trisodium phosphate; $FeSO_4$ co-precipitation; other iron compounds (also used for oxidation/reduction); other inorganic complexing agents; and organic complexing agents. Sulfides and silicates are used more than carbonates, phosphates, and iron co-precipitation.

8.2.2.1 Sulfides, Sulfur, and Organo-Sulfur Compounds

Metal sulfides are not amphoteric and, as such, treatment of metal cations with sulfide (usually Na_2S or $NaHS$, occasionally CaS) or sulfide-producing reagents can reduce the amount of leachable metal contaminants significantly below that possible with simple hydroxide precipitation. Figure 8.1 shows typical solubility behavior of some metal sulfide constituents of concern in environmental work. It should be understood, however, that these curves were determined for the pure metal sulfides in water. The actual solubility minima for individual metals will vary somewhat in complex systems. However, metal sulfides can resolubilize in an oxidizing environment.[4] And the pH must be kept above 8 to prevent possible evolution of hydrogen sulfide gas. The sulfide should be added before any S/S reagents because the calcium, magnesium, and iron present in these reagents will compete for the soluble sulfide ion.

Elemental sulfur, sulfides, and organic sulfides have been used to reduce the solubility of mercury compounds through the formation of insoluble mercuric sulfide. The most common way to reduce mercury solubility is by the use of various sulfur compounds, typically sodium or calcium sulfide. Both the mercury hydroxide and oxide are too soluble (see Figure 8.2 generated from published data).[5] In an alkaline system, sulfur and OH^- react to form polysulfide ion, S_x^{2-}, that in turn reacts with polyvalent metal ions to form very low-solubility compounds. Mercury solubility does increase with the concentration of excess dissolved sulfide anions, similar to increasing metal solubility with excess hydroxide anions at high pH, so it is important to not use a large excess of sulfide.

Organo-sulfur compounds are primarily used for wastes containing mercury. Organo-sulfur compounds such as hexadecyl mercaptan also appear to work well.[6,7] Other reagents that have been used include thiourea[8] and polydithiocarbamates.[9]

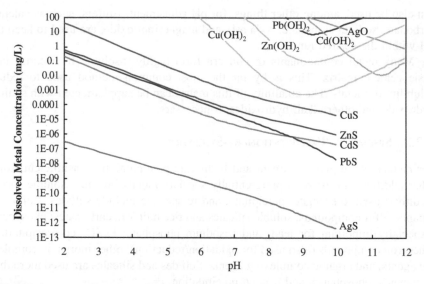

FIGURE 8.1 Comparing the solubility of metal sulfides and hydroxides.

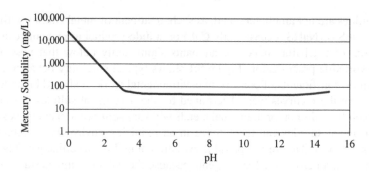

FIGURE 8.2 Mercuric oxide solubility versus pH from published data in Ref. 5.

8.2.2.2 Soluble Silicates and Rice Hull Ash

Numerous metals can be treated using silicates. Soluble silicates are especially effective at precipitating the following metals (listed in order of decreasing effectiveness): copper, zinc, manganese, cadmium, lead, nickel, silver, magnesium, and calcium. Metal silicates exhibit low solubility over a wide pH range of 2 to 11 or greater.[10] In addition, soluble silicates are also effective at reducing the permeability of the stabilized grout. By forming precipitates in the matrix that block pores,[11] the movement of any mobile species through the matrix into the environment is slowed. If the metals to be treated are already present as insoluble hydroxides, either the waste must be pretreated by reducing the pH to resolubilize the metal hydroxide or silicate must be slowly released or produced over a period of time to react with metal hydroxide that is slowly dissolving and being re-speciated.

Rice hull ash is an amorphous biogenetic silica.[12] Because of its sorptive and alkali-reactive nature, rice hull ash has some unusual properties. Under alkaline conditions, the amorphous silica reacts slowly to produce soluble silicates, which can then react with toxic metals ions to form low-solubility metal silicates. At the same time, the soluble silicate can react with available calcium from cement, or other polyvalent metal ions, to set and harden the system in a controlled manner. The advantage of this method over that of most soluble silicate processes is that the slow, continuous generation of soluble silicate provides a reserve capacity analogous to the action of buffers in a pH-control system. As metal hydroxides and other species slowly dissolve in the alkaline environment of the waste form, they can then become re-speciated as the "silicate." The process is patented in the U.S.[13] and elsewhere.

8.2.2.3 Carbonates

Precipitation by the addition of carbonates has not been widely used for S/S. Carbonates decompose at low pH to release carbon dioxide and only a few metal carbonates (e.g., cadmium, barium, and lead) are less soluble than their corresponding hydroxides.

8.2.2.4 Phosphates and Trisodium Phosphate

Although many metals can form low-solubility phosphates with PO_4^{3-}, the results solubility-wise are quite variable due partly to the complexity of phosphate chemistry. In recent years, considerable work has been done on the use of phosphates for metal fixation. Phosphates can consist of monomeric PO_4^{3-}, and various polymeric species having the potential to sequester the metals as water-soluble species. Phosphates alone (without any S/S reagents) can be effective in the stabilization of lead in some wastes, and this is the most common application for phosphate.[14] The resulting products are different from those in most stabilization processes in that they exhibit very low solubility over a wide pH range. Cementitious materials can be added to improve the physical characteristics of the treated waste if desired, in which case the process is usually considered to be a cement- or pozzolan-based process with phosphates as additives.

8.2.2.5 $FeSO_4$ Co-Precipitation

Co-precipitation of metals with hydrous oxides formed from iron salts has been used in S/S.[15] The co-precipitation reduces metal solubility to levels below those obtained with hydroxide precipitation. Optimum results are obtained when the Fe(II)/Fe(III) ratio is between 1:1 and 1:2.

8.2.2.6 Other Inorganic Complexing Agents

Some metals or metalloids can be present as anions and require a different treatment approach. For example, arsenic can be present as the arsenate ion, AsO_4^{3-}. Addition of either calcium or ferric ion results in the formation of highly insoluble inorganic arsenates: calcium arsenate, $Ca_3(AsO_4)_2$ and ferric arsenate, $FeAsO_4$. In the unusual instances where barium leaching is a concern, the waste can be treated with sodium or calcium sulfate to form insoluble barium sulfate.

8.2.3 OXIDATION/REDUCTION

Oxidation/reduction can be used to treat metals, non-metals, organics, and organometallic compounds. Reduction is usually performed prior to S/S, but adding reducing agents directly to the grout slurry has also been used.[16]

The presence of strong oxidants or reductants can change the valence state of a number of metals, affecting their chemical speciation and often drastically changing their solubility. Metals with more than one oxidation state possible in aqueous solutions include arsenic, chromium, iron, mercury, nickel, selenium, and technetium. And even though they only have one oxidation state in aqueous systems, lead, silver, copper, cadmium, and zinc can also be strongly influenced by redox processes.[17] Ferrous sulfate, sodium bisulfite ($NaHSO_3$), sodium metabisulfite ($Na_2S_2O_5$), metallic iron (zero valent iron), sodium hydrosulfite ($Na_2S_2O_4$), sodium hypochlorite, and potassium permanganate are commonly used to reduce or oxidize metals to the desired valence state.

Ferrous sulfate is probably the most widely used. It is safe and inexpensive, and often co-precipitates other metal contaminants. Its main drawbacks are that it can require a low pH for acceptable treatment times and it can result in a large volume increase. However, low pH does not always appear to be necessary. Rapid reduction of Cr^{6+} in solution up to pH 10 has been reported.[18]

Sodium bisulfite and sodium metabisulfite (the anhydrous form of sodium bisulfite) are effective reducing agents for Cr^{6+} and require much less acid and alkali than ferrous sulfate, and therefore generate less sludge. However, bisulfites can generate sulfur oxides on contact with acids, and this can be a disadvantage to their use. In general, shipping, handling, storage, and air pollution considerations make it impractical for most treatment processes.[19]

Ground granulated blast furnace slag, as described in detail in Chapter 4, is an effective reducing agent because of the iron sulfide naturally present in the slag[20] (see Figure 8.3 regenerated from the data of Angus and Glasser[21-23]). A slag:cement combination of 75:25 virtually eliminates calcium hydroxide as a hydration product; i.e., the presence of excess slag prevents buildup of this cement hydration product.[24] Oak Ridge National Laboratory (ORNL) tested various mixtures of slags in combination with portland cement and flyash for the reduction and stabilization of radioactive wastes containing technetium and nitrates.[25] This natural reducing capability of slag has proven effective in stabilizing technetium by reducing the soluble +7 pertechnetate anion to the more insoluble +4 cation.[26] Westinghouse Savannah River Company has been using slag formulations since 1984 for reduction of Tc^{+7} and Cr^{+6}, as well as for improved nitrate retention and better durability.[27] Ferrous sulfide decreased pertechnetate leaching from cement–flyash grouts, but increased leaching of other anions (chromate, selenate, and nitrate).[28] Figure 8.4 gives the E_h–pH diagram for Tc(VII)–Tc(IV) calculated by HSC 5[28b] using NH_4^+ as the counterbalancing cation in aqueous solution.

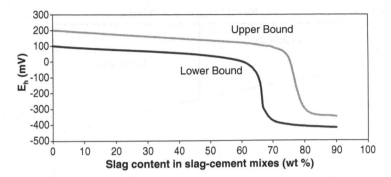

FIGURE 8.3 Matrix oxidation potential with slag content in slag-cement mixes regenerated from the data in References 29 to 31.

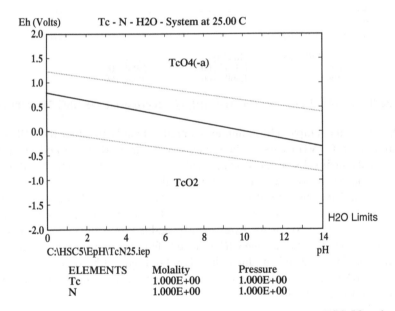

FIGURE 8.4 E_h–pH diagram for Tc(VII)–Tc(IV) (generated using the HSC 5® software).

Sodium hydrosulfite is effective at the alkaline pH of many waste residuals and so does not usually require pH adjustment before addition. And it remains effective even after the highly alkaline solidification reagents are added, allowing both the hydrosulfite and S/S reagents to be added together. The relatively high cost of sodium hydrosulfite limits its use in S/S treatment.

Often treatment may be two-part. For example, re-speciation to free inorganic arsenate (AsO_4^{3-}) is necessary for inorganic arsenic species containing the arsenite (AsO_3^{2-}) ion, for arsenic sulfides containing As(III) (e.g., As_2S_3), and for organoarsenicals. This re-speciation is most commonly accomplished by the use of strong oxidizing agents such as $KMnO_4$, $K_2S_2O_8$, NaOCl or Ca(OCl)$_2$, or H_2O_2, under alkaline conditions.

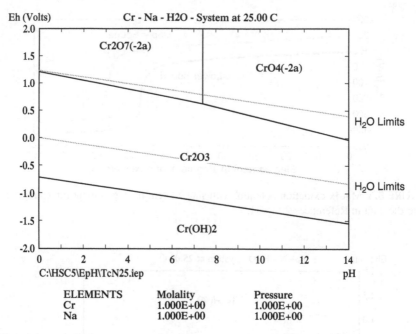

FIGURE 8.5 E_h–pH diagram for Cr(VI)–Cr(III) (generated using the HSC 5® software).

As a second example, chromium exists predominantly in either the +III or +VI oxidation state. Cr(VI) exists as the oxyanions chromate (CrO_4^{2-}) and dichromate ($Cr_2O_7^{2-}$), and their salts, all of which are soluble. The trivalent state, Cr(III), readily forms a relatively insoluble hydroxide and is the desired form for environmental management. The conventional process for management of Cr(VI)-containing waste streams is a two-step process: 1) reduction of Cr(VI) → Cr(III), with a suitable reducing agent, and 2) precipitation as the hydroxide, by addition of a suitable base. Reducing agents commonly used to treat the Cr(VI) include ferrous sulfate, $FeSO_4$, sodium metabisulfite, $Na_2S_2O_5$, and sodium dithionite, $Na_2S_2O_4$. Figure 8.5 gives the E_h–pH diagram for Cr(VI)–Cr(III) calculated by HSC 5 using Na^+ as the counterbalancing cation in aqueous solution.

8.2.4 SORPTION AND ION EXCHANGE

Reagents used to sorb metal contaminants (and organic contaminants) include activated carbon, ion exchange resins, metal oxides, natural materials (such as clays and zeolites), and synthetic materials. Flyash and rice hull ash also have sorptive properties. Some of these sorbents such as the ion exchange resins, zeolites, clays, and flyash both sorb and function as ion exchangers. Reagents that sorb work well for neutral, complexed, or organometallic compounds. The reagents with active ion exchange sites can work well to remove cationic metals from solution.

8.2.4.1 Illite Clay

Illite clay, (OH)4Kx(Al4Fe4Mg4Mg6)(Si8-xAl)O20, has proven to be an effective additive in cementitious waste forms for retarding the release of the soluble radio-isotope ^{137}Cs.[29-32] Illite was shown to be an effective selective sorbent for ^{137}Cs in the early 1960s.[33-35] Illite has a relatively low equivalent exchange capacity among clays, but the gap between illite layers is apparently ideal to allow cesium ions to diffuse between the silicate sheets and essentially irreversibly trap these ions. The cesium ions must remain mobile and not sorbed on the external ion exchange sites to become irreversibly trapped in this manner. For example, a concentration of a few molar of sodium ions is enough to replace the cesium ions sorbed on the external ion exchange sites and allow this intercalation diffusion to occur. Illite increases the ANSI/ANS-16.1-2003 cesium leachability index for a slag-cement-flyash grout from about 8 to about 10. This represents a decrease of two orders of magnitude in the effective diffusion coefficient for this waste form.

8.2.4.2 Crystalline Silicotitanate

Crystalline silicotitanate (CST) is a synthetic sorbent that has been developed and marketed recently as a highly specific sorptive agent for ^{137}Cs. CST increases the ANSI/ANS-16.1-2003 cesium leachability index for a slag-cement-flyash grout from about 8 to about 11, a decrease of three orders of magnitude in the effective diffusion coefficient. In addition, CST increased the ^{90}Sr leachability index from 10 to 12. The high cost of CST makes this synthetic sorbent less attractive for S/S than the natural alternatives, despite its high loading capacity and specificity for ^{137}Cs.

8.2.4.3 Clinoptilolite and Other Zeolites

The cation exchange capabilities of zeolites may make them an effective barrier to radionuclide migration.[36] For the same reason, they have been proposed as a treatment agent to sorb/extract cations from waste,[37-41] to immobilize cations by mixing directly with waste,[42-46] and to immobilize cations by mixing in a waste grout.[28] Natural zeolites form from weathering volcanic glass, such as clinoptilolite, ideally (Na, K, Ca$_{0.5}$)$_6$Si$_{30}$Al$_6$O$_{72}$•24H$_2$O.[47] The reported cation selectivity for four natural zeolites follow:[40]

Chabazite: Cs > NH$_4$ > Pb > Na > Cd > Sr > Cu > Zn
Clinoptilolite: Cs > Pb > NH$_4$ > Na > Sr > Cd Cu Zn
Mordenite: Pb > Cs > NH$_4$ > Na > Cd
Phillipsite: Cs > Pb > NH$_4$ > Na > Sr > Cd > Zn

These results imply that Cd, Sr, Cu, and Zn may not be effectively immobilized for the Na- form of these zeolites or for high Na-bearing wastes.

Synthetic mordenite and two other synthetic zeolites effectively sorbed uranium from solution, but not natural clinoptilolite and another synthetic zeolite.[41] This relative effectiveness was premised to be a function of the pore dimensions, chemical composition, and cation concentration of each zeolite. In addition, the uranium remained sorbed under redissolving conditions. Clinoptilolite proved effective in sorbing ^{60}Co from solution.[40] Surfactant modification made clinoptilolite effective in sorbing anions (chromate).[37,38] The sorbed chromate was resistant to water extraction.

Clinoptilolite or other zeolites have proven effective in sorbing metals (Cd, Cr, Cu, Fe, Mn, Ni, Pb, Zn) when mixed with contaminated compost, soil, or waste.[42–45] Even though sorption makes the metals less leachable by acids, it has been demonstrated that metals (Pb and Zn) sorbed on clinoptilolite are more bioavailable and biotoxic than the same metals in untreated soil.[42] Clinoptilolite decreased cesium leaching from a cement-flyash grout, but increased anion leaching.[28]

The results of Colella[40] and Serne et al.[28] are in agreement on clinoptilolite increasing cesium leach resistance.[28,40] The strontium leach resistance did not increase upon adding clinoptilolite; but it is not clear this is a confirmation, since the cement-flyash grout gave such a high baseline leach resistance for strontium. In summary, clinoptilolite is expected to increase the leach resistance of cesium, lead, and cobalt, but not strontium, cadmium, copper, and zinc cations, or any anions (chromate, pertechnetate, selenate, and nitrate) without surfactant modification of the surface. Zeolites do increase the leach resistance of uranium, but apparently their effectiveness depends on the zeolite pore size.

8.2.4.4 Vermiculite and Other Sorbents

Vermiculite strongly sorbed cesium and strontium.[48] More strontium sorbed as the pH increased, but the sorbed strontium readily extracted into 0.1 N EDTA.[48] The same source reported cesium sorption being unaffected by pH and the EDTA solution,[48] whereas another source reported that the cesium distribution ratio increased from about 10 to 400 as the pH increased from about 2 to 12.[49] The reason for the discrepancy is not clear, but there are differences between the two studies; e.g., source of vermiculite. Both sources indicated high affinity of vermiculite for cesium at high pH, which may be important agreement for the purpose of cement stabilization. The sodium form vermiculite was superior to both the potassium form and the calcium form for cesium sorption.[49] Incomplete conversion of vermiculite into organophilic clay allowed the simultaneous sorption of both anionic radioiodine and cationic radio-cesium and strontium, but the vermiculite distribution ratio significantly decreased for both cations.[50]

Cesium was displaced from vermiculite saturated with cesium by the following cations, presented in the order from greatest to least displacement:[51]

K > Ce ≈ Y ≈ La > Sr ≈ Ca ≈ Ba > Li
K > Na > Cs (isotopic exchange) ≈ Li

Potassium displaced about 40% of the cesium loaded on vermiculite, greatest of all the ions studied. Only about 10% was displaced during isotopic exchange, with more being displaced if the vermiculite was not saturated with cesium. Vermiculite columns removed 99.5% of $^{137/134}$Cs and ^{60}Co from aqueous solution, but removed little or no [131]I.[52] The sorbed radioisotopes were not eluted by water or 4 N ammonia.

Vermiculite ranked first for cesium sorption in the following ranking of natural clays and minerals:[53]

Vermiculite > phillipsite > mordenite ≈ montmorillonites ≈ clinoptilolite > illite ≈ salona shale > labradorite basalt ≈ mica > conasauga shale ≈ kaolinite > chlorite

Under hydrothermal conditions, vermiculite, tobermorite, and zirconium phosphate were best for cesium immobilization. Tobermorite and hydroxyapatite were best for strontium, lead, and cadmium immobilization.[54]

Vermiculite from South Africa was used as a comparative base for sorption of Cs, Sr, and Ca by Polish minerals.[55] The relative rankings of Cs distribution ratios (from 900 for vermiculite to 1.6 for shales) follow:

Vermiculite > bentonites > silica > sand > iron sands > iron and silty sands > phosphorites > halloysite ≈ bog iron ore > chlorite detritus > shales

Another study ranked the effect of additives on the cesium effective diffusion coefficient in the following order:[56]

Ca-bentonite (Karlich) > Na-bentonite (U.S. & Geisenheim) ≈ H-bentonite > Ca-bentonite (Neiderbayern) ≈ Na-bentonite (Karlich) ≈ vermiculite ≈ kaolin ≈ illite > zeolite > attapulgite

The equivalent leachability indices ranged from 11.2 for the Karlich Ca-bentonite down to 7.4 for attapulgite, with a value of 8.6 for illite, significantly below that measured at ORNL (> 10). The value for vermiculite was 8.8. As the bentonite increased from 0 to 20 wt% in steps of 5 wt%, the effective diffusion coefficient decreased with each addition from about 10^{-1} to about 10^{-5} cm^2/d (leachability index of 6 to 10) for cesium, but slightly increased with each addition around a value of 10^{-3} cm^2/d (leachability index of 8) for strontium.

Higher salt concentrations of sodium or calcium nitrate suppressed sorbent selectivity for ^{137}Cs and [85]Sr.[57] Of the agents tested, ferrocyanides had the highest selectivity for [137]Cs (distribution ratio > 10,000 cm^3/g). At low salt concentrations, high distribution ratios for ^{137}Cs were also observed for natural aluminosilicates (vermiculite and clinoptilolite), synthetic zeolites, and phosphates. The lowest ^{137}Cs ratios were observed for ion exchange resins, hydrous metal oxides, and carbon adsorbents. Synthetic shabazite had the highest selectivity for [85]Sr in sodium nitrate solutions (distribution ratio > 43,000 cm^3/g). Efficient Sr sorption also occurred for ion exchange resins, some synthetic zeolites, and modified manganese dioxide.[85] Vermiculite, clinoptilolite, other synthetic zeolites, phosphates, ferrocyanides, and carbon adsorbents showed relatively low selectivity for [85]Sr. Clinoptilolite and modified manganese dioxide had the highest selectivity for [85]Sr in calcium nitrate solutions (distribution ratio > 500 cm^3/g), though synthetic shabazite was still efficient

(250 cm³/g). Vermiculite selectivity for ^{85}Sr was relatively low in both sodium and calcium nitrate solutions.

The calcined waste at the Idaho National Engineering and Environmental Laboratory (INEEL) can be reacted to make hydroceramic, an insoluble ceramic of sodium alumino silicates.[58] Vermiculite was added to enhance the cesium leach resistance of this waste form. Illite was destroyed in the hydrothermal reaction to produce hydroceramic and did not enhance the cesium leach resistance.

Heat-treated vermiculite had aqueous distribution ratios of 420, 52, and 11 for ^{137}Cs, ^{90}Sr/^{90}Y, and ^{239}Pu, respectively.[59] The ratio ranges for several natural and synthetic materials were $< 5 - 1800$, $< 5 - 1800$, $< 5 - 680$, respectively, with the following ranking for ^{137}Cs ratios:

Ferricyanide NZA > ferricyanide FS-2 > synthetic zeolite > modified mordenite > mordenite ≈ clinoptilolite > modified glauconite > glauconite > caoline > vermiculite > cambria clay > cation exchange resin > modified charcoal > charcoal > anion exhange resin

Although the synthetic materials exhibited high sorption efficiency, the natural materials were favored because of their longer working cycle and lower costs.

Combining copper ferricyanide (CFC) with porous media (silica gel, bentonite, vermiculite, and zeolite) removed cesium from aqueous solutions.[60] The treated vermiculite and zeolite removed cesium better than treated silica gel or bentonite. A compound sorbent of CFC and vermiculite was used to increase the cesium leachability index of solidified borate wastes from around 8 to around 10, with CFC receiving the most credit for the enhanced leach resistance.

8.3 NON-METALS AND ORGANICS

Simple complexes such as cyanides can often be destroyed by alkaline chlorination, for example, by using sodium or calcium hypochlorite. Destruction may be pH sensitive.

The most common application has been for non-metals such as cyanide, hydrogen sulfide, and phenol.[18] If organics such as oil and grease are present they will also consume the oxidizing agent. This may require large quantities of oxidizing agent, making treatment costly or practically prohibitive.

8.4 ORGANIC STABILIZATION

For waste streams containing low levels of hazardous organic contaminants, useful sorbents include activated carbon, organoclays (natural clays treated with quaternary ammonium compounds), rubber particulates, rice hull ash, and other natural minerals. In such instances, the purpose of treatment is to sorb and reduce the leaching of organic contaminants to below regulatory levels. Activated carbon, organoclays, and rubber particulate are useful with specific contaminants in specific test methods, but none works best for all. Carbon is effective overall for reduction in TCLP leachability, but not for reduction in trichloroethane (TCA).[61] Particulate rubber is not as effective in TCLP reduction, but is the only additive that was broadly useful for TCA reduction, especially for the low-volatility compounds. Organoclays are effective with select contaminants.

For solidification of moderate levels of non-hazardous organic contaminants, surfactants have been used to disperse an organic phase in an aqueous phase before the emulsion is solidified. Organic reagents tend to retard pozzolanic reactions and may be easily leached from the resulting solidified waste form.[62] Organic modified clays can overcome these limitations.[63] The modified organophilic clays are prepared by exchanging ammonium ions for metal ions in the clay. The modified clay has both inorganic and organic properties. The ammonium ions increase the distance between the layers of alumina and silica in the clay, and this allows organics to penetrate. Studies indicate that chemical bonding may occur between the organo-philic clay binder and certain organic wastes.[64]

For solidification of higher levels of non-hazardous organic wastes, a mixture of lime and flyash or other cementitious additives is commonly used. The lime makes oils and tars miscible with solidification/stabilization reagents. Lime/flyash processes are able to accommodate large quantities of organics as well as the more common inorganic sludges. Therefore, they are useful in the solidification of oily wastes[65] and other water-insoluble organic materials at organic levels of 20% or more. In fact, this is a major use for these processes. Combining lime and flyash with water forms a cementitious material. The reaction product formed is initially a non-crystalline gel, but eventually becomes calcium silicate hydrate, a compound found in hydrated portland cements. In general, however, these reactions are slower than those of cement and do not produce exactly the same products in terms of chemical and physical properties. Flyash used in lime/flyash processes is a by-product of coal-burning power plants, and its composition, and thus its reactive property, is dependent on not only the composition of the coal burned, but also how the plant is operated.

8.5 ADDITIVES AND TREATMENT METHODS FOR ORGANOMETALLIC AND COMPLEXED COMPOUNDS

Additive and treatment methods for metal/organic compounds are dependent on the type of compound present. Simple complexes such as cyanides can often be treated by alkaline chlorination. However, many industrial waste streams contain soluble metal complexes that are very difficult to treat because of their stability.

Simple complexes such as metallo-cyanides, e.g., $Cd(CN)_4^{-2}$, can be treated using alkaline chlorination (e.g., sodium or calcium hypochlorite). Nickel cyanide complexes can be broken with alkaline chlorination, leaving a nickel species that can be stabilized to low leaching levels. In cases where a stable, soluble silver complex is present in large amounts, several techniques are available. Magnesium sulfate and lime can be used to precipitate a mixed sulfate-oxide. Alkaline chlori-nation can be used to break the complex and precipitate silver chloride. Sulfides and hydrosulfites are also used to treat silver complexes.

More stable complexes such as citrates, EDTA complexes, oxalates, gluconate, and a wide variety of non-chelate organometallics can sometimes be sorbed onto activated carbon, clays, or other similar sorbents. This approach has worked for nickel and chromium and probably arsenic and cadmium chelates.[66] The sorbed organometallic compounds may then be solidified using portland cement or another solidification reagent.

If neither of these approaches work then the metal must be released from the organic so that typical treatment processes for metals can be applied. This usually involves oxidative destruction of the organic molecule using strong oxidants such as potassium permanganate, potassium persulfate, or hypochlorite, and the possible use of elevated temperature. A plausible reaction for the hypochlorite oxidation/destruction of the EDTA ligand, necessary for precipitation/removal of $Ni^{2+}(aq) \rightarrow Ni(OH)_2(s)$, is depicted below:

$$NiEDTA^{2-} + 12\ ClO^- \rightarrow H_2NCH_2CH_2NH_2 + 8\ CO_2 + Ni^{2+}(aq) + 12\ Cl^- + 4\ OH^-$$
$$\downarrow$$
$$Ni(OH)_2(s)$$

With either oxidizing or reducing agents, species other than the target compound may compete for the reagent. As a result, large additions of oxidizing agent may be necessary, and this becomes very expensive and time consuming. Therefore, a non-oxidative method for handling complexed metals, such as sorption, is preferred. Other problems can complicate the oxidative destruction of organic complexes. For example, if chromium is present, it will be oxidized to Cr^{+6} and must then be reduced before precipitation.

8.6 PROCESSING AND ANTI-INHIBITION AIDS

The first additives used probably came from the concrete industry. More and more, additives are being used in chemical systems for purposes other than stabilization. Cement-based systems may contain many additives, such as given in Table 8.1. The purposes of additives, in addition to those already discussed, are many and varied. Among the more common purposes are control of setting, retarders, water-reducing agents, increase or control viscosity, and physical property development.

Pozzolan/portland cement processes consist primarily of silicates from pozzolanic-based materials like flyash, kiln dust, pumice, or blast furnace slag and cement-based materials like portland cement.[67] These materials chemically react with water to form a solid cementitious matrix, which improves the handling and physical characteristics of the waste. They also raise the pH of the water, which may help precipitate and immobilize some heavy metal contaminants. Pozzolanic and cement-based binding agents are typically appropriate for inorganic contaminants. The effectiveness of this binding agent with organic contaminants varies.

TABLE 8.1

Properties of Some Processing and Anti-Inhibition Aids

Additive	Control of Setting	Retarders	Water Reducers	Control Viscosity	Physical Property Development	Control Free Water Content
Lime	X				X	X
Soluble Silicates	X				X	
Alumina	X	X				
Surfactants			X			
Clays				X		
Polymers				X		
Flyash		X		X	X	X
Kiln Dust				X	X	X
Silica Fume				X		
Diatomaceous Earth						X
Vermiculite						X
Rice Hull Ash	X				X	X
Mineral Sorbents						X

8.6.1 CONTROL AGENTS

Cement-based systems, especially, are sensitive to many waste components discussed previously in their setting and strength development. Additives such as lime and sodium silicate are often used to counter the inhibition effects of certain metals and allow proper set and hardening to be achieved.[68] Research with soluble silicates indicates that these materials are beneficial in reducing the interference from metal ions in the waste solution.[69] Cement/lime and mixtures of cement, lime, and ferrous sulfate have been used to produce solid, low-leaching products with lead, chromium, and other heavy metal-containing wastes. Another additive commonly used along with cement and lime is aluminum sulfate or alumina, which reportedly counters the retarding effect of organic constituents in the waste.[70] Calcium chloride can be used to accelerate setting in portland cement systems and overcome retardation by organic compounds.[71] Clay, diatomaceous earth, flyash, and activated carbon can remove interfering ions and organics.

8.6.2 RETARDERS

Some waste components cause premature setting and require the use of set retarders commonly used in making concrete. Control of setting is especially important for *in situ* S/S.

8.6.3 WATER-REDUCING AGENTS

Water-reducing agents are usually surfactants that are used to lower the viscosity. Flyash also reduces viscosity, but results in significant volume increase.

8.6.4 Viscosity Control Agents

Small amounts of clays or polymers are sometimes employed to control grout viscosity and reduce available water as well as to sorb organics and metals.[72] Another technique is to use sorbents such as diatomaceous earth, vermiculite, rice hull ash, zeolites, and various expanded mineral absorbents to control free water content.[73] Kiln dust (lime and cement) also increases grout viscosity and reduces the available water. One reason for increasing viscosity is to prevent phase separation.

Gel clays have been used for decades in geotechnical applications, e.g., construction (slurry walls and clay caps) and drilling (drilling muds and cement mixes), to resist solids segregation (suspension aid), prevent bleed water, and act as an engineered hydraulic barrier to water penetration (into a construction zone, waste disposal site, and so on). The most commonly used clay for these purposes is bentonite, sodium montmorillonite, " ... a colloidal clay mined in Wyoming and South Dakota. It imparts viscosity and thixotropic properties to fresh water by swelling to about 10 times its original volume. Bentonite (or gel) was one of the earliest additives in oilwell cements to decrease slurry weight and to increase slurry volume."[74,75] The individual clay particles of bentonite are plate-shaped. The particle faces are positively charged, while the edges are negatively charged. When mixed with water the platelets separate and disperse throughout the fluid. When mixing ceases, the clay particles form a multilayered colloidal gel structure due to the attraction of opposite charges. However, the electrostatic double-layer forces are lessened with increasing ionic strength.[76] Consequently, high-salt solutions (notably chloride, sulfate, and phosphate salts as well as acids and bases) collapse these gels, lessening their dispersive effectiveness and releasing the large volume of water collected around the clay particles (i.e., free water can form if salt solutions are grouted).[77]

This susceptibility compromised the use of bentonite in off-shore oil drilling in salty waters. For this reason, attapulgite was adapted as the gel clay used in such salty applications. Attapulgite clay particles carry no charge and are not affected by high salt content. The individual attapulgite particles resemble needles, rather than platelets. When mixed with water, these needles are dispersed throughout the fluid and become aligned along shear planes. When mixing ceases, the random entanglement of these particles, referred to as a "brush-heap effect," forms a gel structure. Attapulgite is commercially available from northern Florida and southern Georgia. Attapulgite has been adopted as the gel clay of choice for salty wastes. Note that although several forms of attapulgite have been tested for DOE salty wastes, only attapulgite 150 (Attagel® 150) proved effective.[78]

8.6.5 Additives that Increase Strength or Improve Other Physical Properties

Using flyash as part of a grout recipe typically results in a solidified product that has increased strength and reduced permeability (see detailed description below). In one study the leaching of chromium decreased as more portland cement was substituted with silica fume. The incorporation of silica fume into the cement matrix minimized the detrimental effects of organic materials on the cement hydration reaction and contaminant leachability.[79] Silica fume is a by-product of producing silicon metal or ferrosilicon alloys. Silica fume consists primarily of amorphous (non-crystalline) silicon dioxide (SiO_2). The individual particles are extremely small, approximately 1/100th the size of an average cement particle. Because of its fine particles, large surface area, and the high SiO_2 content, silica fume is a very reactive pozzolan. Cement containing silica fume can have very high strength and can be very durable.[80]

8.7 MULTITALENTED OR MULTIPURPOSE REAGENTS

There are a few additives that have wide usage. This is due in part to the multiple positive characteristics that they impart to the grout or the resulting stabilized waste form. These select additives include flyash, soluble silicates, kiln dust, and ferrous sulfate. Chapter 4 describes the use of flyash, soluble silicates, and cement kiln dust as cementing components.

The characteristics of flyash are described in detail in Chapter 4. The use of flyash with portland cement has many advantages, including increased viscosity, preventing phase separation, acting as a pozzolan, binding additional water, decreasing the pore pH, adsorbing metal ions, and maybe also retarding setting of the cement. The resulting product typically has increased strength, decreased permeability, and increased durability in tests such as freeze-thaw and wet-dry resistance. Thus, a cement–flyash combination is a popular system where critical physical properties are required.

The greatest potential disadvantage of the cement–flyash process is the weight and volume increase associated with large additions of flyash. The flyash:cement ratio, by weight, is typically 2:4, with total weight increases of 50 to 150% corresponding to volume increases of 25 to 75%.[81] This is especially true with low solids waste streams, where the flyash acts as a bulking agent to increase viscosity and prevent phase separation until the mass sets. Where the increase is not important, as in some remedial action projects, portland flyash cement may be the optimum choice. Waste components, which interfere with cement setting and hardening, will likely also interfere with those functions in the portland flyash cement system.

Soluble silicates combined with portland cement act as immobilization agents for metals,[18] as anti-inhibition agents for cement setting,[18] and to reduce permeability.[82]

Lime kiln dust and cement kiln dust have been used on hundreds of remedial solidification projects in the United States over the last 25 years. They have also been used extensively at central hazardous waste management facilities (TSDFs). Attractive characteristics or features of kiln dust include functioning primarily as an absorbent or bulking agent and removing free water as additional water is consumed in their pozzolanic reactions. In addition, lime kiln dust with its high calcium oxide content is used to neutralize acidic wastes. Kiln dust is used for stabilization of organics and inorganics. Systems with kiln dust usually set slower that portland cement systems and have less strength. The chemical properties for kiln dusts vary widely from source to source because of how these materials are produced.[83]

The kiln dusts are effective because of their calcium oxide content. This gives them high alkalinity and the ability to remove free water by the hydration of CaO to $Ca(OH)_2$, an advantage over inert sorbents. They also use up water in their pozzolanic reactions. Often, kiln dust and flyash solidification techniques result in friable, even granular, products, which are usually desirable from the operational point of view at the landfill. The relatively large volume increase associated with these materials — the weight addition percentage ranges from 50 to 200% — makes them unattractive in some applications. However, there are instances, especially with highly organic waste streams, where these systems are the most efficient for solidification, even on a weight-to-weight basis.

Cement kiln dust has been successfully used to treat numerous types of waste materials. Everything from coalmine waste effluents[84] and industrial wastewater to sewage and oil sludges[85] have been stabilized using cement kiln dust.

REFERENCES

1. Mackay, M., and J.J. Emery. Practical Stabilization of Contaminated Soils, *Land Contamination and Reclamation* , 1, 3, pp. 149–155 (1993).
2. Conner, J.R. and S.L. Hoeffner. A Critical Review of Stabilization/Solidification Technology. *Critical Reviews in Environmental Science and Technology.* 28(4):397–462 (1998).
3. Wilk, C.M. *Stabilisation of Heavy Metals with Portland Cement: Research Synopsis.* Waste Management Information, Public Works Department, Portland Cement Association, Skokie, IL (1997).
4. Moore, J.N., W.H. Ficklin, and C. Johns. Partitioning of Arsenic and Metals in Reducing Sulfide Sediments. *Environ. Sci. Technol.* 22:432–437.
5. W.F. Linke, *Solubilities: Inorganic and Metal-Organic Compounds.* Volume I, Fourth Edition, pp. 1245–1248, American Cyanamid Co., D. Van Nostrand Company, Inc., Princeton, NJ, 1958.
6. Chemfix Technologies Inc., Kenner, LA (1987).
7. Lo, P.C. et al. *Chemical Stabilization — More than a Fixation Process,* Chemfix Technologies, Kenner, LA (1986).
8. Yagi, T. *Japan Kokai* 75 105,541 (Aug. 20, 1975).
9. Nakaaki, O. et al. *Japan Kokai* 75 99,962 (Aug. 8, 1975).
10. Gowman, L.P., Chemical Stability of Metal Silicates vs. Metal Hydroxides in Ground Water Conditions. In *Proc. 2nd National Conference on Complete Water Reuse.* (1975).
11. Falcone, J.S. *Soluble Silicates.* New York: Reinhold (1982).
12. Durham, R.L. and C.R. Henderson. *U.S. Patent 4,460,292* (July 17, 1984).
13. Conner, J.R. and R.S. Reber. *U.S. Patent 5,078,796* (Jan. 7, 1992).
14. O'Hara and Surgi. *U.S. Patent 4,737,356* (April 12, 1988).
15. Pojasek, R.B. *Toxic and Hazardous Waste Disposal.* Vols. 1–4. Ann Arbor, MI: Ann Arbor Science Publishers. (1980).
16. Spence, R.D., T.M. Gilliam, and A. Bleier. Cementitious Stabilization of Chromium, Arsenic and Selenium in a Cooling Tower Sludge. Presented at the *88th Annual Meeting of the Air and Waste Management Association,* San Antonio, TX, June 18–23 (1995).

17. Dragun, J. The Fate of Hazardous Materials in Soil. *Hazardous Materials Control:*41–65 (May/June 1988).
18. Eary, L.E. and D. Rai. Chromate Removal from Aqueous Wastes by Chromium Reduction with Ferrous Ion. *Environ. Sci. Technol.* 22(8):972–977 (1988).
19. Conner, J.R. *Chemical Fixation and Solidification of Hazardous Wastes,* New York: Van Nostrand Reinhold (1990).
20. US EPA. *Stabilization/Solidification Processes for Mixed Waste.* EPA 402-R-96-014. US Environmental Protection Agency. June (1996).
21. Angus, M.J. and Glasser, F.P. The Chemical Environment in Cement Matrixes, *Materials Research Society Symposium*, MRS, Boston, 50: 547–556 (1986).
22. Glasser, F.P. Chemistry of Cement-Solidified Waste Forms, Chapter 1, *Chemistry and Microstructure of Solidified Waste Forms*, Spence, R.D., Editor, Lewis Publishers, Boca Raton, FL (1993).
23. Spence, R.D., Bostick, W.D., McDaniel, E.W., Gilliam, T.M., Shoemaker, J.L., Tallent, O.K., Morgan, I.L., Evans-Brown, B.S., and Dodson, K.E. Immobilization of Technetium in Blast Furnace Slag Grouts, *3rd International Conference on the Use of Fly Ash, Silica Fume, Slag & Natural Pozzolans in Concrete*, Trondheim, Norway (1989).
24. International Atomic Energy Agency (IAEA). *Improved Cement Solidification of Low and Intermediate Level Radioactive Wastes*, Technical Reports Series No. 350, International Atomic Energy Agency, Vienna (1993).
25. Gilliam, T.M., L.R. Dole, and E.W. McDaniel. *Waste Immobilization in Cement-Based Grouts.* Philadelphia: ASTM Special Technical Publication (1986).
26. Spence, R.D. et al. Immobilization of Technetium in Blast Furnace Slag. Presented at the *Third International Conference on the Use of Flyash, Silica Fume, Slag & Natural Pozzolans in Concrete.* Trondheim, Norway. June 19–24 (1989).
27. Langton, C., EPA/ORIA Conference on Treatment Technologies for Low-Level Radioactive and Mixed Wastes. Crystal City, VA (1995).
28a. Serne, R.J., C.W. Lindenmeier, V.L. LeGore, P.F.C. Martin, L.L. Ames, and S.J. Phillips. *Leach Testing of in situ Stabilization Grouts Containing Additives to Sequester Contaminants*. PNL-8492. Pacific Northwest National Laboratory (1993).
28b. Roine, A. *Outokumpu HSC Chemistry® for Windows: Chemical Reaction and Equilibrium Software with Extensive Thermochemical Database*, User's Guide Version 5.0, 02103-ORC-T, Outokumpu Research Oy, Information Service, P.O. Box 60, FIN-28101 PORI, Finland (2002).
29. Moore, J.G., H.W. Godbee, A.H. Kibbey, and D.S. Joy. *Development of Cementitious Grouts for the Incorporation of Radioactive Wastes. Part 1: Leach Studies*. ORNL-4962. Oak Ridge National Laboratory. Oak Ridge, TN (August 1975).
30. Moore, J.G. *Development of Cementitious Grouts for the Incorporation of Radioactive Wastes. Part 2: Continuation of Cesium and Strontium Leach Studies*. ORNL-5142. Oak Ridge National Laboratory. Oak Ridge, TN (September 1976).
31. Gilliam, T.M. and J.A. Loflin. *Leachability Studies of Hydrofracture Grouts.* ORNL/TM-9879. Oak Ridge National Laboratory, Oak Ridge, TN (November 1986).
32. Gilliam, T.M. Leach testing of hydrofracture grouts containing hazardous waste. *Journal of the Underground Injection Practices Council*. 1:192–212 (1986).
33. Tamura, T. Cesium sorption reactions as indicator of clay mineral structures. *Clays and Clay Minerals, Proc. Natl. Conf. Clays and Clay Minerals*, 10:389–398 (1961).
34. Tamura, T. Cesium sorption reactions as indicator of clay mineral structures. *Intern. Clay Conf., Proc. Conf. Stockholm*, 1:229–237 (1963).

35. Tamura, T. and D.G. Jacobs. Structural implications in cesium sorption. *Health Phys.* 2:391–398 (1960).

36. Ogard, A.E., K. Wolfsberg, W.R. Daniels, J. Kerrisk, R.S. Rundberg, and K.W. Thomas. Retardation of radionuclides by rock units along the flow path to the accessible environment. *Materials Research Society Symposium Proceedings.* 26:329–336 (1984).

37. Ingram, C.W., R. Szostak, K. Cleare, M.M. Smith, D.A. Cook, C.D. Parker, and P. Abrahams. Zeolite shape selectivity in the uptake of uranium from solutions. *Nucl. Hazard. Waste Manage.* Proceedings of an International Meeting. 2:1098–1105 (1996).

38. Li, Z. and R.S. Bowman. Counterion effects on the sorption of cationic surfactant and chromate on natural clinoptilolite. *Environ. Sci. Technol.* 37(8):2407–2412 (1997).

39. Li, Z. Chromate extraction from surfactant-modified zeolite surfaces. *J. Environ. Qual.* 27(1):240–242 (1998).

40. Colella, C. Environmental applications of natural zeolitic materials based on their ion exchange properties. *Natural Microporous Materials in Environmental Technology.* P. Misaelides et al. (eds.) NATO Sci. Ser., Ser. E. Kluwer Academic Publishers. Netherlands. 207–224 (1999).

41. Hernandez-Barrales, E. and F. Granados-Correa. Sorption of radioactive cobalt in natural Mexican clinoptilolite. *J. Radioanal. Nucl. Chem.* 242(1):111–114 (1999).

42. Lewis, M.A., D.F. Fischer, and L.J. Smith. Salt-occluded zeolites as an immobilization matrix for chloride waste salt. *J. Am. Ceram. Soc.* 76(11):2826–2832 (1993).

43. Greene, J.C. and J.J. Barich. Biological and chemical evaluation of remediation performed on metal bearing soils. *Tailings & Mine Waste'94* . Proceedings of the 1st International Conference. Balkema, Rotterdam. 157–166 (1994).

44. Crawford, P. and J. Gafford. Method for the stabilization and detoxification of waste material. US patent 5,484,533 (1996).

45. Zorpas, A.A. and M. Loizidou. The use of inorganic material such as zeolite for the uptake of heavy metals from the composting process. *Harzard. Ind. Wastes* . 31st. 611–620 (1999).

46. Zorpas, A.A., E. Kapetanios, G.A. Zorpas, P. Karlis, A. Vlyssides, I. Haralambous, and M. Loizidou. Compost produced from organic fraction of municipal solid waste, primary stabilized sewage sludge and natural zeolite. *J. Haz. Mater.* 77(1–3): 149–159 (2000).

47. Bowers, T.S. and R.G. Burns. Activity diagrams for clinoptilolite: Susceptibility of this zeolite to further diagenetic reactions. *American Mineralogist* . 75:601–619 (1990).

48. Jha, J.C., A.R. Chinoy, and K.T. Thomas. Studies on caesium sorption properties of some Indian vermiculites. *Nucl. Radiat. Chem. Symp.* Proceedings. 40–45 (1966).

49. Lee, S.H., Studies on the sorption and fixation of cesium by vermiculite. *J. Korean Nucl. Soc.* 6(2):97–111 (1974).

50. Bors, J., A. Gorny, and St. Dultz. Studies on the interaction of radionuclides with organophilic clays. *Radiochim. Acta* . 74:231–234 (1996).

51. Levi, H.W. and N. Miekeley. Studies on ion diffusion in vermiculite. *Disposal Radioact. Wastes Ground* . Proceedings. 161–168 (1967).

52. Sebastian, T.A., I.S. Bhat, and P.R. Kamath. Vermiculite decontamination of atomic power station effluents. *Environ. Pollut.* Proceedings. 95–97 (1973).

53. Komearneni, S. and R. Roy. Interactions of backfill materials with cesium in a bittern brine under repository conditions. *Nucl. Technol.* 56(3):575–579 (1982).

54. Komearneni, S. and R. Roy. Low-temperature materials for waste disposal: I, Hydroxylated phases. *Advances in Ceramics: Nuclear Waste Management II* . American Ceramic Society. 20:199–206 (1986).
55. Tymochowicz, S. Sorptive properties of mineral deposits occurring in Poland. *Nukleonika*. 26(4–6):595–599 (1981).
56. Vejmelka, P., G. Rudolph, W. Kluger, and R. Koster. *Conditioning of radioactive waste solutions by cementation* . KfK 4800. Kernforschungszent Karlsruhe (1990).
57. Gelis, V.M. and E.A. Kozlitin. Application of inorganic adsorbents and ion exchange resins for decontaminating solution from cesium and strontium radionuclides. *Technol. Programs Radioact. Waste Manage. Environ.* 2:1839–1841 (1993).
58. Gougar, M.L.D., D.D. Siemer, and B.E. Scheetz. Disposal of INEL spent nuclear fuel reprocessing waste using a glass-forming cement. *Embedded Top. Meet. DOE Spent Nucl. Fuel Fissile* . 359–366 (1996).
59. Doilnitsyn, V.A., A.F. Nechaev, A.S. Volkov, and S.N. Shibkov. Purification of slightly contaminated low-salt water from the long-lived radionuclides. *Radioact. Waste Manage. Environ. Rem.* Proceedings International conference. ASME. 527–528 (1997).
60. Huang, C.-T. and G. Wu. Improvement of Cs leaching resistance of solidified radwastes with copper ferrocyanide (CFC) – vermiculite. *Waste Manage.* 19(4):263–268 (1999).
61. Lear, P.R. and J.R. Conner, Immobilization of Low-Level Organic Compounds in Contaminated Soil. *Sixth Annual Conference on Hydrocarbon Contaminated Soils.* Amherst, MA. (September 23–26, 1991).
62. Bishop, P.L., Contaminant leaching from Solidified-Stabilized Wastes. Chapter 15 in *Emerging Technologies in Hazardous Waste Management II.* D.W. Tedder and F.G. Pohland (eds.). American Chemical Society, Washington, D.C. (1991).
63. Alther, G., J. Evans, and S. Pancoski, No Feet of Clay, *Civil Engineering,* 60, pp. 6–61 (1990).
64. Soundararajan, R., E. Barth, and J. Gibbons. Using an Organophilic Clay to Chemically Stabilize Waste Containing Organic Compounds. *Hazardous Materials Control* , 3, pp. 42–45 (1990).
65. Chestnut, R., J.J. Colussi, D.J. Frost, W.E. Keen, Jr., and M.C. Raduta, *U.S. Patent 4,514,307* (Apr. 30, 1985).
66. Conner, J.R. Private study (1988).
67. Remediation Technologies Screening Matrix and Reference Guide, Version 4.0, Federal Remediation Technologies Roundtable. http://www.frtr.gov/matrix2/section1/toc.html (April 2002).
68. Iffland, N. et al. *U.S. Patent 4,122,028* (Oct. 24, 1978).
69. Colombo, P. and R.M. Neilson. *Properties of Wastes and Waste Containers. Progress Report No. 7. BNL-NUREG 50837.* Brookhaven National Laboratory. Upton, NY (1978).
70. Uemura, T. and E. Hirotsu. *Japan Kokai* 80 47,251 (Apr. 3, 1980).
71. Valls, S. and E. Vazquez. Leaching Properties of Stabilised/Solidified Cement-Admixtures-Sewage Sludge Systems. *Waste Management* , 22, pp. 37–45 (2002).
72. Conner, J.R., *U.S. Patent 4,518,508* (May 21, 1985).
73. US EPA. *Handbook for Stabilization/Solidification of Hazardous Wastes.* EPA/540/2-86/001. US Environmental Protection Agency. (June 1986).
74. Smith, D.K. 1951. Physical properties of gel cements. *Pet. Eng.* , B7-B12 (April 1951).
75. Smith, D.K. 1990. *Cementing* , Monograph Volume 4, SPE Henry L. Doherty Series, Society of Petroleum Engineers Inc., New York (1990).

76. Grim, R.E. *Applied Clay Mineralogy*, McGraw-Hill, New York (1962).
77. American Colloid Co. Technical Bulletin Data No. 201, American Colloid Co., Chicago (1945).
78. de Laguna, W., T. Tamura, H.O. Weeren, E.G. Struxness, and W.D. McClain. *Engineering Development of Hydraulic Fracturing as a Method for Permanent Disposal of Radioactive Wastes*. ORNL-4259. Oak Ridge National Laboratory, Oak Ridge, TN (1968).
79. K.-S. Jun, B.-W. Hwang, H.-S. Shin, and Y.-S. Won. *Chemical characteristics and leachability of organically contaminated heavy metal sludge solidified by silica fume and cement.* Water Science & Technology Vol 44 No 2–3 pp. 399–407 © IWA Publishing 2001, http://www.iwaponline.com/wst/04402/wst044020399.htm.
80. Silica Fume Association, http://www.silicafume.org/general-silicafume.html.
81. Conner, J.R. and S.L. Hoeffner. *The History of Stabilization/Solidification Technology.* Critical Reviews in Environmental Science and Technology. 28(4):325–396 (1998).
82. Davis, E.L., J.S. Falcone, S.D. Boyce, and P.H. Krumrine. *Mechanisms for the Fixation of Heavy Metals in Solidified Wastes Using Soluble Silicates.* The PQ Corp.: Lafayette Hill, PA (1987).
83. Collins, R.J. *DOT/DOE Evaluation of Kiln Dust/Flyash Technology.* Presented at the 6th Annual International Conference on Economic and Environmental Utility of Kiln Dust and Kiln Dust/Flyash Technology. Feb. 19–21 (1985).
84. Haynes, B.W. and G.W. Kramer. Characterization of U.S. Cement Kiln Dust. *Bureau of Mines Information Circular 8885*, U.S. Department of the Interior, Pittsburgh (1982).
85. Morgan, D.S., J.I. Novoa, and A.H. Haliff. Oil Sludge Solidification Using Cement Kiln Dust. *Journal of Environmental Engineering*, vol. 110, no. 5, pp. 935–948 (1984).

9 Microstructure and Microchemistry of Waste Forms

Amitava Roy and Frank Cartledge

CONTENTS

9.1 INTRODUCTION

Microstructural investigations of materials trace their history to the development of light microscopy in the 1600s. The premise that detailed structural information can lead to understanding of their history, function, macroscopic properties, and so on

1-56670-444-8/05/$0.00+$1.50
© 2005 by CRC Press

has proven to be enormously productive and was extended to investigations at the atomic and molecular level in the 20th century. There is thus an enormous literature that is relevant to microstructural investigations of S/S waste products. This chapter summarizes a wide range of experimental techniques and their applications for S/S waste forms.

Microstructural studies of solidified wastes help explain the efficacy of the S/S process by assessing the microstructure of the binder and its interactions with the wastes, and predict its long-term stability. Among the kinds of information of interest are

- Efficiency of mixing of the waste and binder
- Macroscopic structure and the presence of cracks or micro-cracks
- Microscopic pore structure including size and distribution
- Mineralogy and morphology of hydration products of the binder
- Interactions between the hydration products and wastes
- Stability of the mineral phases

A common waste form is cement-stabilized soil, which illustrates very well the complications involved in acquiring structural information. Both cement and soil consist of several mineral phases in various concentrations, and while some minerals are easy to identify because of their properties, others are not. For example, a few percentages of quartz can be easily identified by X-ray diffractometry (XRD) because of strong reflection of X-rays from its atomic planes. In contrast, clay minerals are not as easy to identify because these reflections are weak, and the chemical composition is variable. Multiple analytical techniques are necessary for positive identification of clay minerals. It is highly desirable to be able to identify small quantities of clays, because clay in even small amounts can have an inordinately large influence on the chemical and physical properties of a soil.

9.2 MICROSTRUCTURAL CHARACTERISTICS OF CEMENT PASTES AND CEMENT-BASED WASTE FORMS

Chapters 4 and 5 discussed hydraulic cement systems and organic binders used for S/S of hazardous, radioactive, and mixed wastes. A hardened cement paste is a heterogeneous multi-phase system. At room temperature, a fully hydrated portland cement paste consists of 50 to 60% calcium-silicate-hydrate (C-S-H) gel, 20 to 25% $Ca(OH)_2$, 15 to 20% AFt and AFm by volume. These minor hydration products, such as $Ca(OH)_2$, $3CaO \cdot Al_2O_3 \cdot 6H_2O$, and AFt, form in small quantities depending on the composition of the cementing material and hydration conditions. AFt dominates the early hydration matrix and initial set of portland cement as the gypsum ground with the cement clinker hydrates and interacts with the hydrating cement minerals. Ettringite (AFt) converts to monosulfate (AFm) as the sulfate anion becomes depleted from continued hydration reactions and C-S-H replaces AFt as the dominant phase in the hardened cement pastes.

C-S-H gel is the main binding component and responsible for the hardened pastes. Its chemical composition varies with hydration conditions. C-S-H is a microporous material and has a very high surface area of about 100 to 700 m²/g, depending on the measurement technique used. The structure of C-S-H is regular within a particular sheet and resembles a clay structure;[1] however, there is no long-range order. The hydrated paste has a more massive inner structure (inner C-S-H) different from the more porous outer rim, the outer C-S-H, which is somewhat fibrous.[1,2]

Calcium hydroxide is crystalline and is easily identified in the scanning electron microscopy (SEM) by its platy hexagonal outline. The crystals are usually a few μm in diameter, but the size depends on many variables. In mature systems, "massive" calcium hydroxides are found where these crystals are much larger and the crystalline outline is absent. Ettringite can be easily identified in the optical microscopy (OM) or SEM by its needle-like crystals. The morphology and dimensions of AFm are similar to that of calcium hydroxide. Thus morphology alone cannot be used to differentiate between AFm and calcium hydroxide; energy dispersive X-ray spectrometry (EDX) is necessary. AFm is quite common in fly ash containing waste forms. Some of the contaminants may also enter its crystal structure.

Hardened cement pastes are porous materials and contain three types of pores: gel pores, capillary pores, and air voids. In portland cement pastes, gel pores constitute about 28% of the total C-S-H gel volume and have a size of 1.5 to 2.0 nm, which is the size of a water molecule and will not permit the flow of water. Gel porosity cannot be resolved by SEM and would be included in the volume occupied by C-S-H. Capillary pores are the portion of the space that was originally filled by water in the fresh cement pastes, and has not been filled by cement hydration products. Air voids are the result of incomplete consolidation or entrapped air, or both. Capillary pores are usually tortuous and tubelike, while air voids are shorter but much larger. Table 9.1 describes the classification of gel and capillary pores and their effects on the properties of hardened cement pastes.[1] Physical encapsulation plays a very important role in immobilizing contaminants within solidified waste forms. Thus it is critical to reduce or eliminate the capillary pores to control the movement of contaminants within solidified waste forms.

The inclusion of these wastes, in most cases, interferes with the hydration of hydraulic cements as discussed in Chapter 7. Of course, it also affects the microstructure of waste forms. However, in most cases, these wastes have no effect on the structural characteristics of organic binders, except they are physically encapsulated.

9.3 ANALYTICAL TECHNIQUES

A variety of analytical techniques have been developed and used for the microstructural characterization of materials. Mollah et al.[41] classified the techniques for characterization of S/S waste forms into three categories: molecular information, structural information, and surface information, as summarized in Figure 9.1.

An analytical technique usually probes a certain narrow property of a material. Applications of spectroscopy in materials characterization have been reviewed in a recent publication.[3] Most techniques also require specimens in a certain specific physical form, and a specimen for one can rarely be used for another. Sometimes

TABLE 9.1

Gel Pores and Capillary Pores in Hardened Cement Pastes[1]

Designation	Diameter	Description	Role of Water	Paste Properties Affected
Capillary Pores	10 μm – 50 nm	Large capillaries	Behaves as bulk water	Strength, permeability
	50 – 10 nm	Medium capillaries	Moderate surface tension generated	Strength, permeability, shrinkage at high humidity
Gel Pores	10 – 2.5 nm	Small (gel) capillaries	Strong surface tension generated	Shrinkage to RH = 50%
	2.5 – 0.5 nm	Micropores	Strongly adsorbed water, no menisci form	Shrinkage, creep
	< 0.5 nm	Micropores "interlayer"	Structural water involved in bonding	Shrinkage, creep

conductive coatings are also necessary for insulating materials. The analytical techniques differ in whether or not they allow *in situ* experimentation. For example, many X-ray-based techniques can be conducted in air or in a wet state, and thus "real world" conditions can be simulated. In contrast, electron beam-based techniques usually require high vacuum, and consequently moist specimens cannot be observed. However, new developments in instrumentation have often blurred these boundaries. Environmental electron microscopes allow observation of moist samples without any conductive coating.

9.4 MICROSCOPIC CHARACTERIZATION OF S/S WASTE FORMS

The microstructure of a waste form can be defined as the spatial relationship between its different constituents, including pores. This information often cannot be resolved with the naked human eye and can be obtained only with the aid of a microscope. Among the properties that need to be documented are the grain size, morphology of the individual phases, relationship between the different phases, whether or not it is surrounding another (encapsulation, for example), pore spaces, and pore-filling phases.

At an elementary level, the overall color of the waste form can be an indicator of the degree of mixing of the waste and the binder. If a waste is solid or paste, mechanical mixing is not usually achieved at the microscopic level. The waste and the binder can easily be identified as regions of different colors. At low magnifications up to 100 or so, a hand lens or an optical stereomicroscope may suffice. At this stage no sample preparation is necessary. The microstructure can be studied at magnifications up to 1000X with a transmission optical microscope. The color, an important optical property, can still be observed with such a microscope. Thin

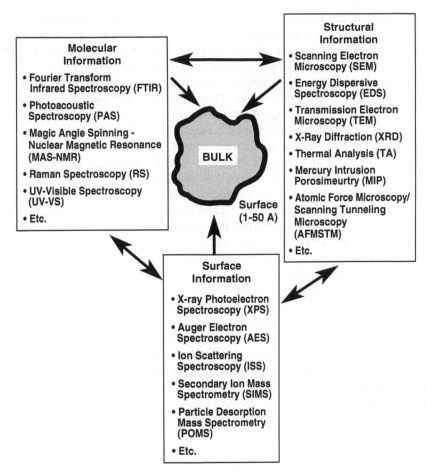

Molecular Information

• Fourier Transform Infrared Spectroscopy (FTIR)

• Photoacoustic Spectroscopy (PAS)

• Magic Angle Spinning - Nuclear Magnetic Resonance (MAS-NMR)

• Raman Spectroscopy (RS)

• UV-Visible Spectroscopy (UV-VS)

• Etc.

Structural Information

• Scanning Electron Microscopy (SEM)

• Energy Dispersive Spectroscopy (EDS)

• Transmission Electron Microscopy (TEM)

• X-Ray Diffraction (XRD)

• Thermal Analysis (TA)

• Mercury Intrusion Porosimeurtry (MIP)

• Atomic Force Microscopy/ Scanning Tunneling Microscopy (AFMSTM)

• Etc.

BULK

Surface (1-50 A)

Surface Information

• X-ray Photoelectron Spectroscopy (XPS)

• Auger Electron Spectroscopy (AES)

• Ion Scattering Spectroscopy (ISS)

• Secondary Ion Mass Spectrometry (SIMS)

• Particle Desorption Mass Spectrometry (POMS)

• Etc.

FIGURE 9.1 Summary of techniques used for characterization of S/S waste forms.[41]

sections (30-μm thickness), however, need to be prepared. Observations at higher magnifications require electron microscopes. SEM, which allows much higher depth of field than transmission microscopes, has improved our understanding of the microstructure of waste forms at magnifications ranging from a few tens to many tens of thousands. For insulators, SEMs usually require that the specimens have a conductive coating and need to be placed in an evacuated chamber. This process may affect the microstructure of moisture-rich specimens. This is particularly true for clay minerals, such as bentonite, which absorb water many times their original volume. Transmission electron microscopes (TEMs) operate very much like transmission optical microscopes, but because electrons are used for observations, instead of light, the resolution is much higher. Magnifications of several hundreds of thousands can be attained with a TEM.

Many attachments are available with the SEM and TEM, which provide information complementary to the microstructure. For example, an EDX spectrometer can be attached to the SEM and the local chemical composition can be analyzed (both qualitative and quantitative) in seconds. The backscattered electrons in SEM can be used to map the compositional contrast. Selected area electron diffraction in the TEM can provide crystal structure information from a spot a few μm in diameter. All of these techniques have been utilized to study waste forms.

Though many forms of microscopy are used to study the microstructure of a waste form, SEM is the dominant one. This is because SEM sample preparation is relatively easy; it provides a high depth of field compared to other techniques. SEM has been widely used to examine the microstructure of S/S waste forms. Some SEM representative photomicrographs of a waste form composed of lime and ASTM Class C fly ash, containing an electroplating waste, with 8% by weight of $Pb(NO_3)_2$, are shown in Figures 9.2 and 9.3. At very low magnification (Figure 9.2, magnified 108X) the waste form appears monolithic with some voids. Some of the voids are due to trapped air, as these are spherical in shape. Cracks of various lengths are common, some of them extending into the voids. Various lines of evidence suggest that the cracks are original and not induced by the high vacuum of the SEM observation chamber. Outlines of fly ash particles are seen in some areas. Some of the voids are encrusted with very fine particles. Figure 9.3 (magnification 5300X) shows one of the encrustations of one of the voids (below letter X in Figure 9.2). The encrustation consists of very small equi-dimensional crystals, between 5 to 10 μm in diameter. Some of the crystal faces are visible. The morphology of the crystals is unlike that of any of the common cement hydration phases. EDX from that spot showed that these crystals are composed of Pb, and showed no other peak so the anion could not be determined. This series of waste form contained $Pb(NO_3)_2$ in various amounts, 0%, 2%, 5%, and 8% by weight. The Pb precipitate could be observed only when the amount of $Pb(NO_3)_2$ reached 8%. In this experiment, microstructural observations were the key to understanding the behavior of $Pb(NO_3)_2$ in this system.

Phosphate-based waste forms have been proposed for both radioactive[4] and non-radioactive wastes[5] as discussed in Chapter 6. Phosphate minerals apatite

FIGURE 9.2 Photomicrograph of electroplating waste with 8 wt% $Pb(NO_3)_2$ stabilized in lime–fly ash (108×).

FIGURE 9.3 Photomicrograph of electroplating waste with 8 wt% $Pb(NO_3)_2$ stabilized in lime–fly ash (5300×) (inside void below × in Figure 9.2).

$(Ca_5(PO_4)_3(F,Cl,OH))$ and monazite $A(PO_4)$ (where A = light rare earth minerals) often naturally contain high levels of radioactive elements. These minerals have also been made to crystallize from radioactive wastes by techniques such as sol-gel or sintering. Eusden et al.[6] reported the microstructure and mineralogy of the waste form generated from old mine tailings in Colorado containing Pb as the principal contaminant. The treatment resulted in the precipitation of several phosphate phases. The phosphate grains that crystallized were of very small size, in the nanometers range, or were amorphous. Many of the reaction products occurred as rinds around the original minerals.

Immobilization of high-level radioactive wastes using ceramics and glasses has recently been reviewed.[7] The waste form can be completely glassy, or crystalline (ceramic), or a mixture of the two. The ceramic phases are such that they can contain most of the waste elements in solid solution, i.e., replacing some of the original elements in the crystal structure of the ceramics. The microstructure of such waste forms can be gainfully studied by comparing them with similar textures produced in nature or laboratory. For example, waste vitrification processes produce a microstructure similar to glassy volcanic rocks, where both consist of a glassy matrix with enclosed bubbles. The waste may be completely dissolved in the matrix or can be enclosed by the glass. Similarly, for the ceramic waste forms, crystals will form with triple junction contact, just like slowly cooled igneous rocks or slowly cooled molten metals.

Raman[8] recently reported a study of the microstructure of a partially vitrified glass-ceramic waste form containing simulated high-level radioactive waste produced by hot isostatic waste (1000°C, 138 MPa). The proportion of crystalline phases varied in direct proportion to waste loading. Some relict phases remained depending on the waste loading. The phases that are present or should be present can usually be predicted from phase diagrams.

The use of pure sulfur as a waste form has been suggested.[9] Contaminants are trapped between the crystals of sulfur. The microstructure of such waste forms is similar to those of ceramic waste forms from nuclear wastes.

If the process used to treat a waste is unknown, much can be learned by studying its microstructure and chemistry. Different cementitious materials can be identified by the morphology and chemistry of the reactants and the products. The microstructure of waste forms depends fundamentally on the processing temperature. The low-temperature waste forms produced from portland cement-like material yields a relatively porous material. However, it is possible to reduce the porosity by the addition of pozzolans such as silica fume and fly ash, which will fill up the pore spaces. The high-level radioactive wastes, where the tolerance for failure is low, are mostly treated by high-temperature processing, leading to glass or ceramic formation. These products have much lower porosity and would have less interconnected pores. The presence of radioactive waste in a waste form also poses another special difficulty: radiation damage. Radiation damage can change the structure of glasses and ceramics and ultimately the microstructure.

The information obtained from microscopy is mostly qualitative, but with image processing it can be turned into a quantitative tool. Image processing techniques are now commonly available with the electron microscope. Its application to waste forms may be difficult because of the complexity of its microstructure. However, even semi-quantitative to qualitative observations can provide enough constraints for an understanding of the S/S process.

9.5 MICROCHEMISTRY

Ultimately, the stability of a waste form is controlled by its chemistry at the microscopic or even molecular level. The availability or release of a contaminant to the environment will depend on the speciation of the contaminants. For example, chromium in the trivalent form is relatively insoluble and benign, whereas the hexavalent form is readily soluble and very toxic. A wet chemical analysis of the total chromium in a waste form is no indication of its availability to the environment. The answers about the speciation of a contaminant in a waste form can be obtained by studying its microchemistry.

The contaminants can be present as discrete crystalline phases, as amorphous phases, in solid solution with other elements, or adsorbed onto the surface of certain phases. The degree of crystallinity of a waste form can be highly variable. One of the main hydration products of portland cement, calcium silicate hydrate, is practically amorphous and constitutes about 70% (by mass) of the hydration products. Fly ash, a commonly added constituent to waste form binders, can contain as much as 90% glass.

Typically, regulatory tests are conducted on cement-based waste forms mostly 28 days after the stabilization. Though a substantial amount of the hydration reactions in portland cement are completed by this time,[1] the addition of a waste to a cementitious binder often retards its hydration process. However, only about 20% fly ash in a blended cement is reacted at 28 days.[10] This means that the microstructure and performance of cement-based waste forms continue to develop after that time. The glassy waste forms containing high-level nuclear wastes are definitely metastable. However, they may continue in that state for millions of years, as nature suggests.

Very few long-term studies are available.[11,12] The long-term stability of a waste form can be monitored and, hopefully, predicted by studying its microchemistry.

9.5.1 X-Ray Analysis

At present, an EDX spectrometer is an almost integral part of an electron microscope. Coupled with the microscopic observations, it can vastly improve our understanding of the S/S processes.

With a properly prepared specimen, local quantitative chemical analysis from a μm-diameter volume is possible with this technique. Thus, location of contaminants in a waste form can be easily identified. The detection limit is typically 0.1%. The electron beam in the microscope can also be rastered across the surface of the specimen, and an elemental distribution map at the μm scale can be obtained. The distribution of several elements can be mapped simultaneously. The elemental distribution correlates to the microstructure, as both are obtained without moving the specimen. The technique can show whether a waste is segregating at grain boundaries or is distributed within a phase. The spatial correlation between different elements can give information about the affinity between different elements and the mineralogy at the μm scale. The morphology of a phase sometimes may be insufficient for its proper identification. For cementitious waste forms, calcium hydroxide and AFm have the same morphology. EDX can then be used in their identification. The X-ray fluorescence (XRF) process is used in the TEM for local chemical analysis. The spot size in that case is much smaller, as the specimen is very thin (only a few hundreds of nanometers) and spreading of the electron beam does not occur. The XRF phenomenon is also used for X-ray mapping with synchrotron sources, where specimens can be mostly analyzed in air, analysis can be obtained from a few μm-diameter spot size, and, in addition, the speciation of an element can be determined.

9.5.2 Solid-State NMR Spectroscopy

Solid-state nuclear magnetic resonance (NMR) spectroscopy was applied to the characterization of cementitious materials almost as soon as effective techniques for examining solids were devised.[13-15] Since cement materials are hydrated silicates and aluminates, three NMR-active nuclei — Al, H, and Si — have been utilized as probes in most of the work. H and Si are both spin-1/2 nuclei yielding spectral data that are easier to interpret than that of Al, which is quadrupolar. A significant advantage of NMR spectroscopy is that it is non-destructive; hence, the same sample can be examined at different times after initial mixing of the ingredients, allowing kinetic studies to be carried out. A limitation is that some binders, such as ordinary portland cement (OPC), contain enough paramagnetic iron to interfere with some NMR experiments. Low-iron-content "white" cements are often used in NMR experiments, but the bulk of the evidence available suggests that the conclusions reached in the white cement systems apply equally well to OPC.

The most common application of solid-state magic angle spinning NMR (MAS NMR) to cements and cement-based waste forms is to monitor the development of the silicate matrix by following the hydration reactions over time. The solid-state

^{29}Si spectra contain peaks, usually overlapping ones, that correspond to Si in five different environments.[16] In describing these spectra the symbol Q represents a Si atom surrounded by four oxygen atoms. A superscript following the Q shows the number of other Q units attached to the Si atom under study. The principal transformation occurring during cement hydration, as followed by NMR, is the formation of chains containing Q^1 (chain-terminating) and Q^2 (chain-lengthening) units starting from the orthosilicate ions (Q^0) present in the cement clinker. The loss of Q^0 units as a function of time is a measure of the overall degree of hydration of the cement clinker and agrees well with calorimetric and other measurements of degree of hydration.[17] It is well known that under acidic conditions, silicate species condense to relatively long-chain polymers, while under basic conditions, which is what pertains in cement hydration, the polymers are low-molecular-weight oligomers. This is obvious in the NMR spectra from the relative proportions of Q^1 and Q^2 units. The Si peaks are usually broad and often show shoulders, and the interpretation has sometimes been made that the shoulders represent bonding of the type Si-O-M, where M is a metal or metalloid atom taking the place of Si in the silicate oligomers.

Fly ash, kiln dust, and related pozzolanics undergo hydration reactions similar to those in portland cement, and Si NMR, although more complex, is applicable in these cases as well. Addition of wastes, even in relatively low concentration, can have a major effect on these hydration reactions, altering the proportions of the Q^n components, retarding or accelerating the reactions, or even completely inhibiting silicate matrix formation.[18] In such cases ^{29}Si NMR is an extremely useful diagnostic tool for determining whether effective cement-setting reactions have occurred.

Hydration of aluminate phases occurs more rapidly than that of the silicate phases, and NMR spectroscopy can distinguish the chemical shift differences between tetrahedral 4-coordinate and octahedral 6-coordinate Al atoms, as shown initially for hydration of tricalcium aluminate.[15] In most samples of cement clinker and other pozzolanic binders both 4- and 6-coordinate Al are present initially, and different lots from different sources show wide variations. Nevertheless, hydration almost always converts the mixture essentially completely to 6-coordinate Al. There are, however, exceptions to this. A notable example is the reversion of 6-coordinate Al to 4-coordinate after long curing times in a case of arsenic immobilized in portland cement/fly ash mixtures.[19] The reversion is also associated with increased arsenic leachability.

As curing takes place in the common binders that are formed via hydration reactions, water distributes itself into a number of different environments that are characterized by different nuclear magnetic spin-lattice relaxation times.[20,21] Again, the process can be followed over time as water (or hydroxide) appears in three main phases (C-S-H, calcium hydroxide, and pore waters with freely exchangeable protons), with a small amount of a fourth. This process is sensitive to the presence of waste materials in the curing matrix. For instance, a retardation of matrix formation in the presence of Cd^{2+} is obvious in smaller proportions of the magnetization fraction arising from protons in C-S-H and CH, and larger proportions from exchangeable protons.[22]

Most NMR studies have focused on the effects of wastes on the matrix of a cement or related material, and these effects are relevant to waste containment and expected long-term stability of the waste form. However, NMR can also be used to look directly at the waste within the matrix and obtain information about chemical speciation and binding mechanisms. Most of the reported applications of this type have involved organic wastes. In a variation of the spin-lattice relaxation measurements described in the previous paragraph, measurements of proton relaxation times have been made when the source of the protons is ethylene glycol and the hydrating species is D_2O rather than H_2O. Partitioning of the ethylene glycol into a bound and a dissolved fraction can be followed over time.[23]

^{13}C NMR spectra of phenol and substituted phenols can distinguish phenol from phenolate, since the *ipso*-carbon of the aromatic ring shifts upfield by about 7 ppm in either PhOCaOH or $(PhO)_2Ca$ compared to PhOH.[18] A more elaborate investigation of phenols in cement involved the use of completely deuterium-substituted phenol. The main feature of deuterium NMR spectroscopy is the ability to monitor molecular reorientations over a wide range of reorientation rates. Comparison of the predicted and experimental NMR spectra show that phenol is mainly dissolved in pore waters and thus poorly immobilized in a cement matrix. But the smaller fraction that is matrix-bound shows an activation energy for ring-flipping and hence a lower limit to the bond strength of 5.5 kcal/mol.[24]

A combination of techniques (synchrotron X-ray microtomography, electron probe microanalysis, and solid-state NMR) have been used to suggest that at concentrations up to about 3% of toluene in portland cement, only a small fraction of the toluene is contained in matrix pores of the size of micrometers or larger.[25] A reinvestigation of this and related materials using NMR relaxometry techniques has confirmed the conclusion that toluene exists mainly as a finely dispersed phase, but that other organic solvents behave differently.[26] A nonpolar aliphatic solvent (cyclooctane), and polar organics that are poorly miscible in water (hexanol and a melamine/naphthalene sulfonate superplasticizer), do lead to liquid pocket formation.

9.5.3 X-RAY ABSORPTION FINE STRUCTURE (XAFS) SPECTROSCOPY

XAFS is the oscillations seen in the X-ray absorption spectrum of an element (Figure 9.4). The oscillations are produced by backscattering of the electron ejected from a

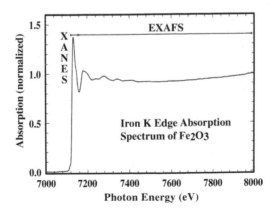

FIGURE 9.4 X-ray absorption spectrum of Fe_2O_3.

central absorber atom by the surrounding atoms. The oscillatory region is divided into two parts: one close to the absorption edge, within 30 to 40 eV from the edge (X-ray absorption near edge structure, or XANES), and the second from a few tens of eV above the edge to about 1000 eV above the edge (extended X-ray absorption fine structure, or EXAFS). The ejected electron is scattered multiple times in the XANES region, whereas single scattering occurs in the EXAFS region.[27] XAFS studies the local structure of an atom, and no long-range order (crystallinity) is necessary, as it is with XRD. It is thus an ideal technique for investigating waste forms where the waste (and also the binder) is often poorly crystalline or amorphous. The XANES oscillations are much stronger than the EXAFS ones. Thus, XANES can be used for elemental concentrations in the hundreds or tens of ppm range. The application of this technique in environmental sciences exploded in the last decade with the wider availability of synchrotron X-ray sources.

One of the best advantages of X-ray-based techniques is that they can be applied in air. In addition, solid, liquid, or gaseous specimens can be studied. In the fluorescence mode, which is used for elements at low concentrations, blocks of materials can also be studied.

X-ray absorption spectroscopy is a direct method of monitoring the oxidation state of a contaminant in a waste form. Figure 9.5 shows the XANES spectra for different oxidation states of iron. As can be seen from the figure, the absorption edge of the spectra (here defined as the first derivative of the spectrum) shifts to higher energy as the oxidation state increases. Thus, ferrous iron can be easily distinguished from the ferric state. Even though hematite and goethite have the same oxidation state and thus a very similar edge position, the spectra are slightly different. For example, the minor peak around 7150 eV is less pronounced in goethite. If these mineral phases are present in a material, they can be easily identified. It is also possible to quantify the amounts of different species of an element from the XANES spectra if suitable standards are available. For example, if a material contains hematite and goethite, their proportions can be determined by least squares analysis. Another advantage of XANES spectroscopy is that species at low concentrations can be detected, which is not easily possible by other techniques. XRD may be able

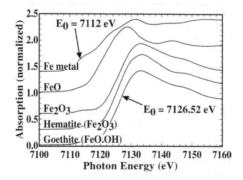

FIGURE 9.5 XANES spectra for different oxidation states of iron.

to detect these phases only up to a several tenths of a percent concentration. XANES can push this detection limit much lower, to several tens of ppm.

Direct S/S of hexavalent chromium in portland cement is difficult because of its high solubility. TCLP tests of such waste forms usually yield very elevated levels of chromium in the leachate. Pre-treatment or additives that can reduce the oxidation state of chromium in the waste form can lead to successful stabilization of hexavalent chromium waste. Zhang[28] studied the effect of three reducing agents on chromium oxidation state when 10% chromate (added as sodium chromate) by weight was stabilized by portland cement. The reducing agents were Fe_3O_4, $FeCl_2$, and $FeSO_4$. The reagents were added in molar ratios ranging from 1:1.5 to 1:4.5. Overdose of iron salts reduced short-term leachability by one to two orders of magnitude. The long-term leachability, however, increased. At low concentrations of hexavalent chromium all reagents were effective in reducing the leachability below the acceptable limit.

XANES measurements of the effect of the reducing agents are shown in Figure 9.6. Hexavalent chromium has a strong pre-edge peak at 5992 eV due to the lack of symmetry in the CrO_4^{-2} tetrahedron. This peak is absent when chromium is present in the trivalent oxidation state. The area under the pre-edge peak can be integrated to obtain a quantitative estimate of the amount of hexavalent chromium present. A calibration curve can be obtained from a mixture of standards. Figure 9.6 shows that by 28 days $FeSO_4$ has reduced hexavalent Cr below the detection limit (in this case several hundred ppm), whereas Fe_3O_4 has reduced very little, if any. The effect of $FeCl_2$ is somewhere in between. The effectiveness of the reducing agent $FeSO_4$ can be seen in Figure 9.7, where little transformation has taken place in 3 hours, but most of the hexavalent chromium is essentially reduced in 28 hours.

Scheidegger et al.[29] recently applied EXAFS to study speciation of nickel in a model nickel-cement system. They observed that the solubility of the nickel in the system is controlled by the formation of nickel-aluminum double hydroxides instead of nickel hydroxide, the commonly assumed mechanism in such systems.

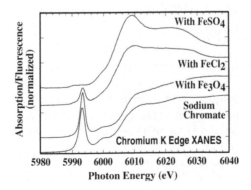

FIGURE 9.6 The effect of three reducing agents on the XANES spectra of chromate (10 wt% sodium chromate) in portland cement after 28 days.

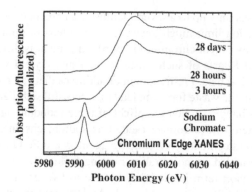

FIGURE 9.7 The effect of the reducing agent $FeSO_4$ on the XANES spectra of chromate (10 wt% sodium chromate) in portland cement.

The use of ettringite for the removal of toxic anions such as borate[30] and selenate[31] has been suggested. It is expected that when ettringite crystallizes from a waste stream containing these anions they will replace the sulfate ions in the crystal structure. However, Myneni et al.[32] showed that more of the waste anions were sorbed on the surface of the ettringite than were incorporated within the crystal structure.

EXAFS and XANES have been traditionally applied to bulk specimens. It is possible to focus the synchrotron X-ray beam to a very small area, a few micrometers across, and obtain speciation information from such an area. Thus, spatial variation in speciation can be studied. It is well established by now that the addition of granulated blast furnace slag (which contains +2 Fe and –2 S) to the waste form can reduce the oxidation state of various toxic elements.[33] For example, the oxidation state of chromium can be reduced from hexavalent to trivalent. Bajt et al.[34] used a synchrotron-based X-ray microprobe to show that this is indeed the case. With the presently available synchrotron sources, the spot size of the focused X-ray beam can be as small as 1 μm, and the detection limit for chromium can be 10 ppm.

9.5.4 THERMAL ANALYSIS

In thermal analysis, a small amount of a material is heated at a programmed rate, and the change in its properties over temperature is monitored.[35] A wide range of thermal analysis techniques, such as thermogravimetry (TG), differential thermal analysis (DTA), and differential scanning calorimetry (DSC), are used by cement chemists. TG, for example, is routinely applied to quantify the amount of calcium hydroxide in hydrated portland cement.

Cement hydration reactions are often affected by the presence of a waste, usually retarded. This process can be monitored by thermal analysis and often quantified. Figure 9.8 shows the pozzolanic reaction of a Class F fly ash when mixed with Type I portland cement in 1:1 proportion, with a water to solids (W/S) ratio of 0.5. A significant amount of calcium hydroxide is present by 7 days in the hydrated mixture. The amount of calcium hydroxide is lower in the year-old specimen, and no calcium

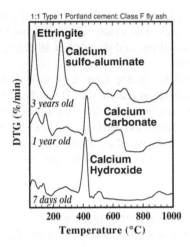

FIGURE 9.8 Differential thermogravimetric (DTG) analysis of a portland cement:Class F fly ash (1:1) matrix (W/S = 0.5).

hydroxide is detected in the 3-year-old specimen. Class F fly ash hydration reactions proceed much more slowly compared to other pozzolans such as Class C fly ash, silica fume, and granulated blast furnace slag. This reaction thus would have been much faster in the presence of the latter pozzolans. Figure 9.9 shows the effect of a sodium arsenite waste on the portland cement:Class F fly ash binder. No calcium hydroxide is detected in specimens containing sodium arsenite up to 1 year of age, and it is detected only in specimens 3 years of age. Sodium arsenite is thus clearly retarding the hydration process of the binder. The peaks corresponding to other phases such as ettringite and gypsum are less pronounced in the binder in the

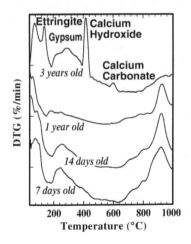

FIGURE 9.9 The effect of a sodium arsenite waste on a portland cement:Class F fly ash (1:1) matrix (W/S = 0.5).

presence of the waste. Evidence obtained from thermal analysis corroborates those obtained by other techniques. For example, XRD would show in such cases the presence of more unhydrated tri- and di-calcium silicates. MAS NMR would also show more of these phases.

The pH of a cement-based waste form may well depend on the free calcium hydroxide. If calcium hydroxide is present the pH will be buffered at around 12.6. The amount of calcium hydroxide can be monitored by TG or XRD, the former being used more. Apart from calcium hydroxide, ettringite (pH ca. 10.7) and the calcium silicate hydrate (pH 9) can also buffer the pH of a cementitious system. The amount of ettringite can be potentially quantified and its effect on the pH can be measured.

9.5.5 XRD

XRD is a widely used analytical tool in many branches of science and engineering. It is typically used for phase identification. With careful sample preparation and proper experimental conditions, however, the phase proportions can be quantified and important information about particle size (diffracting domain size) can be obtained. XRD is a relatively rapid technique and is widely available.

A finely powdered sample of a waste form can be scanned over a large angle to obtain a graph of angle (2θ) versus diffraction intensity. A huge data base, published by the Joint Committee on Powder Diffraction Standards-International Center for Diffraction Data (JCPDS-ICDD) and containing XRD patterns of tens of thousands of pure phases, can be used to match the peaks in an unknown pattern.

Figure 9.10 shows the diffraction patterns of a waste form. The basic waste form contains an electroplating metal sludge (consisting of nickel, chromium, cadmium, and mercury) in a Class C fly ash and calcium hydroxide matrix. An increasing amount of zinc nitrate was added to the waste form. The diffraction pattern of the sludge is also included in the figure for comparison. Aluminum powder was added to all samples as an internal standard to correct for 2θ errors.

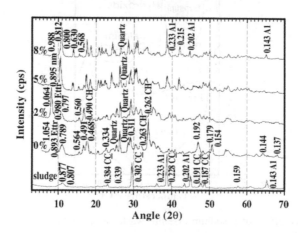

FIGURE 9.10 XRD of electroplating sludge (Ni, Cr, Cd, Hg) in a Class C fly ash–Ca(OH)$_2$ matrix with 0, 2, 5, and 8 wt% zinc nitrate (the untreated sludge XRD is shown for comparison).

The sludge is a synthetic one, prepared by adding calcium hydroxide to a mixture of nitrates of the metals nickel, cadmium, chromium, and mercury.[36] None of the calcium hydroxide could be presently detected in the XRD pattern of the sludge, whereas several calcium carbonate peaks could be found. Thus, any remaining calcium hydroxide had carbonated on exposure to air. The remaining peaks could not be matched with simple hydroxides of the metals in the sludge. A careful analysis of the XRD pattern in the region 10° to 13° 2θ shows that there are really two highly overlapping peaks. These peaks, corresponding to 0.877 and 0.807 nm d-spacings, are very broad. Below 1 μm grain size, the peak width increases with decreasing grain size.[37] The widths of these two peaks suggest that these particles are much less than 1 μm across. Considering the chemistry of the sludge, these peaks appear to be from the very fine-grained or poorly crystalline heavy metal precipitates in the sludge. The very small grain size of the precipitates would indicate that they are probably quite unstable and would also dissolve easily.

The XRD pattern of the control sample (0% zinc nitrate) indicates the presence of calcium hydroxide, gypsum, ettringite, quartz, calcium carbonate, and iron oxide. In addition, there is a broad peak corresponding to 0.789 nm d-spacing, and another unidentified peak corresponding to 1.054 nm. Calcium hydroxide came from the binder, and ettringite is a reaction product between the Class C fly ash and calcium hydroxide. Quartz, gypsum, and iron oxide were present in the fly ash. A comparison of the XRD patterns with increasing zinc nitrate content shows some systematic changes. No calcium hydroxide is observed when 5% or more zinc nitrate is present. Ettringite is present irrespective of zinc nitrate concentration. Zinc nitrate reacted with calcium hydroxide, but no simple zinc hydroxide could be detected in the XRD patterns. The broad sludge peak could be seen in the 0% and 2% zinc nitrate samples but not at higher concentrations. A stronger, sharper peak corresponding to 0.812 nm is seen in 5% and 8% samples, but it is difficult to assign this peak to any particular phase. The peak at 1.054 nm, found in the control, is diminished in the 5% sample and is absent in the 8% sample. This peak could be from a calcium silicate hydrate type phase from the binder.

The above discussion shows that the XRD patterns of waste forms can be very complex. A waste form is very likely to be a multi-phase material. When cementitious materials are present, poorly crystalline phases, impure phases, and solid solutions between phases can make phase identification difficult. Peak overlap of various degrees makes this task even more difficult. Some prior chemical knowledge about the binder can make phase identification somewhat easier. Ideally, three or more peaks of a phase need to be identified in a pattern for its definite detection. Prior knowledge may allow identification of a phase with one peak. For example, very often one and at the most two peaks of ettringite are found in the XRD pattern of a cementitious material. Since ettringite is expected in such a system, its identification by one or two peaks is easier.

Even from the complex XRD patterns in Figure 9.10, where several peaks remain unidentified, important conclusions can be reached. The absence of calcium hydroxide at zinc nitrate concentration 5% and higher indicates that the pH value of the waste form was lowered by its presence. The pH, however, remained alkaline as seen by the presence of ettringite. The presence of ettringite also indicates that the sludge did not strongly retard the pozzolanic reactions between calcium hydroxide and the Class C fly ash in the binder. The gradual disappearance and appearance of some peaks suggest that zinc nitrate is strongly modifying the hydration process of the binder. The very small grain size of the sludge precipitate would also make it unstable.

The grain size is not very critical for qualitative phase identification. Strictly speaking, however, the XRD pattern is not representative if the grain size is too coarse (> 15 to 20 μm). For qualitative data, the grinding should be done at least with an agate mortar and pestle. The sample powder should be further ground with a mechanical mill for quantitative analysis. Quantitative XRD analysis requires proper grain size (5 to 1 μm). The powder also needs to be mounted in the specimen holder in such a way that preferred orientation is minimized. Sample spinning during data collection reduces the effect of preferred orientation and coarser grain size. Modern quantification software also allows correction for preferred orientation. The data should be collected over a wide 2θ range and for a longer time so that statistically valid proportions are obtained. If all the phases in a powder are crystalline it is possible to quantify them without any internal standard. If amorphous phases are present then an internal standard is needed. Quartz, which is usually present in soil or fly ashes used as binders, can be used as an internal standard in simple experiments. Quantification would require a better characterized standard with proper grain size. NIST Standard Reference Material 676 Alumina (corundum) is often used in such instances.

If more than one diffraction pattern is obtained, it is important that uniform parameters are employed in data collection. Strict comparison is otherwise difficult. Even for the same diffractometer, the conditions may subtly (for example, the intensity of the X-ray tube) change over time. In particular, the widths of the slits used before and after the specimen should be noted.

9.5.6 FOURIER TRANSFORM INFRARED (FTIR) SPECTROSCOPY

Infrared (IR) spectroscopy, particularly Fourier transform infrared (FTIR) spectroscopy, is commonly used in the characterization of stabilized/solidified waste. The technique is rapid, does not require extensive sample preparation, and can provide an overall characterization of the specimen. In conjunction with other techniques, such as XRD and TG, it can provide important constraints in sample characterization. Since IR spectroscopy focuses on the molecular bonds in a phase, both amorphous and crystalline phases can be studied by this method. In the absence of more expensive, time-consuming, and difficult-to-access characterization data, such as NMR, many qualitative but important constraints can be determined from IR spectroscopy.

IR has been widely used for characterization of clay minerals (sheet silicates); other types of silicate structures can be easily identified by IR. Cement hydration can be followed by IR. The nature of water in a mineral structure, as hydroxyl or molecular water, can be easily resolved by IR. Standard spectra for minerals necessary for comparison are not always easily available, but can be obtained from a number of sources.[38-40] The standard spectra for organics are more easily found.

Most IR observations are made in transmission with potassium bromide pellets; the diffuse total reflectance method requires less difficulty in sample preparation, however. The moisture content of the powder should be as low as possible, which otherwise makes KBr (potassium bromide) pellet preparation difficult and dominates the spectrum. Powdered samples are usually dried in an oven at 105°C. In cementitious materials, this process would dehydrate ettringite and gypsum and will also affect other low-thermal-stability hydrated phases. An alternative to this drying process would be drying over P_2O_5 and silica gel in a desiccator.

Figure 9.11 shows the FTIR spectra of a series of lime–Class C fly ash (LFA) mixtures containing a heavy metal (Ni, Cr, Cd, and Hg) sludge and increasing amounts of an interference, copper nitrate. The quartz (from the Class C fly ash) bands are strong in the lime–fly ash mixture without the sludge and any interference. The 796- to 781-cm^{-1} doublet of quartz can be usually traced through all the samples even after the sludge and interference were added. In hydrated cements, the silicate (Si-O-Si antisymmetric stretching,$_{v_3}$) band is typically present as seen around 970 cm^{-1}. Such a band is present in the LFA, with sludge but with or without copper nitrate at 968 cm^{-1}. The position of the band stays constant to up to 5% copper nitrate concentration, indicating that the copper nitrate is not affecting the degree of polymerization of the calcium silicate hydrate. The band cannot be observed in the presence of 8% by weight of copper nitrate. There is a band at 943 cm^{-1}, which, if it corresponds to the Si-O-Si band in the C-S-H, would indicate a remarkable change in its structure, getting less polymerized. Even after drying at 105°C, the broad water band around 3500 cm^{-1} stays strong with 8% copper nitrate, indicating that moisture is strongly sorbed to the binder. In all other specimens this band is significantly reduced. The peak for molecular water around 1640 cm^{-1} is prominent when the sludge and the copper nitrate are present. This peak maintains the same intensity with up to 5% copper nitrate, but is particularly strong for 8% copper nitrate by weight.

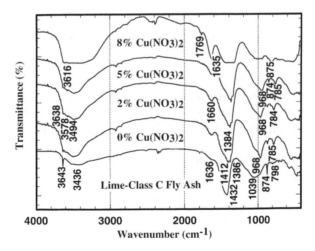

FIGURE 9.11 FTIR spectra of a heavy metal (Ni, Cr, Cd, Hg) sludge in lime–Class C fly ash for 0, 2, 5, and 8 wt% $Cu(NO_3)_2$ (the spectra for lime–fly ash alone is shown for comparison).

The OH^{-1} band (3643.8 cm^{-1}) from calcium hydroxide is very strong in the lime–fly ash mixture. The OH^{-1} peak persists in the presence of sludge without any copper nitrate and even when 2% and 5% by weight of copper nitrate has been added.

Many strong bands from calcite (e.g., 1432 cm^{-1} and 874 cm^{-1}) are present in the LFA mixture. In the presence of the sludge, with or without copper nitrate, the carbonate band around 1400 cm^{-1} is broader. It is the broadest with 8% by weight copper nitrate. Vaterite has a very wide band in this region, and it appears that the presence of the waste is aiding vaterite formation. Since copper is a significant component of the system, formation of copper carbonate phases, such as malachite or azurite, is a possibility. FTIR did not show any match with the spectra of these minerals, and the green color of malachite, commonly found on copper objects due to weathering, was not observed in the waste form. The peaks around 2400 cm^{-1} are from the atmospheric carbon dioxide in the sample compartment.

FTIR is a widely used complementary analytical tool for waste form characterization. Mollah et al.[41] observed that curing of a waste form in a carbon dioxide-rich atmosphere polymerized the silicate matrix and also altered the nature of the waste species. Van Jaarsveld and van Deventer[42] used FTIR to show that Cu and Pb are incorporated in the silicate (geopolymer) binder of a waste form. Jing et al.[43] used FTIR to monitor the fate of adsorbed arsenic on iron hydroxide sludge. Arsenic was adsorbed on the sludge before treatment, but precipitated as calcium arsenate when mixed with portland cement. The FTIR band for arsenic moved from 830 cm^{-1} to 860 cm^{-1}.

9.6 EXAMPLES OF SOLIDIFICATION/STABILIZATION

The application of several complementary techniques for characterization of a waste form is much more powerful than any technique alone, and some typical examples are discussed in this section.

9.6.1 Behavior of Chromium in a Simulated Waste Form

Omotoso et al. and others[44-46] studied the behavior of chromium, in different valence states, in a simple matrix (tricalcium silicate). The premise for choosing tricalcium silicate is that it is the most important component of portland cement. A range of analytical techniques, including quantitative XRD, SEM, TEM, and EDX, were used to monitor the hydration process of tricalcium silicate in the presence of Cr^{III} and Cr^{VI} solutions.

Contrary to some earlier reports, they found that Cr^{III} was precipitating as some Ca-Cr complexes. These complexes recrystallized into calcium hydroxide and a different Ca-Cr complex when abundant moisture was present. Otherwise, it remained unaltered. Leaching of this model waste form led to the removal of large quantities of Ca but little Si or Cr. Apparently, the high pH of the waste form converted the Cr to an oxide or hydroxide form, which kept the Cr solubility low.

The behavior of Cr^{VI} was very different. With its mixing with tricalcium silicate, a calcium chromate hydrate phase ($Ca_2CrO_5 \cdot 3H_2O$) precipitated. The calcium chromate hydrate phase is highly soluble, and most of it leaches out easily. FTIR indicated that some of the Cr^{VI} entered the C-S-H structure and increased its degree of polymerization. NMR independently corroborated that the degree of polymerization of the C-S-H increased. Electron microscopy showed that the presence of Cr^{VI} also changed the physical characteristics of the C-S-H, making it more permeable.

9.6.2 STRUCTURAL CHARACTERIZATION OF A SOLIDIFIED SYNTHETIC WASTE FORM

A synthetic waste containing arsenic, cadmium, chromium, lead, and phenol was stabilized/solidified in a suite of unknown binders. The principal aim of the study was to assess the efficacy of different binders by leaching tests. Microstructural and phase characterization was a secondary objective. Still, such studies can provide additional constraints about the behavior of the waste forms. EDX microanalysis of some of the waste forms are shown in Figure 9.12. All the analyses came from waste form specimens, which were coated with carbon, not gold, so that the gold M peak did not interfere with the sulfur K peak. The analysis shows that the principal elements present in a μm-volume region are sulfur and calcium. A minor amount of silicon is detected, while a trace amount of chromium is also visible. The major elements and their proportions indicate that the analysis came from a calcium sulfate. Since EDX does not detect water, whether the calcium sulfate is gypsum or anhydrite could not be confirmed. The XRD pattern of the same waste form is shown in Figure 9.13D. The pattern matches with that of gypsum, and no other peaks are present. A gypsum-based binder was thus used to stabilize/solidify the waste. The other EDX patterns shown in Figure 9.12 have calcium and silicon as the principal elements, which is typical of cementitious material-based patterns. SEM of the waste form from which the analysis B was obtained had abundant diatom. Diatomaceous earth was thus used in this waste form. The XRD pattern of the waste form indicated the presence of opal. The analysis B also has more silicon than calcium, which is unlike portland cement. This observation also points to the fact that there is an additional

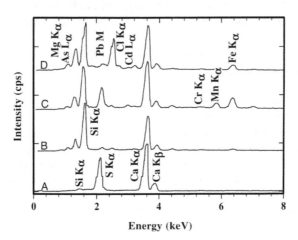

FIGURE 9.12 EDX microanalysis of waste forms of synthetic waste (As, Cd, Cr, Pb, phenol).

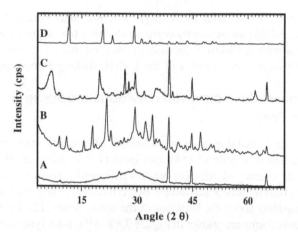

FIGURE 9.13 XRD pattern of waste forms of synthetic waste (As, Cd, Cr, Pb, phenol).

source of silicon in this waste form. Figure 9.12D shows a significant amount of chlorine. Since this was not present in the waste, it is coming from the binder.

EDX analyses of the waste show varied distribution of the waste elements. Chromium was quite uniformly distributed in these waste forms, with some occasional locally high concentrations. Arsenic and lead were rarely detected, but some areas had high concentrations. These elements were possibly precipitated, whereas chromium was adsorbed on the surface. The XRD patterns indicate a wide variation in the presence of crystalline phases. (All these patterns include aluminum as an internal standard.) Figure 9.13A has very few peaks and a broad hump. The pattern matches that of granulated blast furnace slag. Melilite, typically found in slag, can be identified in the pattern. No new peaks due to hydration could be identified in the pattern. Figure 9.13B also has a broad hump but several other peaks are present, suggesting a significant amount of crystalline mineral content. Figure 9.13C has several broad peaks, particularly at low two thetas. This is typical of clay minerals like smectite. A clay was thus used in this waste form in binding the waste elements.

9.6.3 STRUCTURAL CHARACTERIZATION OF FIELD CEMENT-SOLIDIFIED CONTAMINATED SOIL

The samples come from a site where the soil was contaminated from polynuclear aromatic hydrocarbons, volatile organic compounds and cyanide, and was remediated by *in situ* solidification in 1992 by injection of a bentonite/portland cement slurry using a rotating auger. A total of 16 samples were taken 10 years after the solidification with the aim of examining long-term effectiveness of the remediation. Several techniques, including XRD, SEM, DTG, DSC and FTIR, were used to examine the structural characterization of these samples.

As shown by XRD in Figure 9.14A, the most dominant mineral in all XRD patterns is quartz, which is the mineral form of sand. The amount of quartz in these soils could be easily 60% or higher. All the major clay minerals, smectite, illite, and kaolinite, are present in the soil samples. Sometimes all three are present in the same

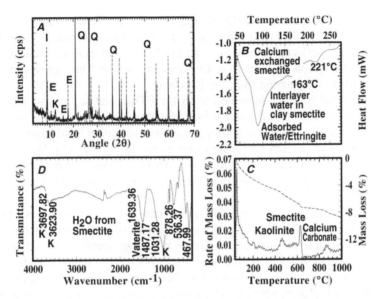

FIGURE 9.14 XRD, DTG, DSC, and FTIR on field samples of contaminated soil 10 years after remediation.

soil, in some cases two of the three. The basal reflection peaks for illite and kaolinite are often strong while that of smectite is weak. XRD patterns of some soils showed that two peaks were present in the 4° to 8° 2θ range. One of these corresponds to smectite, and the other could be chlorite. The principal calcium carbonate phase is vaterite. Calcite, another calcium carbonate mineral, was detected in only one or two soils. Ettringite, a mineral characteristic of set portland cement, was detected in many soils, and often more than one peak of ettringite is present. A peak from unhydrated cement clinker, tricalcium silicate, is tentatively identified in some soils, as the peaks are weak, and only one peak of tricalcium silicate is observed (many peaks of C3S can be detected by XRD very easily; we are afraid that it is not evident enough to confirm the presence of C3S with only one peak of C3S). However, previous studies of cement hydration has indicated which peak of unhydrated tricalcium silicate is likely to remain. This peak is present only in those soils containing ettringite. The presence of ettringite also allows us to infer the minimum pH (> 10.7) of the waste form.[47,48]

TG of the waste forms shows mass loss ranging from 5 to 12% (for example, it is 8% for the sample shown in Figure 9.14C). In some instances the mass loss occurs at less than 100°C, which is due to loss from adsorbed moisture and ettringite dehydration, if the latter is present. In some waste forms the dehydration events of the individual minerals are separated, whereas in others, they are highly overlapping. It is easier to identify kaolinite by DTG because of its well-defined dehydroxylation characteristics. Dehydroxylation for smectite and illite occurs over a wider temperature range and is somewhat diffuse. The peak location can be used to identify these clay minerals. For smectite, desorption from surface and dehydration from interlayer spaces can be observed by DTG.[49] Calcium carbonate decarbonation can be easily identified by DTG. It occurs at a slightly higher temperature than smectite. The DSC curves of the soil usually have a strong peak at temperatures less than 100°C followed by some weaker peaks at higher temperatures. The low-temperature peaks are associated with desorption of water from the smectite surface. The higher temperature peaks are due to dehydration from the interlayer.

Kaolinite and illite are identified from the FTIR patterns in most waste form samples. The location of the molecular H_2O vibration corresponds to that of smectite. Quartz is also easily identified. A strong carbonate vibration band is present in the FTIR pattern. The calcium carbonate phase is vaterite (μ-$CaCO_3$). Vaterite is rare geologically, but is often a carbonation product of portland cement. However, not all waste form samples show an equal degree of carbonation, and not all cement-containing soils show carbonation.

In the SEM, the clay minerals and quartz can be identified in the waste forms by their morphology. The quartz grains are angular and often have smooth crystallographic faces. In some cases the hexagonal plate-like structure of kaolinite can be observed. Kaolinite typically occurs in nature as stacked hexagonal crystals and thus can be differentiated from calcium hydroxide or calcium aluminates, the typical cement hydration phases, which also have hexagonal shape. Some thicker hexagonal plates are observed in the SEM; however, they are not stacked like kaolinite and EDX indicates them to be calcium aluminate.

Some typical photomicrographs of the waste form are shown in Figure 9.15. One of the products of cement hydration, calcium aluminate hydrate (Figure 9.15A), can be identified in the photomicrograph. Illite from the soil can be identified (Figure 9.15B) by its layer structure and chemical composition. Ettringite is locally seen in these waste forms (Figure 9.15C). When C-S-H is present, it is intimately mixed with other phases of the soils. Foil-like morphologies corresponding to the C-S-H phase in hydrated cement pastes are also often noted. Figure 9.15D shows the growth of C-S-H next to the illite crystals.

In the SEM ettringite is sometimes seen, but not as commonly as XRD suggested. Since XRD is a bulk technique and SEM can show spatial variation, the occurrence of ettringite then must be localized. EDX spectrometry in the SEM shows a high amount of calcium in these same samples. Clearly, after 10 years of ageing, these samples show the presence of hydrated cement phases as well as products from the carbonation of hydrated cement phases.

9.7 CONCLUDING REMARKS

Even a cursory literature search on the subject of S/S shows that multiple analytical techniques are commonly used in the characterization of the waste forms. Since the waste form is a multi-element, multi-phase, complex material, a better understanding of its properties can be obtained only through the application of multiple techniques. The total concentration of a contaminant has little correlation to its leachability, which depends on the chemical species present. Very few techniques yield direct evidence about the nature of the species of a contaminant, particularly at low concentrations. XANES is one such technique.

FIGURE 9.15 SEM photomicrographs of field samples of contaminated soil 10 years after remediation (upper: A and B; lower: C and D).

The sample preparation process for some techniques is not very straightforward. Though qualitative data can be easily obtained, getting quantitative data from a particular technique may require more extensive sample preparation. However, meaningful results can be obtained from a technique only from a properly prepared specimen. Before any major investigation is initiated, the sample requirements should be properly studied.

ACKNOWLEDGMENT

This chapter has drawn on research conducted at Louisiana State University over the last two decades. The research was supported by various funding agencies, such as the Waterways Experimental Station of the U.S. Army Corps of Engineers, the U.S. Environmental Protection Agency, Environment Canada, the National Science Foundation, the Louisiana Board of Regents, and a number of industrial supporters, most notably Southern Company Services and the Electric Power Research Institute.

GLOSSARY OF TERMS

AFm:	Monosulfate
AFt:	Ettringite
BFS:	Granulated Blast Furnace Slag
CH:	Calcium Hydioxide
C-S-H:	Calcium Silicate Hydrate. The dashes in between means that there is no definite stoichiometric relationship between the three constituents.
DSC:	Differential Scanning Calorimetry
DTG:	Derivative Thermogravimetry
EDX:	Energy Dispersive X-ray Spectrometry
eV:	Electron volt
EXAFS:	Extended X-ray absorption fine structure
TG:	Thermogravimetry
FTIR:	Fourier Transform Infrared Spectroscopy
μm:	Micrometer
nm:	Nanometer
OM:	Optical Microscopy
OPE:	Ordinary portland cement
PDF:	International Center for Crystal Diffraction, Powder Diffraction File
SCM:	Supplementary Cementing Material
SEM:	Scanning electron microscopy
TEM:	Transmission electron microscopy
W/S:	Water to solid ratio
XAFS:	X-ray Absorption Fine Structure
XANES:	X-ray Absorption Near Edge Structure
XRD:	X-ray Diffractometry
XRF:	X-ray Fluorescence

REFERENCES

1. Mindess, S. and Young, J. F., *Concrete*, Prentice-Hall, New York, 1981.
2. Taylor, H., *Cement Chemistry*, Thomas Telford, London, 1997.
3. Hagamand, E., *Spectroscopy for Materials Scientist*, John Wiley & Sons, New York, 2003.
4. Ewing, R. C. and Wang, L., Phosphates as Nuclear Waste Forms, in *Phosphates - Geochemical, Geobiological, and Materials Importance*, Kohn, M. J., Rakovan, J., and Hughes, J. M. Mineralogical Society of America, Washington, D.C., 2002, pp. 673–699.
5. Eighmy, T. T., Crannell, B. S., Butler, L. G., Cartledge, F. K., Emery, E. F., Oblas, D., Krzanowski, J. E., Eusden, J. D., Shaw, E. L., and Francis, C. A., Heavy metal stabilization in municipal solid waste combustion dry scrubber residue using soluble phosphate, *Environmental Science & Technology* 31 (11), 3330–3338, 1997.
6. Eusden, J. D., Gallagher, L., Eighmy, T. T., Crannell, B. S., Krzanowski, J. R., Butler, L. G., Cartledge, F. K., Emery, E. F., Shaw, E. L., and Francis, C. A., Petrographic and spectroscopic characterization of phosphate-stabilized mine tailings from Leadville, Colorado, *Waste Management* 22 (2), 117–135, 2002.

7. Donald, I. W., Metcalfe, B. L., and Taylor, R. N. J., The immobilization of high level radioactive wastes using ceramics and glasses, *Journal of Materials Science* 32 (22), 5851–5887, 1997.
8. Raman, S. V., Microstructures and leach rates of glass-ceramic nuclear waste forms developed by partial vitrification in a hot isostatic press, *Journal of Materials Science* 33 (7), 1887–1895, 1998.
9. Fuhrmann, M., Melamed, D., Kalb, P. D., Adams, J. W., and Milian, L. W., Sulfur polymer solidification/stabilization of elemental mercury waste, *Waste Management* 22 (3), 327–333, 2002.
10. Shi, C., Studies on several factors affecting on the strength and hydration of lime-pozzolan cements, *Journal of Materials in Civil Engineering, ASCE* 13 (6), 441–445, 2001.
11. Roy, A. and Cartledge, F. K., Long-term behavior of a Portland cement-electroplating sludge waste form in presence of copper nitrate, *Journal of Hazardous Materials* 52 (2–3), 265–286, 1997.
12. Akhter, H., Cartledge, F. K., and Roy, A., Solidification/stabilization of arsenic salts: effects of long cure times, *Journal of Hazardous Materials* 52 (April), 247–264, 1997.
13. Lahajnar, G., Blinc, R., Rutar, V., Smolej, V., Zupancic, Z., Kocuvan, I., and Ursic, J., On the use of pulse NMR techniques for the study of cement hydration, *Cement and Concrete Research* 7, 385–394, 1977.
14. Lippmaa, E., Magi, M., Tarmak, M., Wieker, W., and Grimmer, A.-R., A high resolution 29Si study of the hydration of tricalcium silicate, *Cement and Concrete Research* 12, 597–602, 1982.
15. Muller, D., Rettel, A., Gessner, W., and Scheler, G., An application of solid-state magic-angle spinning 27Al NMR to the study of cement hydration, *Journal of Magnetic Resonance* 57, 152–156, 1984.
16. Lipmaa, E., Magi, M., Samoson, A., Engelhardt, G., and Grimmer, A.-R., Structural studies of silicates by solid-state high-resolution 29Si NMR, *Journal of the American Chemical Society* 102, 4889–4893, 1980.
17. Barnes, J. R., Clague, A. D. H., Clayden, N. J., Dobson, C. M., and Hayes, C. J., Hydration of Portland cement followed by 29Si solid-state NMR spectroscopy, *Journal of Materials Science Letters* 4, 1293–1295, 1985.
18. Butler, L. G., Cartledge, F. K., Eaton, H. C., and Tittlebaum, M. E., Microscopic and NMR spectroscopic characterization of cement-solidified hazardous wastes, in *Chemistry and Microstructure of Solidified Waste Forms*, Spence, R. D., Lewis Publishers, Boca Raton, FL, 1993, pp. 151–168.
19. Akhter, H. C. F., Roy, A., and Tittlebaum, M. E., Solidification/stabilization of arsenic salts: effects of long cure times, *Journal of Hazardous Materials* 52 (2–3), 247–264, 1997.
20. Blinc, R., Lahajnar, G., Merljak, P., and Zupancic, I., The determination of surface development in cement pastes by nuclear magnetic resonance, *Journal of the American Ceramic Society* 65, 25–31, 1982.
21. MacTavish, J. C., Miljkovic, L., and Pintar, M. M., Hydration of white cement by spin grouping NMR, *Cement and Concrete Research* 15, 367–377, 1985.
22. Butler, L. G., Cartledge, F. K., Chalasani, D., Eaton, H. C., Frey, F., Tittlebaum, M. E., and Yang, S., Immobilization mechanisms in solidification/stabilization using cement/silicate fixing agents, in *2nd Annual Symposium on Hazardous Waste Research*, Baton Rouge, LA, 1988, pp. 42–61.

23. Cartledge, F. K., Butler, L. G., Tittlebaum, M. E., H., A., Chalasani, D., Janusa, M. A., and Yang, S., Solid-state NMR characterization of organics in cement, in *1st International Symposium on Cement Industry Solutions to Waste Management,* Canadian Portland Cement Association, Calgary, Canada, 1992, pp. 289–305.

24. Janusa, M. A., Wu, X., Cartledge, F. K., and Butler, L. G., Solid-state deuterium NMR spectroscopy of d5-phenol in white Portland cement: a new method for assessing solidification/stabilization, *Environmental Science and Technology* 27, 1426–1433, 1993.

25. Butler, L. G., Owens, J. W., Cartledge, F. K., Kurtz, R. L., and Byerly, G. L., Synchrotron X-ray microtomography, electron probe microanalysis, and NMR of toluene waste in cement, *Environmental Science and Technology* 34, 3269–3275, 2000.

26. Nestle, N., Zimmermann, C., Dakkouri, M., and Niessner, R., Action and distribution of organic solvent contaminations in hydrating cement: time-resolved insights into solidification of organic waste, *Environmental Science and Technology* 35 (24), 4953–4956, 2001.

27. Brown, G. E., Jr. and Sturchio, N. C., An overview of synchrotron radiation applications to low temperature geochemistry and environmental science, in *Applications of Synchrotron Radiation in Low-Temperature Geochemistry and Environmental Sciences,* 2002, pp. 1–115.

28. Zhang, A., *Characterization of solidification/stabilization of chromium wastes in cement matrix,* PhD, Louisiana State University, 1999.

29. Scheidegger, A. M., Wieland, E., Scheinost, A. C., Dahn, R., and Spieler, P., Spectroscopic evidence for the formation of layered Ni-Al double hydroxides in cement, *Environmental Science & Technology* 34 (21), 4545–4548, 2000.

30. Csetenyi, L. J. and Glasser, F. P., Borate Substituted Ettringites, in *Proceedings of the 16th International Symposium on the Scientific Basis for Nuclear Waste Management,* Materials Research Society, Boston, 1993, pp. 273–278.

31. Hassett, D. J., McCarthy, G. J., Kumarathasan, P., and Pflughoeft-Hassett, D., Synthesis and characterization of selenate and sulfate-selenate ettringite structure phases, *Materials Research Bulletin* 25 (11), 1347–1354, 1990.

32. Myneni, S. C. B., Traina, S. J., Waychunas, G. A., and Logan, T. J., Vibrational spectroscopy of functional group chemistry and arsenate coordination in ettringite, *Geochimica Et Cosmochimica Acta* 62 (21–22), 3499–3514, 1998.

33. Kindness, A., Macias, A., and Glasser, F. P., Immobilization of chromium in cement matrices, *Waste Management* 14, 3–11, 1994.

34. Bajt, S., Clark, S. B., Sutton, S. R., Rivers, M. L., and Smith, J. V., Synchrotron X-ray microprobe determination of chromate content using x-ray-absorption near-edge structure, *Analytical Chemistry* 65 (13), 1800–1804, 1993.

35. Wendlandt, W. W., *Thermal Analysis,* 3rd ed., John Wiley & Sons, New York, 1986.

36. Roy, A., Eaton, H. C., Cartledge, F. K., and Tittlebaum, M. E., Solidification stabilization of hazardous-waste — evidence of physical encapsulation, *Environmental Science & Technology* 26 (7), 1349–1353, 1992.

37. Jenkins, R. and Snyder, R. L., *Introduction to X-ray Powder Diffractometry,* John Wiley & Sons, New York, 1996.

38. Marel, H. W. V. D. and Beutelspacher, H., *Atlas of Infrared Spectroscopy of Clay Minerals and their Admixtures,* Elsevier Scientific Publishing Company, Amsterdam, 1976.

39. Farmer, V. C., Infrared Spectroscopy, in *Data handbook for clay materials and other non-metallic minerals: providing those involved in clay research and industrial application with sets of authoritative data describing the physical and chemical properties and mineralogical composition of the available reference materials*, Van Olphen, H. and Fripiat, J. J. Pergamon Press, Oxford, 1979, pp. 285–288.

40. Farmer, V. C., *The Infrared Spectra of Minerals*, Mineralogical Society, London, 1974, pp. 539.

41. Mollah, Y. M., Hess, T. R., Tsai, Y.-N., and Cocke, D. L., FTIR and XPS investigations of the effects of carbonation on the solidification/stabilization of cement based system-Portland type V with zinc, *Cement and Concrete Research* 23 (4 Jul), 773–784, 1993.

42. van Jaarsveld, J. G. S. and van Deventer, J. S. J., The effect of metal contaminants on the formation and properties of waste-based geopolymers, *Cement and Concrete Research* 29 (8), 1189–1200, 1999.

43. Jing, C. Y., Korfiatis, G. P., and Meng, X. G., Immobilization mechanisms of arsenate in iron hydroxide sludge stabilized with cement, *Environmental Science & Technology* 37 (21), 5050–5056, 2003.

44. Omotoso, O. E., Ivey, D. G., and Mikula, R., Hexavalent chromium in tricalcium silicate: Part II. Effects of CrVI on the hydration of tricalcium silicate, *Journal of Materials Science* 33 (2 Jan 15), 515–522, 1998.

45. Omotoso, O. E., Ivey, D. G., and Mikula, R., Hexavalent chromium in tricalcium silicate: Part I. Quantitative X-ray diffraction analysis of crystalline hydration products, *Journal of Materials Science* 33 (2 Jan 15), 507–513, 1998.

46. Omotoso, O. E., Ivey, D. G., and Mikula, R., Quantitative X-ray diffraction analysis of chromium(III) doped tricalcium silicate pastes, *Cement and Concrete Research* 26 (9 Sep), 1369–1379, 1996.

47. Gabrisova, A., Havlica, J., and Sahu, S., Stability of calcium sulphoaluminate hydrates in water solutions with various pH values, *Cement and Concrete Research* 21 (6), 1023–1027, 1991.

48. Myneni, S. C. B., Traina, S. J., and Logan, T. J., Ettringite solubility and geochemistry of the $Ca(OH)(2)-Al-2(SO4)(3)-H2O$ system at 1 atm pressure and 298 K, *Chemical Geology* 148 (1–2), 1–19, 1998.

49. Bish, D. L. and Duffy, C. J., Thermogravimetric analysis of clay minerals, in *Thermal Analysis in Clay Science*, Stucki, J. W., Bish, D. L., and Mumpton, F. A. The Clay Minerals Society, Boulder, CO, 1990, pp. 96–157.

10 Leaching Processes and Evaluation Tests for Inorganic Constituent Release from Cement-Based Matrices

Andrew C. Garrabrants and David S. Kosson

CONTENTS

1-56670-444-8/05/$0.00+$1.50

10.1 INTRODUCTION

The earliest waste treatment techniques similar to current stabilization/solidification (S/S) processes were descended from industrial practices and aimed at improving the handling of industrial and radioactive process waste sludge through dewatering and encapsulating the waste in a cementitious matrix.[1,2] Leaching performance and environmental impact were rarely considered until the advent of environmental regulations in the early 1970s.

Specific to S/S wastes, the primary goal of leaching characterization is to estimate environmental impact under assumed release conditions. Secondary goals may be to develop more effective treatment recipes, evaluate the efficacy of a selected treatment process, or meet waste acceptance criteria for disposal. At some phase of the treatment development and disposal or utilization process, S/S waste forms are typically subjected to one or more leaching tests in order to predict environmental impact of trace contaminants (e.g., heavy metals, organics, pesticides, radionuclides). However, the development of S/S treatment recipes durable enough for long-term waste containment demands that additional attention be paid to major matrix components (e.g., calcium, hydroxide, aluminum), as well as infiltration of environmental components affecting the durability of the waste material (e.g., acids, carbon dioxide, sulfate or chloride ions, chelating agents).

This chapter provides an overview of the leaching mechanisms, factors that control constituent leaching from cement-based materials, and leaching evaluation tests applicable to S/S materials. The discussion is limited primarily to inorganic species in S/S waste placed in a disposal release scenario; however, with few limitations, similar release mechanisms and environmental factors may apply for organic contaminants in S/S waste forms, alternative applications of waste-cement matrices (e.g., above-ground placements, beneficial reuse of wastes), and the broader class of cementitious materials in general.

10.2 LEACHING MECHANISMS

In its most basic form, leaching involves contact of a solid matrix with a liquid phase (leachant) into which constituents are released and transport, producing a liquor (leachate).[6] The rate of constituent release to the environment depends on the physical and chemical mechanisms of leaching including (i) geochemical control of equilibrium, (ii) sorption and desorption, and (iii) mass transport of constituents into the environment.

10.2.1 GEOCHEMICAL CONTROL

Geochemical control of leaching occurs at the interface between solid and liquid where the dissolution of mineral phases and solubility of constituents in the liquid phase are coupled, pH-dependent processes. The release of a constituent in equilibrium with a liquid bath of fixed volume may be limited by the solid through dissolution of solid mineral phases or by the liquid phase through constituent solubility under the chemical conditions of the liquid phase (e.g., pH, temperature, or the presence of other ions).

Although solubility may be affected by the rates of dissolution, local equilibrium over short intervals of time is often assumed between solid phases and liquid constituents. The magnitude of solubility is influenced by liquid properties including pH and the presence of dissolved organic matter, complexing agents, or other ions. For example, it is typically assumed that dissolution of portlandite, $Ca(OH)_2$, provides the high alkalinity of cement pore solution, although cement pore expression analysis indicates that the pore water pH is higher than the dissolution pH of portlandite. The presence of alkalis, regardless of their speciation (e.g., NaCl, KOH), raises the pore water pH and depresses the solubility of calcium.[7,8]

When the solid phase limits the amount of constituent leached into the liquid bath, the release is "availability-limited."[5,9,10] Here availability is the maximum labile, or leachable, fraction of a constituent under a particular set of chemical conditions. Since this parameter is operationally defined (i.e., based on the determining chemical conditions), it is usually determined at release conditions designed to maximize release of oxyanions (e.g., AsO_4^{2-}, SeO_3^-) and cations (Pb^{2+}, Na^+, Ca^{2+}, Cd^{2+}, Cs^+). The conditions under which availability is determined should reflect the field conditions that would realistically result in maximal release in order to best represent leaching potentials.[5,9]

10.2.2 SORPTION AND DESORPTION

Some cement hydration products (e.g., C-S-H gel, sulfoaluminates) exhibit large surface areas with charges such that metal ions become sorbed to their surface. The reversibility of these surface reactions is influenced, according to surface layer theory, by the presence of Na^+, K^+, Ca^{2+}, and H^+ ions, which compete with metal ions for negatively charged surface adsorption sites. Changes in pH may affect the surface charge of hydration products such as C-S-H gel, disrupting the balance of adsorbed metal ions. In addition, an increase in dissolution of one hydration product may lead to precipitation or adsorption of another. For example, $BaSO_4$ dissolution is typically controlled by the release of other, relatively more soluble SO_4-bearing minerals.[11]

10.2.3 MASS-TRANSPORT CONTROL

When the mechanisms of chemical binding are reversed (e.g., by contact with aggressive leachants) and mobilization begins, the rate of release may be limited by the mass transport of constituents to the environment. The mass-transport pathways for porous materials include (i) diffusion of constituents in liquid-filled pores, (ii) advection or bulk liquid transport, (iii) dispersion of constituents due to the porous nature of the solid material, (iv) crack flow, and (v) diffusion through solid phases. Only the first four pathways will be described here.

10.2.3.1 Diffusion Within Pores

Dissolution of relatively soluble mineral phases results in a pore solution of high ionic strength. Upon contact with infiltrating rainwater or groundwater, a chemical potential gradient is established between the pore solution and the surrounding liquid phase. Leaching is initiated due to the net movement of ions via diffusion from areas of high ionic activity to areas of low ionic activity. In turn, diffusion disrupts the equilibrium between the pore liquid and solid, leading to further dissolution of mineral phases and desorption from mineral surfaces.

10.2.3.2 Advection

Constituents dissolved in the pore water may be transported when pressure gradients induce physical movement of the pore solution. In the disposal scenario, a hydraulic head causes pressure gradients. The velocity of the fluid flow is related to the d'Arcy's permeability of the solidified matrix, the viscosity of the fluid phase, and the pressure head of the infiltrating liquid. Constituents that dissolve or transport into the liquid phase are carried away by the fluid flow. Convective mass transport typically results in increased mass-transport rates compared to pure diffusional systems.

10.2.3.3 Dispersion

Dispersion is the spreading of a concentration front during advective mass transport due to variations in the pore water flow. The coefficient of proportionality between fluid viscosity and constituent spreading is "dispersivity." The variations in pore water velocity can result from the tortuosity of the pore structure, the connectivity of the pore network, or variation in fluid properties. For example, tight pore systems with high tortuosity and low connectivity may be more dispersive than open pore systems. In scenarios where fluid flows through a solid, a hydrodynamic dispersion coefficient representing both diffusion and hydraulic dispersion within the matrix is used to describe mass transport.[12]

10.2.3.4 Crack Flow

Microcracks are commonly found in cement-based material caused by induced stresses that exceed the material tensile[13,14] or facture strength.[15,16] Such stresses may be induced by combinations of (i) mechanical loads (e.g., fatigue cracking),[17] (ii) environmental interactions (e.g., expansive reactions, desiccation),[16,17] and (iii) thermal effects (thermal gradients or thermal incompatibilities between matrix components).[17,18] A continuous crack network may form through progression and bridging of discrete microcracks,[19] resulting in a matrix with increased permeability,[16] enhanced diffusion properties,[15] and reduced strength characteristics.[19,20]

The presence of discrete microcracks (< 50 μm in diameter) seems to have an insignificant influence on permeability;[21] however, when through-cracks of sufficient diameter exist, perched water above the solid matrix may travel through the continuous crack network under the influence of hydraulic gradients. The increased hydraulic conductivity facilitates infiltration of deleterious agents from the environment and further accelerates deterioration of the matrix.[16,21]

When the velocity of infiltrating liquid is high, the rate of leaching and solid phase dissolution are increased by a combination of advective and diffusive mass transport.[20] Constituents in narrow or dead-end pores may diffuse through a stagnant pore fluid toward an intersection with a crack, where advective mass transport carries the constituent with the liquid flow to the environment. Mechanical dispersion, or spreading, due to variations in fluid velocity across the crack width and fracture wall roughness influence diffusion of constituents within the through-crack.[20]

10.3 FACTORS INFLUENCING CONTAMINANT RELEASE

Leaching is a complicated process, requiring knowledge of mineral chemistry, thermodynamics, reaction kinetics, and mass-transport theory to make accurate estimates of release in the long term. The physical and chemical factors that can influence the leaching process as well as the primary degradation pathways for solidified materials are shown in Figure 10.1. The most important factors influencing leaching of inorganic contaminants from S/S waste forms come from (i) matrix characteristics (e.g., matrix mineralogy, permeability, acid neutralization capability (ANC)), (ii) environmental conditions (e.g., infiltration, leachant composition), and (iii) reactions between the waste form and the surrounding environment (e.g., acid attack, carbon dioxide uptake, sulfate attack).

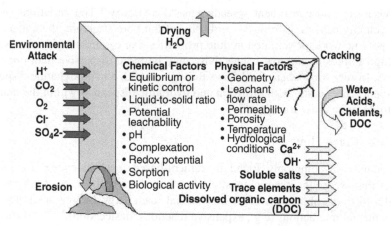

FIGURE 10.1 Factors affecting the rate of leaching from solid materials including S/S waste forms.

10.3.1 S/S MATRIX PROPERTIES

During the hydration and setting of a cementitious matrix, aqueous species are "stabilized" in more thermodynamically stable forms prior to "solidification" into a monolithic waste form with significant compressive strength and minimal hydraulic conductivity. Low hydraulic conductivity is primarily associated with retention of constituents in a highly tortuous, disconnected pore structure. The stabilization aspect of S/S treatment has been referred to as "chemical fixation,"[22] where contaminants may be fixed as (i) precipitated oxides or hydroxides, (ii) adsorbed species onto the surface of hydration products, or (iii) substituted ions within hydration products.[23,24] These binding processes are discussed in Chapter 7.

10.3.1.1 Hydration Products

The retention properties that make S/S treatment so attractive (e.g., favorable pH for contaminant fixation, high buffering capacity, low permeability, significant strength) depend on the stability and dissolution of hydration products. Simple S/S matrices primarily based on portland cement are common; however, supplementary cementing materials and other additives are often used together with portland cement to modify the hydration products and enhance the S/S efficacy. The hydration of portland cement and microstructure of portland cement pastes are described in Chapters 4 and 9. The major hydration products of portland cement are (i) calcium hydroxide, (ii) calcium silicate hydrate (C-S-H), and (iii) calcium sulfoaluminates.[25,26] If the hydration process is incomplete, clinker minerals such as tricalcium silicate (C_3S) and dicalcium silicate (C_2S) may also be found in the cement matrix.[7,25-28]

A saturated solution of calcium hydroxide has a pH of approximately 12.4 at room temperature.[7] The presence of highly soluble alkaline salts, common in cement clinker and many treated wastes, suppresses the dissolution of calcium hydroxide and elevates pore water pH towards a value of 14.[7,8,26] As shown in Chapter 7, the solubility of amphoteric metal hydroxide cations passes through a minimum in the pH range 9 to 11, rising sharply with pH thereafter. Thus, pore water conditions dominated by dissolution of pure $Ca(OH)_2$ are not optimal for the fixation of many priority pollutants.

C-S-H consists of amorphous gel-like phases, which comprise more than 60 wt% of cement paste,[29] and has a variable stoichiometry with different calcium to silica ratios depending on the cement compositions and the hydration conditions.[32-35] Due to its high surface area and amorphous properties, several researchers report that C-S-H is a principle mineral phase for adsorption of metal cations and anions.[29,32,35-37] However, the retention properties of C-S-H gel vary with Ca/Si ratio in that the surface charge of C-S-H gel goes through a point of 0 net charge at a Ca/Si ratio of approximately 1:2.[7] Therefore, anions adsorbed to the positively charged surface of C-S-H with a high Ca/Si ratio may be released in favor of cation adsorption to the negative surface charge of C-S-H with a low Ca/Si ratio during decalcification of C-S-H gels.

Calcium sulfoaluminates, such as ettringite ($3CaO \cdot Al_2O_3 \cdot 3CaSO_4 \cdot 32H_2O$) and monosulfate ($3CaO \cdot Al_2O_3 \cdot CaSO_4 \cdot 12H_2O$), coexist to varying degrees at different stages of the hydration/aging process.[38,39] Chapter 7 discusses the incorporation of cations[29,35,40] and oxyanions[7,29,32,41-43] as well as the stability of calcium sulfoaluminates.

Supplemental cementing materials such as blast furnace slag, fly ash, and silica fume are often used in S/S treatment recipes to modify hydration products and matrix properties. Blended cement matrices exhibit finer, less continuous pore structure,[44,45] higher compressive strength,[46-49] and increased capacity for contaminant adsorption and incorporation.[45,50,51] Chapter 4 discusses the hydration and microstructure of blended cements containing blast furnace slag, fly ash, and silica fume.

10.3.1.2 Hydraulic Conductivity

Hydraulic conductivity is a material property representing the ability of a fluid to flow through a porous material. The hydraulic properties of a material primarily depend on the capillary porosity, pore-size distribution, and connectivity of the pore structure.[52-55] In turn, the pore structure is a function of the water/cement ratio, the presence of hygroscopic admixtures, and curing conditions.[56] The Katz–Thompson model[57] uses percolation theory to correlate water permeability to the pore structure; however, conflicting research has found this model is either valid[58,59] or invalid[58,60] for predicting measured permeability in cement-based materials.

Experimentally determined values of water permeability for mature portland mortar range from 10^{-12} to 10^{-13} m/s, while blended cements tend to exhibit even lower values.[26,44,58,61] Stegemann and Côté[12] state that for low permeability materials ($< 10^{-9}$ m/s), infiltration of water is negligible and the rate of constituent release is expected to be limited by the diffusion of species to the surface. Thus, it can be expected that the rate-limiting release mechanism for undamaged monolithic S/S materials in most disposal scenarios will be mass transport of constituents through the solidified material.

10.3.1.3 Acid Neutralization Capacity

ANC, a material parameter representing the ability of a matrix to resist a decrease in pH, is a key retention characteristic of S/S materials. Resistance to pH change is important, since the solubility of inorganic constituents and the kinetics of mineral dissolution often are pH dependent. Typically, ANC of a material is evaluated by plotting the steady-state pH response* as a function of predetermined acid additions. Plateaus in the curve indicate zones of pH buffering due to mineral phase dissolution.

The pH stability of pure mineral solids[62,63] and single-component cement systems[64] have been resolved from ANC plateaus. Pure calcium oxide-silica-water systems exhibit variable neutralization behavior with plateaus at approximately 12.3, 11.9, and 9.9, depending on their Ca/Si ratio.[65] Calcium hydroxide is often considered the primary contributor to the ANC of cement-based materials, and the ANC plateau at pH 12.3 is typically associated with neutralization of free lime. However, differential acid neutralization analysis[66,67] indicates the calcium hydroxide only accounts for about 25% of the ANC to a pH of 10.[67] Above pH 11.6, C-S-H, ettringite, monosulfate, hydrogarnet, brucite, and hydrotalcite also may contribute to ANC, but below a pH of 9, the solid matrix is generally degraded to the point that these solid phases do not exist.[68] In comparison to pure systems, the addition of waste or mineral additives lowers the levels of pH plateaus or changes the number and magnitude of the buffering zones.[65] Thus, analysis of ANC curves for cements with mixed mineral phases and waste additives is quite complex, and accurate direct resolution of mineral phases has not been possible.[67]

A significant portion of the high acid resistance of cement materials is due to the presence of low solubility C-S-H rather than the relatively more soluble calcium hydroxide. C-S-H gel does not dissolve in the same manner as crystalline minerals like calcium hydroxide, maintaining acid resistance by decalcifying to calcium hydroxide and a siliceous gel of a lower Ca/Si ratio according to the stoichiometry presented by Bonen and Sarkar.[69]

$$x\mathrm{CaO}{\cdot}\mathrm{SiO_2}{\cdot}n\mathrm{H_2O} + y\mathrm{H_2O} \;\rightarrow\; (x{-}y)\mathrm{CaO}{\cdot}\mathrm{SiO_2}{\cdot}n\mathrm{H_2O} + y\mathrm{Ca(OH)_2} \quad (10.1)$$

The decalcification of C-S-H typically occurs slowly, either co-currently or after depletion of the calcium hydroxide phase,[69,70] and provides both buffering capacity and long-term acid resistance.

10.3.2 Release Conditions

Several release scenario parameters influence the rate of constituent leaching from solidified materials in disposal placements. These factors include (i) rate and flow regime of infiltrating fluid, (ii) the composition of the liquid phase in contact with the solidified waste, (iii) the geometry and physical properties of the applied S/S

---

* Since the dissolution reactions within cement materials are irreversible, "steady-state pH response" refers to the conditions during which no net change in pH occurs.

material (e.g., unit size, porosity, permeability), and (iv) environmental or matrix-generated temperature.

10.3.2.1 Flow Regime

Water contact in the environment may occur by a combination of leachant flowing around the material or flowing through the material. When a solid material is contacted by water in a flow-around mode, mass transport of ions through the solid material to the surface is expected to limit constituent release,[9] while release due to flow-through leaching may be mass-transport-limited or equilibrium-limited, depending on the infiltration rate.

The critical parameters that determine the mode of water contact are the hydraulic gradient of the infiltrating water and the relative hydraulic conductivities of the waste form and surrounding material. Barth et al.[71] suggest that infiltrating water will flow around an S/S monolith if the hydraulic conductivity of the solidified matrix is a factor of 100 less permeable than the surrounding fill, while Baker and Bishop indicate that this factor is 1000, reporting that "diffusion through a solid represents a maximal contaminant loss rate when the waste permeability is less than 10^{-3} times that of the geological media."[72]

10.3.2.2 Infiltration Rate

Although leaching rates are rarely controlled by a single limiting mechanism, the rate of subsurface infiltration influences leaching rates in that the limiting release mechanisms are dependent on the liquid-to-solid (L/S) ratio (leachant volume), contact time (infiltration rate), and geometry of the solid phase (material type). Two bounding release scenarios may be identified, roughly defined by the Peclet number (i.e., ratio between the infiltration seepage rate and diffusional mass transport):

$$Pe = \frac{L^2}{D^{obs}} \left(\frac{v}{H} \right) \qquad (10.2)$$

where:
L is the characteristic length for diffusion
D^{obs} is the observed diffusivity of a constituent through the porous material
H is the height through which infiltration percolates
v is the infiltration seepage velocity

If the liquid-to-solid ratio is low (i.e., the volume of the leachant is small in comparison to the surface area of the solid phase), then the Peclet number tends to be less than 1,[73] and a finite bath with a local equilibrium end-state may be assumed. These conditions exist in the following two cases: (1) water percolating through a granular fill, and (2) groundwater flowing around a monolith at low infiltration rates. Equilibrium-based mechanisms, which dominate at low L/S ratios (e.g., sorption and geochemistry), control the extent of release.

When L/S ratios are large or infiltration rates are rapid, the Peclet number may be greater than 10 and the rate of release is assumed to be controlled by mass transport through a solid matrix. This scenario occurs with monolithic materials or when granular material is overlain by low permeability covers or surrounded by material with much greater hydraulic conductivity. Typically release is estimated by evaluating mass transport into an infinite bath with long-term release limited by the available content.

10.3.2.3 Groundwater Composition

In near-surface or above-ground applications, the solid material is most likely to be contacted by rainwater or surface runoff. Although rainwater may contain sulfuric acid due to industrialization, the total acidity of infiltrating rainwater is typically much lower than the ANC of S/S materials. Other components (e.g., CO_2, organic matter, metals, salts) may be dissolved in the leachate as it percolates through the soil zone. For subsurface application scenarios, the commonly anticipated leachant is groundwater, which tends to be mineralogically "soft" in comparison to the high ionic strength of the pore solution. In addition, biological activity, co-application with other materials, reduction of Fe- and S-bearing species in soils or wastes, or interaction with the surrounding environment may result in groundwater that contains aggressive components.

10.3.2.3.1 Acids

The effects of acid attack on the chemical and morphological characteristics of cement pastes have been well documented[69,74-77] and result in the formation of a calcium depletion zone near the exposed surface surrounding a kernel of unreacted cement paste. In the calcium depletion zone, porosity was significantly increased, the Ca/Si was very low, and the pH showed a gradient between the surface and the boundary of the re-mineralization zone.

Exposure of S/S materials to acids (e.g., through carbonation, chloride ingress, or biological activity) reduces the buffering capacity and pH of the S/S matrix. In turn, the solubility of metal ions in solution may be increased or decreased.[25] Organic acids may contribute to uncertainty of the leaching test through complexation with contaminants (e.g., lead acetate) or incomplete dissociation.

ANC should not be confused with acid resistance.[65,67] The kinetics of mineral dissolution are significant in acid resistance and are not accounted for in the ANC. The highest ANC materials (e.g., portland cement) usually contain large amounts of free lime that are easily dissolved in acidic solutions, while matrices with a high content of C-S-H and a low Ca/Si ratio (e.g., activated blast furnace and fly ash/lime) have a lower ANC but exhibit greater resistance to acid attack over time by creating a siliceous protective layer on mineral surfaces.[65]

10.3.2.3.2 Chelation with Organic Molecules

Leachants containing organic acids may have a secondary effect on the local solubility of constituents in that ligands introduced by the acid dissociation may form highly stable and soluble metal chelants within the pore solution. Ethylenediaminetetraacetic acid

(EDTA), nitrilotriacetic acid (NTA), citrate, and oxalate are common chelating agents found in industrial waste streams, applied to remediate metals-contaminated soils or sometimes detected in leachates of hazardous waste landfills.[78-81] In addition, dissolved organic matter incorporated into the S/S materials containing certain waste types (e.g., sludges, soils) or transferred to the matrix from the surrounding fill may act as chelating agents.[82]

Large organic molecules surround and covalently bond with metal ions, forming a stable complex and effectively withdrawing certain metal ions from solution.[83] The result of chelation is that the apparent liquid solubility limit relaxes, total elemental solubility increases (sometimes by several orders of magnitude), and the solid–liquid partitioning is driven toward dissolution of a solid mineral phase. Although most metal species are affected, lead, cadmium, and copper are particularly susceptible to chelation.[84]

A study on chelation of cement-solidified spent reactor decontamination resins concluded that mobilization of cationic metal/radionuclides is significantly enhanced at the highly alkaline pH of cement waste forms.[85] In this study, concentrations of $> 10^{-4}$ M EDTA were shown to mobilize divalent transition metal Ni and Co, while higher concentrations ($> 10^{-3}$ M) of picolinate mobilized Sm^{3+}, Th^{4+}, NpO_2^+, UO_2^+, and oxides of Pu. S/S treatment using highly binding chelants with cationic metals/radionuclides was discouraged.

10.3.2.3.3 Anions

Anions in solution may influence the chemistry of the constituents in the pore solution through competitive adsorption, complexation, or common ion effects. For example, the presence of sulfate and phosphate ions may result in precipitation of metal ions such as lead, barium, and calcium, lowering the activity of these constituents in the pore solution.[11,33] Conversely, the introduction of acetate, ammonium, and chloride ions may increase solubility of some contaminants (e.g., copper, lead) through formation of coordination complexes.[83] The increase in metal solubility at high pH is attributed to the formation of complex metal hydroxide anions.

10.3.2.3.4 Redox Potential

Environmental reducing conditions and gradients in redox potential can alter the chemistry of waste systems over long periods of time, leading to precipitation or dissolution of some contaminants.[86] For example, Cr, Tc, and Mn retention is increased under low E_h conditions,[7] while retention of arsenic decreases as As(V) is reduced to As(III).[33] Redox gradients and reducing conditions may result from waste material characteristics, biological activity, or external sources.

The redox potential of pure portland cement by nature is mildly oxidizing, with E_h measurements between +100 and +200 mV. The few electro-active components inherent to cement paste include dissolved oxygen and trace amounts of sulfite SO_3^{2-} in pore water and small quantities of insoluble Fe^{2+} and Mn^{2+}. Blended cements, especially those with significant replacement of cement with blast furnace slag, may exhibit overall reductive capabilities.[54]

A significant fraction of saturated disposal scenarios are likely to have characteristically low E_h[7] in the range of −200 to −500 mV.[87] Decomposition of organic matter and other biologically active groundwaters tend to be depleted of oxygen content and are likely to remain either reducing or only weakly oxidizing.[7,88] Furthermore, reduced waste materials (e.g., anaerobic sediments, mining wastes, slags) may also control the redox potential of S/S materials. This implies that a high pH, low E_h environment may be both obtainable and sustainable (i.e., re-oxidation prevented) for many disposal scenarios.[7]

10.3.2.4 Particle Size of the Disposed Fill

For granular materials that may be treated using S/S technologies (e.g., soils, ash residues, slags), the average particle size and particle size distribution of the waste material play important roles in determining the limiting release mechanism. Smaller waste particles have a large surface area exposed to a leachant and a characteristic distance (i.e., particle radius) for mass transfer. Thus, with all other conditions equal, smaller particles should release constituents more rapidly than larger particles.

One significant advantage of S/S material is that encapsulation in a monolithic matrix minimizes the surface area of the waste that is exposed to the environment. Therefore, particle size of the waste itself is usually only a concern for physical (e.g., aggregate-sized particles, rather than chemical reasons. Of course, the monolithic attributes of S/S materials decrease with age-induced degradation of the physical structure and are only valid as long as the material durability holds. In one study, aged S/S materials were broken down into a granular or soil-like consistency due to exposure to an aggressive environment.[89] In this scenario, the surface-to-volume ratio of the S/S material (i.e., the particle size distribution of the degraded zone) becomes an important release parameter.

10.3.2.5 Temperature

Often, S/S materials are deposited deep in the subsurface where temperatures are relatively constant at geographically localized values. For example, average temperatures encountered in municipal solid waste landfills range from 10 to 45°C while temperatures for deep-well hydrofracture injection deposition range from 8 to 15°C.[12] Recent initiatives into beneficial reuse of treated wastes require S/S materials to be placed in shallow burial and near-grade applications where temperatures may vary both diurnally and seasonally with ambient conditions.

Under certain conditions, the temperature of the S/S material may be higher than that of the surrounding environment. Sources of internal heat generation come from hydration reactions or waste characteristics. Generation of heat due to cement hydration may be significant for massive concrete elements;[90] however, S/S materials usually are poured on a smaller scale. In addition, the heat of hydration is relatively short-lived in comparison to long-term release assessment intervals. S/S materials containing high concentrations of radioactive wastes may maintain high temperatures (e.g., > 70°C) due to radioactive decay of nuclides retained in the cement.[91,92]

High temperatures influence leaching properties by (i) accelerating ongoing matrix reactions (e.g., hydration, adsorption, and dissolution) via Arrhenius effects at temperatures ~60°C,[93] (ii) increasing the rates of constituent and moisture diffusion within the matrix[94,95] at temperatures $> 20°C$[96], and (iii) decreasing strength.[13,97] The effect of elevated temperature on mineralogy is somewhat ambiguous, since the stability of most mineral hydrates decreases with temperature, whereas portlandite becomes more stable.[98] Blended cements, which typically do not have high portlandite contents, have been shown to demonstrate increased porosity and decreased degree of hydration when cured at high temperatures, especially between 10 and 60°C.[49]

At subzero temperatures, freezing and thawing may cause deterioration of cement materials through (i) hydraulic pressures of capillary pore water freezing, (ii) osmotic pressures in partially frozen pore water containing salts, or (iii) transport of water from C-S-H gels.[16] If the resultant internal stresses exceed the tensile strength[13] or fracture strength[16] of the material, pressures are relieved by cracking.

10.3.3 MATERIAL AGING

Cement-based materials may be considered metastable in the sense that the properties of the material do not remain constant but change as a result of aging. Klich et al.[99] point out that, morphologically, the same physical and chemical mechanisms that degrade structural concretes are responsible for deterioration of solidified waste matrices. Thus, significant advances in the development of S/S systems may be made after consideration of the structural cements durability literature. In general, degradation of cement materials results from internal and external stresses of the matrix. Internal stresses are caused by chemical reactions occurring in the absence of outside influences, whereas external stresses may be induced through interaction between the cement and its surrounding environment.

10.3.3.1 Internal Stresses

Typically, internal reactions are kinetically limited and occur on a geological time scale.[7] Thus, the associated changes in morphology, mineralogy, and leaching properties due to internal stresses are difficult to assess without accelerating the reaction mechanism.[100] Degradation mechanisms that lead to internal stresses include (i) internal chemical reactions, (ii) delayed ettringite formation (DEF), and (iii) alkali–aggregate reactions.

The mineralogical nature of cementitious wastes continues to change beyond the 28 days considered standard for physical hardening of cement materials. Full hydration of cement clinker may take years to complete, especially if reactants must diffuse through siliceous gels surrounding clinker particles. The dissolution of waste materials may limit incorporation into developing C-S-H or calcium sulfoaluminates. The uptake of water, via continued hydration or salt crystallization, can result in cracking due to osmotic pressures,[101] self-desiccation,[26] or autogenous shrinkage.[102,103]

DEF[39] and alkali–silica reactions[26,54,104] are common causes for cracking of concrete products and structures. DEF is related to high-temperature curing of the products, and alkali–aggregate reactions require the presence of alkali-reactive aggregates. Thus, they seldom happen to cement-solidified waste forms, but may need to be considered for higher internal temperatures (e.g., high cement contents, massive pours, and radioisotopic heating) and if the active aggregate phases are present in the waste (e.g., soil remediation or including aggregate for filling and closing empty tanks).

10.3.3.2 External Stresses

Alterations due to external stresses tend to be more dramatic and occur more readily than those from internal stresses. External stresses may be caused by contact with liquids and gases or by changes in environmental conditions (e.g., temperature). Contact of an S/S matrix with a leachant leads to depletion of matrix constituents [70,105-114] and potential degradation through sulfate attack[101,115-123] or carbonation.[8,28,124-142] When S/S materials are stored in unsaturated environments at relative humidity < 100%, additional degradation may occur due to moisture transport.[143-149] The above phenomena have been extensively studied with regard to structural materials or S/S matrices, so only a brief overview is presented with regard to leaching and durability effects.

10.3.3.2.1 Dissolution and Leaching

Dissolution of calcium hydroxide due to leaching results in (i) decreased pore water alkalinity, (ii) reduction in buffering capacity, and (iii) a shift in pH-dependent constituent solubility. Complete depletion of calcium hydroxide in cement paste has been associated with a 13% increase in macroporosity[70,76,106] and a 70% decrease in compressive strength of the matrix.[70,108] Mineral phase depletion is accelerated by direct exposure to acidic leachants (e.g., acetic acid, nitric acid)[75-77] or to acid-forming species (e.g., ammonium nitrate,[105,106] carbon dioxide.[69])

10.3.3.2.2 Sulfate Attack

Sulfate attack is a complex process that primarily is attributed to ingress of sulfate salts; however, the mechanism of degradation is not completely understood and depends on the speciation of sulfate (e.g., Na_2SO_4, $MgSO_4$, $(NH_4)_2SO_4$).[101,150] Two commonly assumed mechanisms of sulfate attack are the expansive reactions: (i) sulfate ions and free lime combine to form gypsum, and (ii) gypsum and hydrated calcium aluminate combine to form ettringite.[101,117-119] In the presence of magnesium ions, these chemical reactions are accompanied by conversion of C-S-H to magnesium silica hydrates.[117,118] Confounding the above mechanism may be other mechanisms typically not considered "sulfate attack," including hydration expansion of sulfate salts and increased osmotic pressures due to sulfate salt crystalization.[101]

The assumed cause of physical deterioration is mechanical pressure resulting from differential expansion of precipitates in pores.[118,151] The presence of gypsum and ettringite in the microcracks implies that the physical damage results from the formation of expansive minerals, but both the physical damage and the crystals may be caused by desiccation.[101] Physical degradation due to sulfate attack may manifest in (i) loss of cohesion and strength, (ii) expansion and cracking, and (iii) scaling and shelling of the surface in successive layers.[122,152,153] Chemical alteration is due to consumption of calcium hydroxide, decalcification of C-S-H, and precipitation of sulfate minerals.[101,118,154,155]

10.3.3.2.3 Moisture Transport

In the absence of thermal gradients, the change in moisture content of hygroscopic, porous materials (e.g., cement, concrete) has been described empirically as non-linear moisture transport[156-160] and also using a two-regime mechanism.[148,161] Liquid water moves under the influence of capillary and osmotic pressures, while water in the vapor phase is transferred due to gradients in relative humidity. The primary physical effect of moisture transport is cracking of the matrix due to local desiccation. Perhaps more importantly, drying facilitates cement degradation by allowing reactive gases (e.g., CO_2, O_2) to penetrate into the pore system through the vapor phase.

10.3.3.2.4 Carbonation

Carbonation is arguably the most common chemical degradation process, with a strong potential to influence physical and chemical properties of S/S materials,[162-164] because of resultant changes in pore water pH, constituent speciation, and pore structure. The carbonation reaction may be caused by attack of carbonic acid under saturated conditions;[69,140] however, changes in physical and chemical properties are more closely associated with atmospheric carbonation under drying conditions.[135,163,165-167] Atmospheric carbon dioxide diffuses into the pore vapor, dissolves into the receding pore solution film, and reacts with aqueous cations, producing carbonate precipitates and water.[26,28] The primary product of the carbonation reaction is calcium carbonate due to the high calcium content of the cement pore water; however, other metal cations have been observed.[8]

The rate of atmospheric carbonation may be limited by the rate of moisture transport or the diffusion of atmospheric CO_2 through the pore vapor space.[28,168] Cement mineralogy plays an additional role in that penetration of a carbonate front depends on the availability of "carbonatable" ions (i.e., slag cements produce less calcium hydroxide, leading to greater carbonate penetration).[169] Thus, the carbonation effect may be controlled by optimization of supplementary cementing materials.[168,169]

The deleterious effects of atmospheric carbonation include (i) neutralization of the system to pH values < 10,[8,132] (ii) decalcification and polymerization of the C-S-H,[26,69] (iii) decomposition of ettringite,[137,167] and (iv) decreased retention of oxy-anionic species.[8,134,170] Conversely, carbonation may benefit S/S treatment through (i) conversion of easily soluble calcium hydroxide into more stable calcium carbonate, (ii) densification of the microstructure (e.g., pore blocking, capping, or diameter reduction),[132,165] (iii) marked increases in compressive strengths,[26,134,135,171] and (iv) enhanced retention of metal cations in some S/S materials.[134,135,165,167]

10.4 OVERVIEW OF LEACHING TESTS

In general, leaching tests measure leaching or release from a representative sample of the subject matrix (sample) during contact with a liquid phase (leachant) under a set of controlled leaching conditions (e.g., contact time, liquid-to-solid ratio, pH, leachant composition). The liquid solution that results from a leaching procedure (leachate) is analyzed for physical and chemical properties in order to determine information on constituent release from the material.

10.4.1 LEACHING TEST APPLICATIONS

Leaching tests, especially regulatory protocols, are designed for use over a wide range of applications and for a diverse set of purposes. Predictions of constituent release may be used for regulatory purposes (e.g., waste classification or delisting applications), development of advanced treatment technologies, preparation of site assessments, bases for selection of risk-based alternative treatments, or fundamental research (e.g., regarding release mechanisms, treatment process, materials science). The objectives for evaluating constituent release from a solid material via leaching protocols may include one or more of the following:

- Screening of a material as "hazardous" or "non-hazardous"
- Development of an effective S/S treatment recipe
- Compliance to waste treatment acceptance criteria
- Characterization of material-specific intrinsic properties
- Determination of a source term for risk management and assessment
- Comparison of alternative management scenarios
- Prediction of potential release under specified release scenario conditions

The goal of leaching evaluation from an environmental perspective is to answer the question, "What is the potential for toxic constituent release by leaching from this waste matrix under this management scenario?"[5] Inherent in the above goal is the additional question, "How is this potential constituent release affected by alteration of the release conditions or long-term interactions with the release environment?"

Due to the complexity of the leaching process and the number of factors influencing release, no single leaching test or single set of leaching conditions is appropriate for the entire range of leach testing objectives and applications. Therefore, researchers have developed many different leaching tests, each designed to address a specific objective or to measure a leaching parameter over a range of leaching conditions. Since leach test development to date has been largely *ad hoc*, there are many slight variations in the parameters of existing leaching protocols, even among tests designed to provide similar evaluations.

For example, leaching tests designed to determine solid–liquid partitioning usually are conducted on particle size-reduced material, although a brief scan of the literature shows that the specific particle-size maximums are observed to range from 125 µm to 9.5 mm. In order to ensure steady-state conditions are obtained, maximum particle-size guidelines are intertwined with contact times (i.e., longer contact times are required to obtain steady-state conditions for larger particles). Additional variations are seen in the number of leaching steps (e.g., a single extraction with one leaching solution or multiple extractions to a series of solutions) and application of agitation during the leaching interval. The composition of the leachant varies among leaching tests; however, the most commonly specified leachants are "reagent grade" water (i.e., deionized, distilled, or demineralized water) and dilute solutions of organic or inorganic acids. Site-specific runoff or groundwater is sometimes used on a more specialized basis.[172]

In recognition of similar, albeit slightly varying, leaching protocols throughout the European community, Comité Européen de Normalisation (CEN) has commissioned the Characterization of Waste Technical Committee TC-292 with the task of harmonization of leach characterization approaches.[173] One focus of the committee is to define a limited set of leaching protocols that may be used to describe inorganic release from a wide variety of materials (e.g., building materials, incinerator residues, solidified wastes) in a broad range of beneficial reuse or disposal scenarios.

The classification system adopted by CEN/TC-292 segregates leaching protocols into three groups: characterization tests, compliance tests, and on-site verification tests based on the level of detail and purpose of the leaching protocol.[174] On-site protocols tend to be simple tests providing the least level of detail, such that the output test information affords only a "pass/fail" comparison to a regulatory limit or a material acceptance parameter. Compliance testing allows for incorporation of regulatory or permissible release limits for specific release scenarios, while highly specific characterization tests are used to determine intrinsic leaching parameters or release mechanisms. Increasing the detail and complexity level of the leaching tests and interpretation protocols allows for elucidation of leaching potentials, retention and release mechanisms, sensitivity analyses, and calculation of systems failure and environmental risk.

A similar effort (i.e., to define leaching tests, optimize leaching test selection, and develop interpretation methods) is ongoing in the United States in coordination with the USEPA Office of Solid Waste. In the proposed assessment framework,[5] the benefits of the more detailed characterization above a basic "pass/fail" of regulatory or compliance level assessment come in terms of a reduced requirement on the detail of subsequent testing (e.g., on-site verification) for large volume producers, more accurate predictions of long-term environmental impact, flexibility for the selection of appropriate waste management solutions, and lower liability associated with accidental release from test materials.

10.4.2 LEACHING TEST CLASSIFICATION BY APPLICATION

Chapter 11 summarizes commonly used standardized leaching (and other) tests in North America and Europe. These leaching tests may be used to (i) screen materials (e.g., classify waste materials as "hazardous" or "non-hazardous"), (ii) mimic field leaching (e.g., to produce a representative field leachate), and (iii) determine intrinsic properties of the waste material. This section describes the applications of these leaching tests. Both this chapter and Chapter 11 are "stand alone" chapters. Although there is some overlap and redundancy on the subject of leach testing between this section and Chapter 11, the objectives are different for the two chapters and they complement each other on this topic.

10.4.2.1 Screening Tests

Screening tests are relatively quick (~24 hours), low-cost evaluations designed to classify a waste material or to ensure that the material meets certain acceptability criteria. Usually, these tests consist of a single-batch extraction using size-reduced material with chemical analysis of a limited set of constituents. Interpretation of the leaching test results is a simple pass/fail comparison to acceptable leachant concentrations for only those constituents critical for performance acceptance. One significant disadvantage of screening tests is that little mechanistic information is provided by most protocols without subsequent comparison to other more detailed characterizations.

An example of a protocol that may serve to screen waste materials is the EN 12457-3 "shake test" for granular waste.[175] This procedure consists of a two-step sequential extraction procedure of a particle-size reduced material in DI water leachant at two levels of liquid-to-solid ratio. The solid phase from the first 3-hour extraction at an L/S ratio of 2 ml/g is extracted for an additional 3 hours at an L/S ratio of 8 ml/g. The concentrations in the leachate may be used to predict pore solution in the fresh material, and infer pore chemistry after soluble species have been leached. The leaching response using this test may be compared to leaching behavior of previously characterized material to check for product consistency or regulatory compliance.

10.4.2.2   Field Mimicking or Simulation Tests

Several hazardous and solid waste regulatory protocols promulgated for waste screening are essentially field mimicking tests.[176,177] This category of leaching tests is not explicitly distinguished by the CEN/TC-292 harmonization committee. Field mimicking tests are designed to provide a representative leachate from a leaching test by simulating specific release scenario conditions.

Simulation protocols may be useful when the exact release scenario, or a limiting case, is known and duplicated in the laboratory. However, implementation of field mimicking leaching tests may be quite complex, as leaching factors (e.g., leachant composition, L/S ratios, particle size, and modes of water contact) may vary by region or by release scenario (e.g., lined or unlined landfills, monofills, or co-disposal). The application of these tests is severely limited or may be completely irrelevant if the proposed field release conditions are uncertain or significantly differ from test conditions. Interpretation of the leaching data applied to release scenarios other than the particular scenario simulated by the test conditions may be misleading at best and can result in significant errors in constituent release prediction (over- and under-estimation).

As an example, the Synthetic Precipitation Leaching Procedure (SPLP)[177] uses dilute nitric/sulfuric acid leachants to simulate contaminant release for scenarios where acidic rainfall percolates through a high permeability fill. The exact composition of the SPLP leachant depends on the location of the disposal site, requiring *a priori* knowledge of the disposal facility location. A higher level of acid in the SPLP leachant is specified for disposal east of the Mississippi River in order to account for the effects of dense industrialization on rainfall composition in the eastern United States.

TCLP[176] is one example of a field mimicking test that is applied as a screening protocol in practice. The field conditions adopted by TCLP are those of an assumed "mismanagement" scenario described as disposal of hazardous waste in a biologically active municipal solid waste landfill. Leachate concentrations resulting from TCLP are compared to the promulgated list of TC permissible limits, and the solid waste is considered "hazardous" by toxicity characteristics if the concentration of any constituents in the leachate exceeds the acceptable value. Technical and scientific limitations of this protocol are well documented[3-5,40,178] and the TCLP has been proven inadequately protective for highly alkaline waste forms.[181-182] Although not promulgated for regulatory purposes other than classification, the TCLP often has been the *de facto* basis for waste delisting applications, development of waste acceptance criteria, prediction of long-term release, and evaluation of waste treatment efficacy.

10.4.2.3 Intrinsic Property Tests

Many leaching tests are designed to provide insight into one or more of the intrinsic leaching properties that characterize the leaching process. For example, the pH Static Test,[183] a published leaching protocol currently in process of CEN/TC-292 standardization as prEN 14429,[184] is an equilibrium-based protocol consisting of parallel challenges of particle-size-reduced material to different-strength acid solutions. The goal of the test is to determine the ANC of the material and the solubility of constituents as a function of solution pH. The pH Static Test does not provide data on the kinetics of release and provides only limited long-term performance information. The results of the pH Static Test, or other equilibrium-based protocols, are commonly combined with those of mass-transport characterization tests to fully characterize waste leachability.[96,185-188]

The results of a properly selected intrinsic property test may be used to create a fundamental leaching profile of the solid material independent of release scenario factors. A minimal fundamental leaching profile may be comprised of the following intrinsic properties:

- Constituent availability or the leachable content under environmental conditions
- ANC of the S/S material
- Solubility and release of constituents as a function of pH and L/S ratio
- Constituent diffusion coefficients for mass transfer through the S/S material

The development of a fundamental leaching profile based on the above parameters requires a combination of both equilibrium and dynamic testing under a range of experimental conditions. Thus, this category of leaching evaluation tends to be relatively complex. Although characterization tests may be time-consuming and expensive for previously uncharacterized materials, the power and cost-savings benefit is evident when developed intrinsic leaching profiles are used to provide comparative release estimates for any number of release scenarios or sets of site-specific information.[5]

10.4.3 Leaching Tests by Procedural Groups

Two broad categories of leaching tests may be defined based on whether or not equilibrium or steady-state is established during the test duration:

- Equilibrium tests
- Dynamic tests

Table 10.1 and Table 10.2 present a selection of commonly used equilibrium and dynamic leaching tests, respectively. In each table, the critical testing parameters (e.g., leachant composition, contact time, L/S ratio) are shown for comparison and comments or criticisms from the literature are recorded.

10.4.3.1 Equilibrium Tests

Extraction tests that are designed to reach a steady-state release are termed "equilibrium tests."[5] Since mineral dissolution reactions in cementitious media are often irreversible, the steady state established during an extraction test is not the same as thermodynamic equilibrium, but rather implies no net transfer of constituents between the solid and liquid phase within a specific time frame.[12] Equilibrium-based leaching protocols typically require particle-size reduction of the subject material in order to reduce the time required to obtain steady-state release via increased surface area and minimized kinetic transport. In agitated extractions, shaking, stirring, or tumbling further accelerates the rate of extraction and ensures continuous solid/liquid contact.

Common equilibrium tests may be single-batch extractions at one set of release conditions (e.g., leachant pH, L/S ratio, contact time), parallel-batch extractions over a range of conditions, or sequential-batch extractions at constant or progressive release conditions. A lesser-used approach is to build up concentration in a single leachate using multiple solid samples. Thus, equilibrium-based leaching protocols may be further divided into four procedural groups based on the relative number of samples and resulting leachates:

- Single-batch extractions (one sample, one leachate)
- Parallel-batch extractions (*n* samples, *n* leachates)
- Sequential-batch extractions (one sample, *n* leachates)
- Concentration buildup extraction (*n* samples, one leachate)

10.4.3.1.1 Single-Batch Extractions

The goal of single-batch extraction tests is to characterize solubility or release of constituents at a single set of release conditions. The general procedure (illustrated in Figure 10.2a) is that a predefined mass of sample is contacted with a volume leachant for set contact time. At the end of the specified contact interval, the solid and liquid phases are separated (e.g., filtration, decanting) and the leachate is analyzed for physical and chemical properties. The solid phase may be discarded or saved for mineralogical analysis typically supplemental to the leaching protocol.

TABLE 10.1
Comparison of Selected Equilibrium Tests

Test Name (reference)	Status, Application, Procedural Group	Leachant	pH Control	L/S Ratio	MPS [mm]	CT [hr]	Output Information	Comment/Criticism
Toxicity Characteristic Leaching Procedure[176] (Method 1311)	U.S. Regulatory Standard Field Mimicking Single-Batch Extraction	Acetic acid (buffered w/1-N NaOH)	Initial leachant acidity a) pH 2.88 b) pH 4.93	20:1 (m/m)	9.5	18	Pass/Fail – leachate conc. [mg/L] vs. TC listing.	Field mimic used for screening. Arbitrary final pH, not reported. Not applicable for S/S alkalinity. Assumes mismanagement of co-disposal with municipal solid waste. Monolith property not credited.
Synthetic Precipitation Leaching Procedure[177] (Method 1312)	U.S. Regulatory Standard Field Mimicking Single-Batch Extraction	Dilute sulfuric/ nitric acid (60/40)[a]	Initial leachant acidity – initial pH 4.2 or 5.	20:1 (m/m)	9.5	18	Concentration. [mg/L]	Field mimic used for screening. Alternative to TCLP for disposal other than MSW landfill. Arbitrary final pH, not reported. Cannot extend data to other scenarios than monofill exposed to acidic precipitation.
ASTM D 3987[189]	U.S. Standard Test Screening Single-Batch Extraction	DI water	—	20:1 (v/m)	–	18	Concentration. [mg/L]	Rapid on-site extraction – not intended to simulate site-specific conditions. Size reduction not required – equilibrium or kinetic control?
DIN 38 414 S4[190]	German Regulatory Std.[b] Screening Single-Batch Extraction	DI water	—	10:1 (v/m)	–	24	Concentration. [mg/L]	Assumes equilibrium with buffering by waste. Size reduction not required – equilibrium or kinetic control?

(continued)

TABLE 10.1 (CONTINUED)
Comparison of Selected Equilibrium Tests

Test Name (reference)	Status, Application, Procedural Group	Leachant	pH Control	L/S Ratio	MPS [mm]	CT [hr]	Output Information	Comment/Criticism
EN 12457[175]	CEN Standard Screening/Intrinsic Property Single/Serial Agitated Batch (1–2 steps)[c]	DI water	—	1) 2:1 2) 10:1 3) 2+8:1 4) 10:1 (v/m)	4	—	Concentration. [mg/L]	Assumes equilibrium with buffering by waste.
NEN 7341[226] – Availability Test	Dutch Regulatory Standard Intrinsic Property Serial-BatchExtraction (2 steps)	Dilute HNO$_3$ or NaOH in DI water	Monitored Automatic acid/base 1) pH 7 2) pH 4	50:1	0.125	3 + 3	Concentration. [mg/L]. Release [mg/kg].	Considered maximum leachable content over range of environmental conditions.
AV002.1[5] – Availability at pH 7.5 with EDTA	Published Test Screening/Intrinsic Property Single Agitated Batch	50 mM EDTA in DI water	Addition of HNO$_3$ for final pH of 7.5 ± 0.5.	100:1	0.3 2 5	18 48 168	Concentration. [mg/L]. Release [mg/kg].	Considered maximum leachable content at extreme conditions. Too aggressive (potential dissolution of FeO, MnO)
PrEN 14429[184]	Proposed CEN Standard[d] Intrinsic Property Parallel-Batch Extraction (8 extracts)	HNO$_3$ or NaOH in DI water	Monitored Automatic acid/base.	5:1 (v/m)	0.3	24	ANC [meq/g]. Concentration. [mg/L] w/pH.	Similar in scope to pH Static Test, SR002.1, and ANC Test.

Test	Description	Leachant	Addition of	L/S (v/m)	MPS	CT	Measurement	Comments
Acid Neutralization Capacity (ANC) Test[12]	Published Test Intrinsic Property Parallel-Batch Extraction (11 extracts)	HNO₃ or NaOH in DI water	Addition of HNO₃ for range of final pH.	5:1 (v/m)	0.3	24	ANC [meq/g]. Concentration. [mg/L] w/pH.	Similar in scope to pH Static Test, SR002.1, and prEN 14429.
SR002.1[5] – Solubility and Release as a Function of pH	Published Test Intrinsic Property Parallel-Batch Extraction (11 extracts)	HNO₃ or NaOH in DI water	Addition of acid/base (2 < pH < 12)	10:1 (v/m)	0.3 2 5	18 48 168	ANC [meq/g]. Concentration. [mg/L] w/pH.	Similar in scope to ANC Test, pH Static Test, and prEN 14429.
SR003.1[5] – Solubility and Release as a Function of L/S	Published Test Intrinsic Property Parallel-Batch Extraction (5 extracts)	DI water	—	0.5:1, 1:1, 2:1 5:1 10:1 (v/m)	0.3 2 5	18 24 168	Concentration [mg/L]. Ionic strength [M].	Indicates ionic strength and competitive adsorption at L/S approaching pore water of porous media. Data similar to column test.

^{a}Leachant composition depends on disposal location: pH 4.2 or pH 5 for east and west of Mississippi River, respectively.

^{b}To be superceded by EN 12457-3.

^{c}Suite of four tests – number of extractions depends on the characteristics (solids content) of the waste material.

^{d}Directly derived from published protocol pH Static Leach Test.[179]

MPS = maximum particle size [mm].

CT = contact time [hr].

TABLE 10.2
Comparison of Selected Mass-Transport Tests

Test Name (reference)	Status, Application, Procedural Group	Leachant Renewal	L/S Ratio	Sample Dimension [mm]	Refresh Interval	Output Information	Comment/Criticism
ANSI/ANS-16.1[204]	U.S. Regulatory Standard; Intrinsic Property; Tank Leach (10 leachates)[a]	DI water Intermittent	10:1 (v/a)	Monolith >40 (min)	2, 5, 17 hr, 1, 1, 1, 14, 28, 43 d	Leachability Index as $\log(-1/D^{obs})$.	Similar in scope to NEN 7345 and MT001.1.
NEN 7345[203] – Dynamic Leach Test	Dutch Regulatory Standard; Intrinsic Property; Tank Leach (8 leachates)[b]	DI water Intermittent	5:1 (v/v)	Monolith >40 (min)	8 hr, 1, 2, 4, 9, 16, 36, 64 d	Observed diffusion coefficient (D^{obs}).	Similar in scope to ANSI/ANS-16.1 and MT001.1.
MT001.1[5] – Mass Transfer from Monolithic Materials	Published Test; Intrinsic Property; Tank Leach (7 leachates)[b]	DI water Intermittent	10:1 (v/a)	Monolith >40 (min)	2, 3, 3, 16 hr, 1, 2, 4 d	Effective diffusion coefficient (D^{eff}). Cumulative release [mg/m²].	Similar in scope to ANSI/ANS-16.1 and NEN 7345.
MT002.1[5] – Mass Transfer from Compacted Granular Materials	Published Test; Intrinsic Property; Tank Leach (7 leachates)[b]	DI water Intermittent	10:1 (v/a)	Granular <2 mm (max)	2, 3, 3, 16 hr, 1, 2, 4 d	Effective diffusion coefficient (D^{eff}). Cumulative release [mg/m²].	Tank leach test for granular materials

ASTM D 4874[208] – Column Extraction Method	U.S. Standard Field Mimicking Flow-Through Column Test (4 void volumes collected)	DI water Continuous	1, 2, 4, 8 void vol	Granular	8 days total (24 hr/vol)	Cumulative release [mg/kg] with L/S. Incremental conc. [mg/L] with L/S.	L/S ratios – short-, medium-, and long-term leaching. Channeling, variable residence time.
NEN 7343[209] – Column Leach Test	Dutch Regulatory Standard Field Mimicking Flow-Through Column Test (7 fractions collected)	DI water or dilute HNO_3 (pH 4) Continuous	0.1:1 – 10:1 (v/m)	Granular < 4 (max)	21 days total	Cumulative release [mg/kg] with L/S. Incremental conc. [mg/L] with L/S.	L/S ratios – short-, medium-, and long-term leaching. Up-flow minimizes channeling and plugging.
PrEN 14405[227] – Up-flow Percolation Test	Proposed CEN Standard Field Mimicking Flow-Through Column Test	DI water or dilute HNO_3 Continuous	0.1:1 – 10:1 (v/m)	Granular < 4 (max)	21 days total	Cumulative release [mg/kg] with L/S. Incremental conc. [mg/L] with L/S.	L/S ratios – short-, medium-, and long-term leaching. Up-flow minimizes channeling and plugging.

[a]Provisions for extended leaching intervals.
[b]Leachate collected in terms of void volumes rather than L/S ratio.

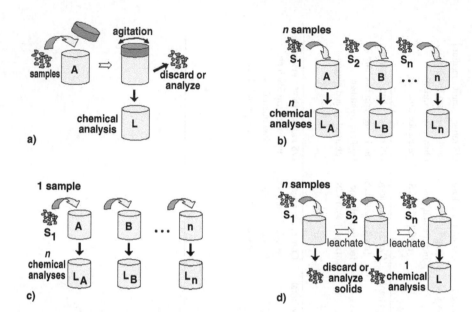

FIGURE 10.2 Schematic representations of extraction tests: a) agitated batch extraction test, b) parallel batch test, c) sequential chemical extraction test, and d) concentration buildup test.

Single-batch extraction tests shown in Table 10.1 include standardized protocols (e.g., TCLP,[176] SPLP,[177] ASTM 3987,[189] DIN 38 414 S4,[190] and EN 12457[175] parts 1, 2, or 4) and published tests (e.g., AV002.1[5]).

10.4.3.1.2 Parallel-Batch Extractions

Parallel-batch tests involve a series of single-batch extractions over a range of release conditions (A, B, C, …, n) as shown in Figure 10.2b. The goal of parallel testing is to represent constituent solubility and release over a range of test conditions typically by varying a single test parameter (e.g., amount of acid in the leachant, L/S ratio, or contact time). Leachate characteristics are usually compared among the n extractions as a function of the test variable. Common uses for parallel-batch extraction tests are to determine the ANC (Acid Neutralization Capacity Test,[12] prEN 14429,[182] SR002.1[5]), constituent solubility over a range of pH values (prEN 14429[184] and SR002.1[5]), and constituent solubility as a function of L/S ratios (EN 12457[175] part 3 and SR003.1[5]).

10.4.3.1.3 Sequential-Batch Extractions

Sequential-batch extraction tests are a family of equilibrium tests in which a single sample is challenged in a serial manner to several different leaching conditions. The basic procedure for sequential-batch extractions, shown in Figure 10.2c, consists of carrying a single solid sample, typically particle-size reduced to minimize mass-transport limitations, through a series of n extractions of specified test conditions. At the end of each extraction interval, the liquid and solid phases are separated via filtration or centrifugation and the solid phase is placed in fresh leachant for the subsequent extraction.

The leaching test variable (e.g., L/S ratio, leachant composition, contact time) may be identical or progressive for each sequential extraction. For example, a published "sequential TCLP"[40] challenges the solid sample to five consecutive batch extractions using TCLP solution at constant release conditions, while EN 12457-3[175] consists of two 3-hour DI water batch extractions at progressive L/S ratios (2 ml/g followed by 8 ml/g). Descriptions and uses of two specific variants on the sequential-batch extraction test approach follow.

10.4.3.1.3.1 L/S Ratio Tests

Concentrations in the leachate may be used to evaluate equilibrium conditions in the short and long term (e.g., to determine which mineral phases control equilibrium before and after soluble salts are released). However, serial batch tests do not provide useful information about the kinetic release rates from granular material during long-term leaching. Leachate concentrations from serial batch tests often are plotted as a function of cumulative L/S ratio and compared to the results of flow-through column tests. The published test, SR003.1,[5] is one example of a sequential extraction test varying L/S ratio at five levels between 10 ml/g and 0.5 ml/g.

10.4.3.1.3.2 Sequential Chemical Extractions

In sequential chemical extraction tests, a solid sample is carried through a series of extractions using increasingly aggressive leaching solutions. The purpose of sequential extraction schemes is to associate fractions of the release constituent (amount released/total content) to a particular extractant.

The most common example of a sequential chemical extraction procedure is that of Tessier et al.[191] designed for soil characterization and modified[40] to account for the high ANC of S/S wastes. Since release of constituents in a sequential extract test is operationally defined (e.g., peroxide extractable fraction), the speciation of the extracted fractions may not be well characterized (e.g., contaminant association with mineral functional groups may not correspond directly to the fractional release) and caution should be taken upon association of a fraction to a particular binding mechanism.

10.4.3.1.4 Concentration Buildup Tests

Concentration buildup tests involve extraction of multiple solid samples in the same leaching solution (Figure 10.2d). These tests are designed to obtain a leachate that is saturated in terms of all constituent concentrations. Thus, samples are typically particle-size reduced to maximize surface area and the volume of leaching solution typically is very small in comparison to other extraction tests. The physical analogy is that of an elemental volume of leachant percolating through a large column of waste material.

10.4.3.2 Mass-Transport Rate Tests

Mass-transport rate tests are used to obtain kinetic information about release prior to steady-state conditions (e.g., release flux, cumulative mass release with time, controlling mass-transport mechanisms). In the general procedure, a specified amount of solid sample is contacted with fresh leachant for a series of n leaching intervals (i.e., a series of distinct, specified contact times) without reaching steady-state release conditions.

At the end of each leaching interval, the solid phase is separated from the leaching solution, the leachate is collected, and the solid is placed in fresh leachant for the subsequent leaching interval. Leachates are analyzed for chemical composition and physical properties to assess the rates and mechanism of kinetic release. The specifics of different mass-transport rate tests vary in terms of the physical mode of water contact (e.g., batches, columns, lysimeters), the duration of leaching intervals, the specified leachant composition, and the method and frequency of leachate renewal.

The defining characteristic of a mass-transport rate test is the periodic renewal of leaching solution in order to maintain a maximal driving force for release from the matrix. The frequency of leachant renewal may be continuous as in a dynamic leaching test (e.g., a one-time pass through or over the solid) or may follow a predetermined schedule of intermittent leachant renewals, as in "semi-dynamic" leaching tests. The ratio of leachant to solid material is much larger than that of the equilibrium leaching test and often is based on the surface area of the subject material exposed to leaching. The physical state of the solid material may be monolithic or granular (e.g., soils, ashes, or particle-size-reduced monoliths), depending on the specifics of the testing protocol.

Within the category of mass-transport rate tests, four groups may be identified based on procedural differences:

- Flow-around tests
- Tank leaching tests
- Flow-through tests
- Soxhlet-type tests

10.4.3.2.1 Flow-Around Tests

Flow-around tests are fully dynamic mass-transport tests in which the leachant is pumped across, or allowed to flow past, the surface of a material and is collected on a continual basis. This dynamic mass-transport rate test may be conducted on either monolithic or granular materials as shown in Figure 10.3a. Although flow-around testing may closely represent the physical contact state for many placement scenarios, practical application of continuous leachant renewal is limited by analytical capabilities. For most S/S materials, the slow rates of diffusion through the solid matrix and high liquid volumes relative to released mass result in leachates that are well below analytical detection or practical quantifiable limits.

10.4.3.2.2 Tank Leaching Tests

Tank leaching tests are "semi-dynamic" mass-transport rate tests that essentially are sequential-batch extractions using large volumes of leachant. The leachant is intermittently renewed at intervals designed to maintain a significant diffusive driving force. In comparison to the dynamic flow-around tests, one advantage of tank leaching testing is that release concentrations tend to be more consistent with analytical capabilities. In addition, procedural simplicity is increased, as no pumps or elaborate gravity-fed dynamic flow systems are required.

Typically, tank leaching tests consist of a series of n leaching periods during which constituents are released from a sample suspended in a large volume of

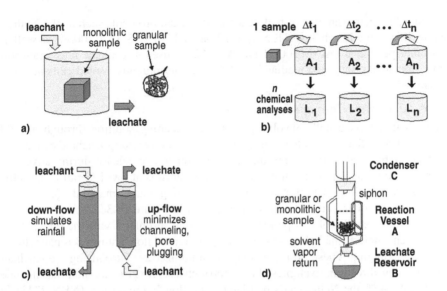

FIGURE 10.3 Schematic representations of mass-transport rate tests: a) flow-around test, b) tank leach test, c) flow-through column test, and d) Soxhlet testing apparatus.

leachant (Figure 10.3b). At the end of each contact time, the sample is removed and placed in a bath of fresh leachant for the subsequent leaching interval. Leachates are analyzed and the results are reported as mean flux (i.e., mass per unit surface area releases over the *i*th release interval) or cumulative release (i.e., total mass release per surface area over a summation of leaching intervals).

All of the tank leaching tests shown in Table 10.2 specify DI water as the leaching solution; however, some published research for S/S materials have used dilute inorganic or organic acids,[140,192-195] buffered DI water using bubbled gases,[196-198] DI water with ionic amendments,[195,199,200] or site-specific groundwater.[172] Although the typical form of the subject material is monolithic, granular materials may be tested in a similar manner after compaction into testing molds using a modified proctor compaction. For example, the MT002.1 (Mass Transport from Compacted Granular Materials Test)[5] has been designed to represent cases where granular wastes may be compressed into consolidated form at lower hydraulic conductivity than the surrounding fill.[9]

Considerable flexibility is offered by tank leaching tests in that leaching periods may be easily interspersed with conditioning outside of the leaching environment (e.g., period of storage or drying in gaseous atmosphere). This technique is especially advantageous for durability assessment of cement-based materials, since many degradation mechanisms, such as drying shrinkage, carbonation, and reinforcement corrosion, are associated with the drying process. In recent research, storage of portland cement mortar under constant atmospheric conditions has been used to assess the process of matrix drying[148] as well as the effects of gradient relaxation[199,200] and carbonation[170] on the release of major constituents (Ca, Na, K) and trace contaminants (As, Cd, Cu, Pb, Zn).

Standardized examples of tank leach tests are the protocols such as the Netherlands Monolithic Leach Test (NEN 7345)[203] and ANSI/ANS-16.1-2003[204] as well as a similar published test (MT001.1 and MT002.1)[5] used to evaluate mass transport for continuously saturated media[187,188,205,206] and intermittently wetted cement materials.[170,199,207]

10.4.3.2.3   Flow-Through Tests

Flow-though tests usually involve passing the leaching solution through a solid material and collecting the leachate after contact. The resulting leachate concentrations may be used to determine the rates of constituent release during advective mass transport and to infer primary release mechanisms at low L/S ratios. Usually, flow-through testing is performed on columns of granular materials (e.g., soils, incinerator residues, crushed cements) as shown in Figure 10.3c.

In column studies of low-permeability materials, problems with the flow pattern may be encountered including wall effects, preferential flow pathways (channeling), and pore plugging. These problems may be minimized by introducing the leachant in an up-flow direction. Examples of flow-through tests for granular materials include ASTM 4874,[208] the Netherlands regulatory "up-flow" column test (NEN 7343),[209] and several published field lysimeter tests using cementitious materials.[210,211]

Flow-through testing for monolithic materials with low hydraulic conductivity is limited to permeameter experiments at pressures up to 300 kPa.[212,213] Regressed diffusion coefficients have been shown to be 7 to 10 orders of magnitude faster in high-pressure flow-through experiments than in the more standard tank leaching protocols. Moreover, the largest reported diffusion coefficient was nearly 2 orders of magnitude higher than molecular ionic diffusion, indicating confusion between regressed advection and diffusion terms. This flow-through experimental approach was allegedly applicable to simulate the leaching process of S/S materials either under the landfill scenario (where the waste is more permeable than the surrounding materials) or when the S/S waste form has degraded to a state that groundwater may pass through the matrix. However, the maximum pressures used in these experiments equate to a pressure head of over 30 m of water at 4°C, which is unrealistic for most placement scenarios for S/S wastes.

10.4.3.2.4   Soxhlet Tests

In a Soxhlet test, a solid sample is continuously contacted with fresh leachant by continuously removing constituents from the leachate. Figure 10.3d shows a Soxhlet test apparatus consisting of a reaction vessel (A), a leachate reservoir (B) where leachate is boiled to produce water vapor, and a condenser (C) where fresh leachant condenses before contact with the solid material. Few published Soxhlet procedures were found in the literature for extraction of inorganic constituents from a synthetic S/S waste.[214,215]

10.4.4   Variables Affecting Leaching Test Results

The testing parameters that most affect the outcome of leaching protocols include (i) leachant composition, (ii) pH control, (iii) sample particle size, (iv) L/S ratio, and (v) open or closed testing conditions.

10.4.4.1 Leachant Composition

In Section 10.3.2.3, the effect of leachant composition on the leaching process has been discussed in detail with respect to groundwater contact. Although the same factors apply to leach testing, the amount of acid added to the extraction solution is the most often varied leaching test parameter in laboratory testing protocols. The intent of acid addition is to create release conditions that neutralize the buffering capacity of the matrix, accelerate dissolution of hydroxide-bearing minerals, and alter the solubility of pH-dependent species. However, the amount of acid specified in field-mimicking leaching tests (e.g., TCLP and SPLP) is too small to control the final pH of the leaching solution. For example, Stegemann et al. report an ANC of 16 meq/g for neat portland cement[65] compared to a maximum of 2 meq/g provided by the TCLP solution.[12] The limited acidity in the extraction fluid results in a near-neutral to slightly alkaline pH, where constituent concentrations of many RCRA-regulated metals are near the minimum of the solubility curve.

Furthermore, the final pH of the extract for several leaching protocols (e.g., TCLP) is not expressly required to be measured or reported. Without a final pH value, little scientific information is provided by the leaching test, as there is no basis for comparison between treatment recipes, no assurance that steady-state conditions have been obtained, and no method of quality control evaluation. Thus, measurement of the final pH and leachate conductivity (as a measure of ionic strength) of all analytical samples is essential in to order to glean useful information regarding solution chemistry from leaching protocols.

10.4.4.2 pH Control

Laboratory tests frequently require strong acids to adjust pH in the leachate to a specified pH. Several tests specify automated pH controllers, which monitor solution conditions and maintain pH to a set value by delivering small quantities of acid or base (e.g., EN 12457, pH Static Test). An alternative method to cumbersome and costly pH controllers is specified in parallel batch extraction tests (e.g., ANC, SR002.1, AV001.1) using predetermined acid additions based on a preliminary titration curve. However, simply dumping in the acid may result in localized areas of low pH, possibly resulting in dissolution of mineral phases and release of some contaminants that otherwise would not have been released using automated pH control.

10.4.4.3 Particle-Size/Contact Time

Most equilibrium-based leaching tests assume that the extraction process reaches steady-state conditions within the time frame specified by the protocol. To approach this end-state, several of these tests require particle-size reduction of samples prior to testing. Since release from larger particles may be mass transfer-limited over the abbreviated time frame for which leaching tests are designed, tests with larger particle-size specifications may require longer contact time to obtain equilibrium or may not reach steady state within the test duration.

The diffusion-limited release of constituents during equilibrium-based tests may be studied using the analytical solution for diffusion from a spherical particle into

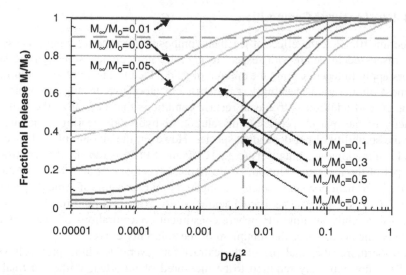

FIGURE 10.4 Fractional release (M_t/M_∞) into a fixed bath as a function of a dimensionless parameter Dt/a^2 and fractional solubility (M_∞/M_o) modified from Garrabrants.[201]

a fixed bath.[216] Figure 10.4 shows the fractional release (i.e., the release with changing time relative to the release at infinite time) for a constituent released into a fixed bath. Parameters that determine the rate of release include the fractional solubility (i.e., the ratio between the amounts of constituent soluble in the fixed bath to the total amount) and a dimensionless parameter determined as the observed diffusivity (D) multiplied by the contact time (t) and divided by the square of the particle radius (a).

If the 90% fractional release ($M_t/M_\infty = 0.9$) is assumed to represent equilibrium and a constituent is known to have a fractional solubility in the liquid phase of 5% ($M_\infty/M_o = 0.05$), the minimum value of the dimensionless parameter Dt/a^2 that will attain equilibrium is 0.005. Thus, extraction of particles sized using the TCLP maximum (2a = 9.5 mm) would require a minimum of approximately 300 hours to establish equilibrium of constituents with a mid-range diffusivity (D = 10^{-13} m²/s). Particle-size reduction to 0.6 mm (2a = 0.6 mm) would accelerate the release process to a minimum of 14 hours. Of course, this first-order analysis assumes that diffusion is the rate-limiting process and that all particles are of the same diameter.

10.4.4.4 Liquid-to-Solid Ratio

The amount of liquid relative to the mass of solid dictates the controlling mechanism of release as detailed in the discussion of infiltration rates (see Section 10.3.2.2). Leaching tests designed to obtain equilibrium or steady-state release usually consist of batch extractions at L/S ratios in the range of 5 to 20 mL/g of dry material (e.g., TCLP, ANC, SR003.1). At these L/S ratios, the release of highly soluble species (e.g., Na^+, K^+, Cl^-) is assumed to be limited by the solid content (availability-limited), while the release of sparingly soluble species may be constrained by solubility in the liquid phase. Equilibrium at L/S ratios below 5 mL/g are used in some leaching tests (e.g., flow-through column tests and L/S ratio batch tests) to infer pore water composition. Several researchers have used high pressures applied to cementitious materials to extract pore solution at L/S ratios much below practical laboratory batch extractions.[178,217-221]

In order to relax solubility constraints (e.g., to measure availability of sparingly soluble constituents), a higher L/S ratio or a shift in solution pH is required. Thus, leaching tests used to measure release flux (e.g., ANSI/ANS-16.1, MT001.1) utilize large L/S ratios, often specified on the basis of liquid-to-surface area ratios.

10.4.4.5 Open/Closed Systems

When tests are conducted in open containers, exchange between the atmosphere and the test system (i.e., solid and liquid) may lead to ambiguous results. For example, tank leaching in vessels that are not air-tight may lead to neutralization of the leachate and precipitation on the solid specimen via carbonation. During equilibrium-based testing, addition of acid aliquots to open vessels may alter the resultant equilibrium by allowing the escape of evolved gases from acid-reactive solids (e.g., carbonated S/S materials). Therefore, control and definition of the testing conditions are essential to accurate interpretation of leach testing results.

10.5 LEACHING DATA PRESENTATION AND INTERPRETATION

Data obtained from leaching tests must be presented and interpreted in the context of the goal of each test and the applied testing conditions. Consistency should be checked between (i) different equilibrium tests and (ii) concentrations in mass-transport rate tests and pH-driven solubility limitations.

10.5.1 EQUILIBRIUM-BASED TESTS

Equilibrium-based data may be presented in terms of leachate concentration (mg/L) (Figure 10.5a) or in terms of the amount of constituent released from the solid phase (mg/kg) (Figure 10.5b). Release is calculated by multiplying the concentration by the L/S ratio of the leaching test. Schematic test data for the following three extraction tests are compared:

1. Solubility/release data as a function of pH (solid line) as might result from the SR003.1 or prEN 14429 protocols
2. Solubility/release data as a function of L/S ratio as determined from sequential extractions for L/S 2 mL/g (○,●) and L/S 8 ml/g (□,■) as might result from EN 12457-3 protocol at natural pH (data shown for solubility-limited release using open symbols and availability-limited release using closed symbols)
3. Available content, or potential leachable amount, determined at pH 4 (Δ) as might be provided by AV001.1 or NEN 7341.

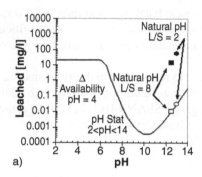 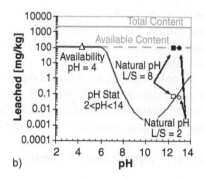

a)   b)

FIGURE 10.5 Schematic leaching data for equilibrium tests at different L/S ratios shown as a) leachate concentration [mg/L] and b) release [mg/kg]. Test data are shown for (i) solubility/release of sparingly soluble constituent as a function of pH (solid line), (ii) solubility/release data as a function of L/S ratio for solubility-limited release (○,□) and availability-limited release (●,■), and (iii) available content plotted at a pH 4 (Δ).

For most sparingly soluble constituents (e.g., most regulated metals and other pollutants), the shape of the solubility curve as a function of pH typically represents availability-limited release at low pH values and solubility-limited release over a controlling solid phase at alkaline pH. Highly soluble constituents (e.g., Na⁺, K⁺, Cl⁻) tend to be availability-limited throughout the entire pH range. As Figure 10.5a indicates, differences in concentration of extractions conducted at different L/S ratios usually are not evident for solubility-limited constituents, unlike that of availability-limited constituents. However, these L/S ratio differences become obvious when solubility-limited leaching is presented on the basis of mass released from the solid phase (Figure 10.5b) and compared to the available content and total elemental content. Leaching results reported as mass released normalize the dilution effect of L/S ratio and allow for comparison of leaching data derived from extractions at different L/S ratios.[174]

Figure 10.6 shows the effects of different retention mechanisms on the equilibrium-based release data. In the alkaline pH range, values of release may be influenced by (i) complexation due to dissolved organic content, chelants, or chlorides, (ii) reduction or oxidation of constituents, or (iii) adsorption to mineral surfaces. At low pH values, the value for availability-limited release is shown as a flattened section of the curve and may be influenced by re-mineralization within the solid material.

10.5.2 Mass-Transport Tests

Kinetic data from mass-transport tests usually are presented in terms of flux [mg/m² s] or cumulative release [mg/m²] as shown in Figure 10.7. The data shown in this figure are common for mass-transport tests such as MT001.1 or ANSI/ANS-16.1-2003. Flux may be defined as a measure of the release rate specific to the interfacial surface area, while cumulative release represents the sum of constituent mass per unit surface area, which has been released from the matrix from the start of leaching.

Figure 10.8 shows schematic cumulative mass-transport release data for S/S materials with different release mechanisms. The conditions of the dynamic test are

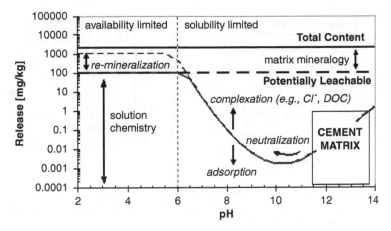

FIGURE 10.6 Factors controlling equilibrium-based leaching of metal constituents in different environments (modified from literature[10]).

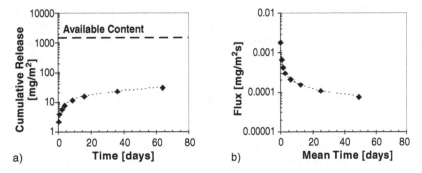

FIGURE 10.7 Schematic data from mass-transport rate tests presented as a) constituent cumulative release [mg/m²] and b) constituent flux [mg/m² s].

designed to study the physical rate-limiting mechanisms of release. As such, a maximum driving force for diffusion with no significant buildup of leachate concentration is assumed. This assumption may be valid only if leachate concentrations are much less than the solubility concentration at the leachate pH (Figure 10.8b). All mass-transport test concentrations in Figure 10.8b are below solubility limits. However, the concentration in Leachate #2 is only slightly different from the solubility concentration, indicating a potentially reduced driving force for leaching during the second leaching interval.

10.5.2.1 Diffusivity in Reactive Media

Typically, the constituent diffusion rate through a matrix is characterized by a diffusivity value; however, this approach is only valid for a limited number of cases for which (i) the observed diffusion coefficient is constant in space and time, (ii) the constituent of concern is not pH-dependent or no large pH gradients exist,

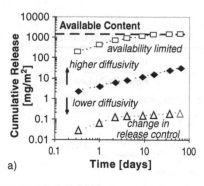
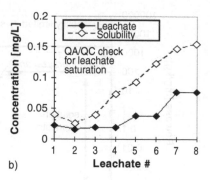

a) Time [days] b) Leachate #

FIGURE 10.8 Mass-transport rate test data consistency: a) the effect of release mechanisms on cumulative release data and b) comparison of leachate concentrations to solubility constraints as a function of leachate pH.

(iii) constituent depletion does not occur, and (iv) matrix degradation does not significantly alter the matrix properties influencing leaching.

The determination of a characteristic diffusivity using release information is somewhat uncertain because of ambiguity in the representation of the diffusion coefficient. The relationship between observed, effective, and molecular diffusivity[218] is shown in the following expression:

$$D^{obs} = \frac{D^{eff}}{R} = \frac{D^{mol}}{\tau \cdot R}$$

(10.3)

Molecular diffusivity, D^{mol}, is the term denoting the rate of change in the concentration gradient for a species diffusing in an aqueous media. Diffusion in a non-reactive porous medium retards the release of constituents in relationship to molecular diffusion through effective surface area and increased path length, or tortuosity τ. The effective diffusivity, D^{eff}, is used to describe the mass transport when physical retardation is present. In chemically reactive systems, the rate of diffusion for a constituent is slowed due to interactions with the surface, or chemical retardation R. One measure of diffusivity in chemically reactive systems is the observed, or apparent, diffusivity D^{obs}.

10.5.2.2 Regression of Diffusivity from Leaching Data

The most common method for calculating diffusivity from leaching data assumes that release of constituents is diffusion-controlled with constant observed diffusivity and that a simple diffusion model may be used to describe constituent release. The diffusion model is characterized by two parameters; the initial concentration of the constituent in the waste form and the diffusivity of the constituent through the solid matrix. The driving force for diffusion is the difference between the constituent concentration in the bulk solid phase and the leachate.

Under assumed simple diffusion controlled release, Fick's second law may be fit to cumulative release data to calculate an observed diffusivity, D^{obs} [m²/s], using either of the following two simple equations for semi-infinite media:

$$D^{obs} = \pi \left[\dfrac{a_n/A_o}{(\Delta t)_n} \right]^2 \left[\dfrac{V}{S} \right]^2 \left[\dfrac{1}{2}\left(\sqrt{t_n} + \sqrt{t_{n-1}}\right) \right]^2 \qquad (10.4a)$$

where a_n/A_o is the fractional release [–] during the time interval $(\Delta t)_n$ [s] from t_{n-1} to t_n;
  V is the volume of the solid phase [m³]; and
  S is the solid surface area in contact with the leachant [m²].

$$D^{obs} = \pi \left(\dfrac{M}{2\rho \cdot C_o\left(\sqrt{t_n} - \sqrt{t_{n-1}}\right)} \right)^2 \qquad (10.4b)$$

where M is the cumulative release [mg/m²] during the time interval from t_{n-1} to t_n [s];
  ρ is the density of the solid phase [kg/m³]; and
  C_o is the leachable content at the start of leaching [mg/kg].

Equation 10.4a was developed in concert with the ANSI/ANS-16.1-2003 protocol for radionuclide release,[204] while Equation 10.4b was developed for heavy metal and matrix constituent release using NEN 7345.[222] Both of these equations are based on the same diffusion model assuming (i) a homogenous solid matrix, (ii) constant release parameters, (iii) constant driving force for diffusion (i.e., zero concentration at the boundary), (iv) no depletion of the constituent at the core of the matrix, and (v) less than 20% of the initial content is leached (allows use for finite media of the solution for semi-infinite media given in the equations).

10.5.2.3 Changes in Mass-Transport Behavior

The condition of a solid matrix after long-term environmental leaching often is simulated by subjecting the matrix to extremely aggressive leachants over a much shorter time interval under the assumption that the same conditions would apply to the matrix in the long term. Thus, the research of acid attack of cement materials showed that the chemical and morphological characteristics are greatly affected by contact with aggressively acidic leachants (see Section 10.2.2.3.1). Several other long-term leaching studies have shown dual release regimes (i.e., a period of fast release kinetic followed by a slower, residual constituent flux), indicating that the behavior of mass-transport-based release from cementitious matrices may change with time or with constituent concentration.[223,224]

In light of these observations and the potential for degradation of the cement matrices in the long term (see Section 10.3.3), the assumptions of the diffusion model, namely constant diffusive properties and a nondepleting matrix, may be concluded invalid in the long term for cementitious matrices. Forcing the cumulative release from S/S matrices to fit the two parameters of the diffusion model, and then projecting those parameters forward in time without regard for chemical or physical changes, is likely to lead to errors in prediction of long-term constituent release, and the simple equations given above cannot be extrapolated beyond a cumulative release of 20%. The solutions for finite media for Fick's second law can be used to correct for a nondepleting media to project forward in time, but do not guarantee that the other assumptions and boundary conditions hold true; i.e., the math can calculate the fraction released for long times for a depleting media, but may deviate significantly from the actual behavior for reasons not obvious from short-term leach testing.

When the diffusion model does not adequately describe contaminant release, more complex models are required to capture the effect of adsorptive or encapsulating mineral phase dissolution on the leaching process. Two such models, the Coupled Dissolution-Diffusion (CDD) model[184] and the Shrinking Un-Reacted Core model,[223] are based on the movement of dissolution fronts within the matrix.

10.6 LEACHING TEST SELECTION

Selection of an appropriate leaching test should be based on the (i) level of release information detail required, (ii) cost of testing including analytical services, data handling, and interpretation, (iii) objective(s) of candidate testing procedures, and (iv) environmental conditions anticipated in the release scenario.

In most cases, more than one leaching test will be required to fully describe release under a particular scenario. Leach characterization and long-term release assessment should also consider changes in chemical and physical properties due to degradation processes via testing of aged and degraded materials. For example, samples may be aged in the laboratory using accelerated carbonation techniques prior to testing in parallel to un-aged materials.[170] In addition to analysis of trace contaminants of the waste material, analytical work on leachates must include (i) leachate pH in order to put solubility, availability, sorption, and complexation processes in perspective, (ii) leachate conductivity as a measure of ionic strength, and (iii) concentrations of major constituents indicative of the structural durability of the S/S matrix. The selection process for leaching test protocols may be illustrated through the two simplified examples for (i) landfill disposal and (ii) near surface placement of S/S materials.

10.6.1 LEACHATE-CONTROLLED MONOFILL EXAMPLE

For an S/S material to be disposed in a monofill scenario with leachate control (e.g., hazardous waste landfill with RCRA-approved caps and liners), the following statements may be considered:

- The infiltration rate will be low due to leachate controls; thus, a low L/S ratio prevails.

- The small amount of leachant that contacts the waste will be highly alkaline (i.e., in the range of the cement pore water) due to percolation through or contact with cementitious wastes above the subject material.
- The concentration in the infiltrating leachant will tend to come to a steady state with the contaminants in the waste as a result of slow percolation rates.

Under these assumed conditions, low L/S ratio leaching tests designed to evaluate equilibrium at the pH dictated by the ANC of the waste would provide an estimate of percolating leachate in the short term. Higher L/S ratio tests would indicate the effect of soluble species depletion on equilibrium chemistry. Suggested tests would include batch tests like EN 12457, SR003.1, and ASTM D 3987 or column tests like NEN 7343, ASTM D 4874, and prEN 14405.

10.6.2 UNCONTROLLED NEAR-SURFACE PLACEMENT

If a solidified material is used in a near-surface placement without leachate controls, several conditions may exist that have significant consequences to constituent release. Examples of such a situation may include beneficial reuse of stabilized wastes in construction (e.g., as roadbase or backfill) or *in-situ* S/S treatment of contaminated soils without leachate diversion. For the above release scenario, the following statements may be considered:

- The cementitious material may be subjected to high volumes of infiltrating water on an intermittent wetting basis (i.e., dependent on local precipitation patterns and volumes).
- The contact mode for infiltrating liquid phases may be either percolation (e.g., loose granulated fill) or flow-around (e.g., compacted granular or monolithic materials).
- Release rates may be controlled by equilibrium or mass transport, depending on infiltration rates, leachant volume, and contact time.
- Degradation via chemical (e.g., carbonation, chloride penetration, sulfate attack) or physical (e.g., cyclic loading, freeze/thaw) mechanisms will change equilibrium- and mass-transport-based release parameters.

The most complete approach for testing in the above scenario would be to combine the results of an equilibrium-based test over a range of pH values with a mass-transport test for a monolithic test sample. Equilibrium testing will allow for consideration of acid neutralization consumption from infiltrating acidic components while the mass-transport tests would provide information on rates of constituent release from the monolithic material. Suggested protocols would include pH-dependent tests (e.g., prEN 14429, SR002.1, and Acid Neutralization Capacity Test) and dynamic tests for monolithic materials (e.g., ANSI/ANS-16.1, NEN 7345, and MT001.1).

REFERENCES

1. Conner, J.R., *Chemical Fixation and Solidification of Hazardous Wastes*, Van Nostrand Reinhold, New York, 1990.
2. Conner, J.R. and Hoeffner, S.L., The history of solidification stabilization technology, *Critical Reviews in Environmental Science and Technology* 28, 325, 1998.
3. USEPA, Leachability Phenomena, EPA-SAB-EEC-92-003, USEPA Science Advisory Board, Washington, D.C., 1991.
4. USEPA, Waste Leachability: the Need for Review of Current Agency Procedures, EPA-SAB-EEC-COM-99-002, USEPA Science Advisory Board, Washington, D.C., 1999.
5. Kosson, D.S. et al., An integrated framework for evaluating leaching in waste management and utilization of secondary materials, *Environmental Engineering Science* 19, 159, 2002.
6. Lewin, K., Leaching tests for waste compliance and characterisation: recent practical experiences, *The Science of The Total Environment* 178, 85, 1996.
7. Glasser, F.P., Chemistry of cement-solidified waste forms, in *Chemistry and Microstructure of Solidified Waste Forms*, Spence, R.D., Ed., Lewis Publishers, Baton Rouge, LA, 1993, p. 1.
8. Garrabrants, A.C., Sanchez, F., and Kosson, D.S., Changes in constituent equilibrium leaching and pore water characteristics of a Portland cement mortar as a result of carbonation, *Waste Management* 24, 19, 2004.
9. Kosson, D.S., van der Sloot, H.A., and Eighmy, T.T., An approach for estimation of contaminant release during utilization and disposal of municipal waste combustion residues, *Journal of Hazardous Materials* 47, 43, 1996.
10. van der Sloot, H.A., Comparison of the characteristic leaching behaviour of cements using standard (EN 196-1) cement-mortar and an assessment of their long-term environmental behaviour in construction products during service life and recycling, *Cement and Concrete Research* 30, 1079, 2000.
11. van der Sloot, H.A. et al., Approach towards international standardization: a concise scheme for testing of granular waste leachability, in *Environmental Aspects of Construction with Waste Materials*, Goumans, J.J.J.M., van der Sloot, H.A., and Aalbers, T.G., Eds., Elsevier Science, Amsterdam, 1994, p. 453.
12. Stegemann, J.A. and Côté, P.L., Investigation of test methods for solidified waste evaluation — a cooperative program, EPS 3/HA/8, 1991.
13. Poon, C.-S. et al., Comparison of the strength and durability performance of normal- and high-strength pozzolanic concretes at elevated temperatures, *Cement and Concrete Composites* 31, 1291, 2001.
14. Piltner, R. and Monteiro, P.J.M., Stress analysis of expansive reactions in concrete, *Cement and Concrete Composites* 30, 843, 2000.
15. Gérard, B. and Marchand, J., Influence of cracking on the diffusion properties of cement-based materials. Part I. Influence of continuous cracks on the steady-state regime, *Cement and Concrete Research* 30, 37, 2000.
16. Basheer, P.A.M., Chidiac, S.E., and Long, A.E., Predictive models for deterioration of concrete structures, *Construction and Building Materials* 10, 27, 1996.
17. Walton, J.C. and Seitz, R.R., Fluid flow through fractures in below ground concrete vaults, *Waste Management* 12, 179, 1992.
18. Fu, Y.F. et al., Thermal induced stress and associated cracking in cement-based composite at elevated temperatures. Part II. Thermal cracking around multiple inclusions, *Cement and Concrete Composites* 26, 113, 2004.

19. Tijssens, M.G.A., Sluys, L.J., and van der Giessen, E., Simulation of fracture of cementitious composites with explicit modeling of microstructural features, *Engineering Fracture Mechanics* 68, 1245, 2001.

20. Mainguy, M., Ulm, F.-J., and Heukamp, F.H., Similarity properties of demineralization and degradation of cracked porous materials, *International Journal of Solids and Structures* 38, 7079, 2001.

21. Wang, K. et al., Permeability study of cracked concrete, *Cement and Concrete Research* 27, 381, 1997.

22. Tittlebaum, M.E., Seals, R.K., and Cartledge, F.K., State of the art on stabilization of hazardous organic liquid wastes and sludges, *Critical Reviews in Environmental Control* 15, 179, 1985.

23. Mollah, M.Y.A. et al., The interfacial chemistry of solidification/stabilization of metals in cement and pozzolanic material systems, *Waste Management* 15, 137, 1995.

24. Johnson, C.A., Metal Binding in the Cement Matrix: An Overview of Our Current Knowledge, Department of Water Resources and Drinking Water, Water-Rock Interaction Group, EAWAG for Cemsuisse, Switzerland, 2002.

25. Lea, F.M., *The Chemistry of Cement and Concrete*, 3rd ed. Chemical Publishing Co., Inc., New York, NY, 1970.

26. Taylor, H.F.W., *Cement Chemistry*, 2nd ed. Thomas Telford Publishing, London, 1997.

27. Mollah, M.Y.A., Parga, J.R., and Cocke, D.L., An infrared spectroscopic examination of cement-based stabilization/solidification systems — portland types V and IP with zinc, *Journal of Environmental Science and Health, Part A* 27, 1503, 1992.

28. Papadakis, V.G., Vayenas, C.G., and Fardis, M.N., Reaction engineering approach to the problem of concrete carbonation, *AICHE Journal* 35, 1639, 1989.

29. Baur, I. et al., Dissolution-precipitation behaviour of ettringite, monosulfate, and calcium silicate hydrate, *Cement and Concrete Research* 34, 341, 2004.

30. Cocke, D.L., The binding chemistry and leaching mechanisms of hazardous substances in cementitious stabilization/solidification systems, *Journal of Hazardous Materials* 24, 231, 1990.

31. Cocke, D.L. et al., Binding chemistry and leaching mechanisms in solidified hazardous wastes, *Journal of Hazardous Materials* 28, 193, 1991.

32. Gougar, M.L.D., Scheetz, B.E., and Roy, D.M., Ettringite and C-S-H Portland cement phases for waste ion immobilization: a review, *Waste Management* 16, 295, 1996.

33. Conner, J.R., Chemistry of cementitious solidified/stabilized waste forms, in *Chemistry and Microstructure of Solidified Waste Forms*, Spence, R.D., Ed., Lewis Publishers, Baton Rouge, LA, 1993, p. 41.

34. Batchelor, B. and Wu, K., Effects of equilibrium chemistry on leaching of contaminants from stabilized/solidified wastes, in *Chemistry and Microstructure of Solidified Waste Forms*, Spence, R.D., Ed., Lewis Publishers, Baton Rouge, LA, 1993, p. 243.

35. Bonen, D. and Sarkar, S.L., The present state-of-the-art of immobilization of hazardous heavy metals in cement-based materials, in *Advances in Cement and Concrete: Proceedings of an Engineering Foundation Conference July 24–29, 1994 New England Center, University of New Hampshire*, Grutzeck, M.W. and Sarkar, S.L., Eds., American Society of Civil Engineers, New York, NY, 1994.

36. Richardson, I.G., The nature of C-S-H in hardened cements, *Cement and Concrete Research* 29, 1131, 1999.

37. Cartledge, F.K. et al., Immobilization mechanisms in solidification/stabilization of Cd and Pb salts using Portland cement fixing agents, *Environmental Science and Technology* 24, 867, 1990.

38. Brown, P.W. and Badger, S., The distributions of bound sulfate and chlorides in concrete subjected to mixed NaCl, MgSO₄, Na₂SO₄ attack, *Cement and Concrete Research* 30, 1535, 2000.

39. Taylor, H.F.W., Famy, C., and Scrivener, K.L., Delayed ettringite formation, *Cement and Concrete Research* 31, 683, 2001.

40. Li, X.D. et al., Heavy metal speciation and leaching behaviors in cement based solidified/stabilized waste materials, *Journal of Hazardous Materials* A82, 215, 2001.

41. Kumarathasan, P. et al., Oxyanion substituted ettringites: synthesis and characterization; their potential role in immobilization of As, B, Cr, Se, and V, in *Fly Ash and Coal Conversion By-Products: Characterization, Utilization, and Disposal VI*, Materials Research Society, Boston, MA, 1990, p. 80.

42. Toyohara, M. et al., Contribution to understanding iodine sorption mechanism onto mixed solid alumina cement and calcium compounds, *Journal of Nuclear Science and Technology* 39, 950, 2002.

43. van der Sloot, H.A., Characterization of the leaching behaviour of concrete mortars and of cement stabilized wastes with different waste loading for long-term environmental assessment, *Waste Management* 22, 181, 2002.

44. Glasser, F.P., Fundamental aspects of cement solidification and stabilisation, *Journal of Hazardous Materials* 52, 151, 1997.

45. Asavapisit, S., Nanthamontry, W., and Polprasert, C., Influence of condensed silica fume on the properties of cement-based solidified wastes, *Cement and Concrete Research* 31, 1147, 2001.

46. Wu, Z. and Naik, T.R., Properties of concrete produced from multicomponent blended cements, *Cement and Concrete Composites* 32, 1937, 2002.

47. Papadakis, V.G., Effect of fly ash on Portland cement systems. Part I. Low-calcium fly ash, *Cement and Concrete Research* 29, 1727, 1999.

48. Papadakis, V.G., Experimental investigation and theoretical modeling of silica fume activity in concrete, *Cement and Concrete Research* 29, 79, 1999.

49. Escalante-García, J.I. and Sharp, J.H., The microstructure and mechanical properties of blended cements hydrated at various temperatures, *Cement and Concrete Research* 31, 695, 2001.

50. Dermatas, D. and Meng, X., Utilization of fly ash for stabilization/solidification of heavy metal contaminated soils, *Engineering Geology* 70, 377, 2003.

51. Thevenin, G. and Pera, J., Interactions between lead and different binders, *Cement and Concrete Research* 29, 1605, 1999.

52. Bágel', L. and Zivica, V., Relationship between pore structure and permeability of hardened cement mortars: on the choice of effective pore structure parameter, *Cement and Concrete Research* 27, 1225, 1997.

53. Ivey, D.G. et al., Electron microscopy characterization techniques for cement solidified/stabilized metal wastes, in *Chemistry and Microstructure of Solidified Waste Forms*, Spence, R.D., Ed., Lewis Publishers, Baton Rouge, LA, 1993, p. 123.

54. Roy, D.M. and Scheetz, B.E., The chemistry of cementitious systems for waste management: the Penn State experience, in *Chemistry and Microstructure of Solidified Waste Forms*, Spence, R.D., Ed., Lewis Publishers, Baton Rouge, LA, 1993, p. 83.

55. Mohr, P. et al., Transport properties of concrete pavements with excellent long-term in-service performance, *Cement and Concrete Research* 30, 1903, 2000.

56. Shafiq, N. and Cabrera, J.G., Effects of initial curing condition on the fluid transport properties in OPC and fly ash blended cement concrete, *Cement and Concrete Composites* 26, 381, 2004.

57. Katz, A.J. and Thompson, A.H., Quantitative prediction of permeability in porous rock, *Physical Review B* 34, 8179, 1986.
58. Christensen, B.J., Mason, T.O., and Jennings, H.M., Comparison of measured and calculated permeabilities for hardened cement pastes, *Cement and Concrete Research* 26, 1325, 1996.
59. El-Dieb, A.S. and Hooten, R.D., Evaluation of the Katz-Thompson model for estimating the water permeability of cement-based materials from mercury intrusion porosimetry data, *Cement and Concrete Research* 24, 443, 1994.
60. Tumidajski, P.J. and Lin, B., On the validity of the Katz-Thompson equation for permeabilities in concrete, *Cement and Concrete Research* 28, 643, 1998.
61. McCarter, W.J., Starrs, G., and Chrisp, T.M., Electrical conductivity, diffusion, and permeability of Portland cement-based mortars, *Cement and Concrete Research* 30, 1395, 2000.
62. Perkins, R.B. and Palmer, C.D., Solubility of ettringite ($Ca_6[Al(OH)_6]_2(SO_4)_3 \cdot 26H_2O$) at 5-75°C, *Geochimica et Cosmochimica Acta* 63, 1969, 1999.
63. Damidot, D. and Glasser, F.P., Investigation of the $CaO-Al_2O_3-SiO_2-H_2O$ system at 25°C by thermodynamic calculations, *Cement and Concrete Research* 25, 22, 1995.
64. Stronach, S.A. and Glasser, F.P., Modelling the impact of abundant geochemical components on phase stability and solubility of the $CaO-SiO_2-H_2O$ system at 25°C: Na^+, K^+, SO_4^{2-}, Cl^-, and CO_3^{2-}, *Advances in Cement Research* 9, 167, 1997.
65. Stegemann, J.A., Shi, C., and Caldwell, R.J., Response of various solidification systems to acid addition, in *International Conference on the Environmental and Technical Implications of Construction with Alternative Materials, WASCON '97*, Goumans, J.J.J.M., Senden, G.J., and van der Sloot, H.A., Eds., Elsevier Science, B.V., Houthem St. Gerlach, The Netherlands, 1997.
66. Glass, G.K. and Buenfeld, N.R., Differential acid neutralisation analysis, *Cement and Concrete Research* 29, 1681, 1999.
67. Glass, G.K., Reddy, G., and Buenfeld, N.R., Corrosion inhibition in concrete arising from its acid neutralisation capacity, *Corrosion Science* 42, 1587, 2000.
68. Stegemann, J.A. and Buenfeld, N.R., Prediction of leachate pH for cement paste containing pure metal components, *Journal of Hazardous Materials* 90, 169, 2002.
69. Bonen, D. and Sarkar, S.L., The effects of simulated environmental attack on immobilization of heavy metals doped in cement-based materials, *Journal of Hazardous Materials* 40, 321, 1995.
70. Carde, C. and François, R., Modelling the loss of strength and porosity increase due to the leaching of cement pastes, *Cement and Concrete Research* 21, 181, 1999.
71. Barth, E.F. et al., *Stabilization and Solidification of Hazardous Wastes,* Noyles Data Corporation, Park Ridge, NJ, 1990.
72. Baker, P.G. and Bishop, P.L., Prediction of metal leaching rates from solidified/stabilized wastes using the shrinking unreacted core leaching procedure, *Journal of Hazardous Materials* 52, 311, 1997.
73. Massry, I.W., The Impact of Micropore Diffusion on Contaminant Transport and Biodegradation Rates in Soils and Aquifer Materials, Ph.D. Dissertation, Rutgers, The State University of New Jersey, New Brunswick, NJ, 1997.
74. Cheng, K.Y. and Bishop, P.L., Sorption, important in stabilized/solidified waste forms, *Hazardous Waste Hazardous Materials* 9, 289, 1992.
75. Cheng, K.Y. and Bishop, P., Metals distribution in solidified/stabilized waste forms after leaching, *Hazardous Waste Hazardous Materials* 9, 163, 1992.

76. Cheng, K.Y. and Bishop, P.L., Morphology and pH changes in leached solidified/stabilized waste forms, in *Stabilization and Solidification of Hazardous, Radioactive, and Mixed Wastes: 3rd Volume, ASTM STP 1240*, Gilliam, T.M. and Wiles, C.C., Eds., American Society for Testing and Materials, Philadelphia, PA, 1996, p. 73–79.

77. Cheng, K.Y. and Bishop, P.L., Property changes of cement-based waste forms during leaching, in *Stabilization and Solidification of Hazardous, Radioactive, and Mixed Wastes, 3rd Volume, ASTM STP 1240*, Gilliam, T.M. and Wiles, C.C., Eds., American Society for Testing and Materials, Philadelphia, 1996, p. 375–387.

78. Allen, H.E. and Chen, P.-H., Remediation of metal contaminated soil by EDTA incorporating electrochemical recovery of metal and EDTA, *Environmental Progress* 12, 284, 1993.

79. Chen, T.C., Macauley, E., and Hong, A., Selection and test of effective chelators for removal of heavy metals from contaminated soils, *Canadian Journal of Civil Engineering* 22, 1185, 1995.

80. Erickson, D.C., White, E., and Loehr, R.C., Comparison of extraction fluids used with contaminated soils, *Hazardous Waste and Hazardous Materials* 8, 185, 1991.

81. Sahuquillo, A., Rigol, A., and Rauret, G., Overview of the use of leaching/extraction tests for risk assessment of trace metals in contaminated soils and sediments, *TrAC Trends in Analytical Chemistry* 22, 152, 2003.

82. Cnubben, P.A.J.P. and van der Sloot, H.A., Leaching characteristics of communal and industrial sludges, in *International Conference on the Environmental and Technical Implications of Construction with Alternative Materials, WASCON '97*, Goumans, J.J.J.M., Senden, G.J., and van der Sloot, H.A., Eds., Elsevier Science, B.V., Houthem St. Gerlach, The Netherlands, 1997, p. 247.

83. Stumm, W. and Morgan, J.J., *Aquatic Chemistry: Chemical Equilibria and Rates in Natural Waters*, 3rd ed. Wiley-Interscience, New York, NY, 1996.

84. Garrabrants, A.C. and Kosson, D.S., Use of a chelating agent to determine the metal availability for leaching from soils and wastes, *Waste Management* 20, 155, 2000.

85. Serne, R.J. et al., Radionuclide-Chelating Agent Complexes in Low-Level Radioactive Decontamination Wastes; Stability, Adsorption and Transport Potential, NUREG/CR-6758, US Nuclear Regulatory Commission, Washington, D.C., 2002.

86. van der Sloot, H.A., Hoede, D., and Comans, R.N.J., The influence of reducing properties on leaching of elements from waste materials and construction materials, in *Environmental Aspects of Construction with Waste Materials*, Goumans, J.J.J.M., van der Sloot, H.A., and Aalbers, T.G., Eds., Elsevier Science B.V., Amsterdam, The Netherlands, 1994, p. 483.

87. van Leeuwen, J. and Ratsma, K., Mine tailings — practical experiences in filling up harbours, in *International Conference on the Environmental and Technical Implications of Construction with Alternative Materials, WASCON '97*, Goumans, J.J.J.M., Senden, G.J., and van der Sloot, H.A., Eds., Elsevier Sciences, B.V., Houthem St. Gerlach, The Netherlands, 1997, p. 175.

88. Crawford, J., Neretnieks, I., and Moreno, L., A generalised model for the assessment of long-term leaching in combustion residue landfills, in *International Conference on the Environmental and Technical Implications of Construction with Alternative Materials, WASCON '97*, Goumans, J.J.J.M., Senden, G.J., and van der Sloot, H.A., Eds., Elsevier Science, B.V., Houthem St. Gerlach, The Netherlands, 1997, p. 501.

89. Fitch, J.R. and Cheeseman, C.R., Characterization of environmentally exposed cement-based stabilised/solidified industrial waste, *Journal of Hazardous Materials* 101, 239, 2003.

90. De Schutter, G., Finite element simulation of thermal cracking in massive hardening concrete elements using degree of hydration based material laws, *Computers & Structures* 80, 2035, 2002.

91. Cho, W.-J., Lee, J.-O., and Kang, C.-H., Influence of temperature elevation on the sealing performance of a potential buffer material for a high-level radioactive waste repository, *Annals of Nuclear Energy* 27, 1271, 2000.

92. Langton, C.A., Direct grout stabilization of high cesium salt waste: salt alternative phase III feasibility study, WSRC-TR-98-00337, Westinghouse Savannah River Company, Aiken, SC, 1998.

93. Eighmy, T.T. et al., Use of accelerated aging to predict behavior of recycled materials in concrete pavements — Physical and environmental comparison of laboratory-aged samples with field pavements, in *Transportation Research Record,* Transportation Research Board National Research Council, Washington, D.C., 2002, p. 118.

94. Cerný, R., Drchalová, J., and Rovnaníková, P., The effects of thermal load and frost cycles on the water transport in two high-performance concrete, *Cement and Concrete Research* 31, 1129, 2001.

95. De Angelis, G., Marchetti, A., and Balzamo, S., Leach studies: influence of various parameters on the leachability of cesium from cemented BWR evaporator concentrates, in *Solidification and Stabilization of Hazardous, Radioactive, and Mixed Wastes, 2nd Volume, ASTM STP 1123,* Gilliam, T.M. and Wiles, C.C., Eds., American Society for Testing and Materials, Philadelphia, PA, 1992, p. 182.

96. Hohberg, I. et al., Development of a leaching protocol for concrete, *Waste Management* 20, 177, 2000.

97. Almusallam, A.A., Effect of environmental conditions on the properties of fresh and hardened concrete, *Cement and Concrete Composites* 23, 353, 2001.

98. Kamali, S., Gérard, B., and Moranville, M., Modelling the leaching kinetics of cement-based materials — influence of materials and environment, *Cement and Concrete Composites* 25, 451, 2003.

99. Klich, I. et al., Mineralogical alterations that affect the durability and metals containment of aged solidified and stabilized wastes, *Cement and Concrete Research* 29, 1433, 1999.

100. Eighmy, T.T. and Chesner, W.H., Framework for evaluating use of recycled materials in the highway environment, FHWA-RD-00-140, Federal Highway Administration, McLean, VA, 2001.

101. Hime, W.G. and Mather, B., Sulfate attack, or is it?, *Cement and Concrete Research* 29, 789, 1999.

102. Bentz, D.P. et al., Influence of cement particle-size distribution on early age autogenous strains and stresses in cement-based materials, *Journal of the American Ceramic Society* 84, 129, 2001.

103. Zhang, M.H., Tam, C.T., and Leow, M.P., Effect of water-to-cementitious materials ratio and silica fume on the autogenous shrinkage of concrete, *Cement and Concrete Research* 33, 1687, 2003.

104. Brown, P.W. and Bothe, J.W., The stability of ettringite, *Advances in Cement Research* 5, 47, 1993.

105. Carde, C., François, R., and Torrenti, J.-M., Leaching of both calcium hydroxide and C-S-H from cement paste: modeling the mechanical behavior, *Cement and Concrete Research* 26, 1257, 1996.

106. Carde, C. and François, R., Effect of the leaching of calcium hydroxide from cement paste on mechanical and physical properties, *Cement and Concrete Research* 27, 539, 1997.

107. Carde, C. and François, R., Aging damage model of concrete behavior during the leaching process, *Materials and Structures* 30, 465, 1997.
108. Carde, C., François, R., and Ollivier, J.P., Microstructural changes and mechanical effects due to the leaching of calcium hydroxide from cement paste, in *Mechanisms of Chemical Degradation of Cement-based Systems*, Scrivener, K.L. and Young, J.F., Eds., E&FN Spon, London, 1997, p. 30.
109. Gérard, B., Le Bellego, C., and Bernard, O., Simplified modelling of calcium leaching of concrete in various environments, *Materials and Structures* 35, 2002.
110. Heukamp, F.H., Ulm, F.-J., and Germaine, J.T., Poroplastic properties of calcium-leached cement-based materials, *Cement and Concrete Research* 33, 1155, 2003.
111. Kuhl, D., Bangert, F., and Meschke, G., Coupled chemo-mechanical deterioration of cementitious materials. Part I. Modeling, *International Journal of Solids and Structures* 41, 15, 2004.
112. Kuhl, D., Bangert, F., and Meschke, G., Coupled chemo-mechanical deterioration of cementitious materials. Part II. Numerical methods and simulations, *International Journal of Solids and Structures* 41, 41, 2004.
113. Mainguy, M. and Coussy, O., Propagation fronts during calcium leaching and chloride penetration, *Journal of Engineering Mechanics* 126, 250, 2000.
114. Bangert, F. et al., Environmentally induced deterioration of concrete: physical motivation and numerical modeling, *Engineering Fracture Mechanics* 70, 891, 2003.
115. Ferraris, C.F. et al., Mechanisms of degradation of Portland cement-based systems by sulfate attack, in *Mechanisms of Chemical Degradation of Cement-based Systems*, Scrivener, K.L. and Young, J.F., Eds., E&FN Spon, London, 1997, p. 185.
116. Santhanam, M., Cohen, M.D., and Olek, J., Sulfate attack research — whither now?, *Cement and Concrete Research* 31, 845, 2001.
117. Santhanam, M., Cohen, M.D., and Olek, J., Mechanism of sulfate attack: a fresh look. Part 1. Summary of experimental results, *Cement and Concrete Research* 32, 915, 2002.
118. Santhanam, M., Cohen, M.D., and Olek, J., Mechanism of sulfate attack: a fresh look. Part 2. Proposed mechanisms, *Cement and Concrete Research* 33, 341, 2003.
119. Taylor, H.F.W. and Gollop, R.S., Some chemical and microstructural aspects of concrete durability, in *Mechanisms of Chemical Degradation of Cement-based Systems*, Scrivener, K.L. and Young, J.F., Eds., E&FN Spon, London, 1997, p. 177.
120. Brown, P.W. and Doerr, A., Chemical changes in concrete due to the ingress of aggressive species, *Cement and Concrete Research* 30, 411, 2000.
121. Brown, P.W., Thaumasite formation and other forms of sulfate attack, *Cement and Concrete Composites* 24, 301, 2002.
122. Al-Amoudi, O.S.B., Attack on plain and blended cements exposed to aggressive sulfate environments, *Cement and Concrete Composites* 24, 305, 2002.
123. Bakharev, T., Sanjayan, J.G., and Cheng, Y.-B., Sulfate attack on alkali-activated slag concrete, *Cement and Concrete Research* 32, 211, 2002.
124. Al-Khaiat, H., Haque, M.N., and Fattuhi, N.I., Concrete carbonation in an arid climate, *Materials and Structures* 35, 2002.
125. Bakharev, T., Sanjayan, J.G., and Cheng, Y.-B., Resistance of alkali-activated slag concrete to carbonation, *Cement and Concrete Research* 31, 1277, 2001.
126. Balayssac, J.P., Detriche, C.H., and Grandet, J., Effects of curing upon carbonation of concrete, *Construction & Building Materials* 9, 91, 1995.
127. Bin Shafique, S.M. et al., Influence of carbonation on leaching of cementitious wasteforms, *Journal of Environmental Engineering* 124, 463, 1998.

128. Chi, J.M., Huang, R., and Yang, C.C., Effects of carbonation on mechanical properties and durability of concrete using accelerated testing method, *Journal of Marine Science and Technology* 10, 14, 2002.

129. Dewaele, P.J., Reardon, E.J., and Dayal, R., Permeability and porosity changes associated with cement grout carbonation, *Cement and Concrete Research* 21, 441, 1991.

130. Dias, W.P.S., Reduction of concrete sorptivity with age through carbonation, *Cement and Concrete Research* 30, 1255, 2000.

131. Freyssinet, P. et al., Chemical changes and leachate mass balance of municipal solid waste bottom ash submitted to weathering, *Waste Management* 22, 159, 2002.

132. Houst, Y.F., Microstructural changes of hydrate cement paste due to carbonisation, in *Mechanisms of Chemical Degradation of Cement-based Systems*, Scrivener, K.L. and Young, J.F., Eds., E&FN Spon, London, 1997, p. 90.

133. Kobayashi, K., Suzuki, K., and Uno, Y., Carbonation of concrete structures and decomposition of C-S-H, *Cement and Concrete Research* 24, 55, 1994.

134. Lange, L.C., Hills, C.D., and Poole, A.B., The effect of accelerated carbonation on the properties of cement-solidified waste forms, *Waste Management* 16, 757, 1996.

135. Lange, L.C., Hills, C.D., and Poole, A.B., Effect of carbonation on properties of blended and non-blended cement solidified waste forms, *Journal of Hazardous Materials* 52, 193, 1997.

136. Maslehuddin, M., Shirokoff, J., and Siddiqui, M.A.B., Changes in the phase composition in OPC and blended cement mortars due to carbonation, *Advances in Cement Research* 8, 167, 1996.

137. Nishikawa, T. et al., Decomposition of synthesized ettringite by carbonation, *Cement and Concrete Research* 22, 6, 1992.

138. Papadakis, V.G., Vayenas, C.G., and Fardis, M.N., Experimental investigation and mathematical modeling of the concrete carbonation problem, *Chemical Engineering Science* 46, 1333, 1991.

139. Roy, S.K., Poh, K.B., and Northwood, D.O., Durability of concrete — accelerated carbonation and weathering studies, *Building and Environment* 34, 597, 1999.

140. Van Gerven, T. et al., Influence of carbonation and carbonation methods on leaching of metals from mortars, *Cement and Concrete Research* 34, 149, 2004.

141. Walton, J.C. et al., Role of carbonation in transient leaching of cementitious wasteforms, *Environmental Science and Technology* 31, 1997.

142. Walton, J.C. et al., Role of carbonation in long-term performance of cementitious wasteforms, in *Mechanisms of Chemical Degradation of Cement-based Systems*, Scrivener, K.L. and Young, J.F., Eds., E&FN Spon, London, 1997, p. 315.

143. Baroghel-Bouny, V. et al., Characterization and identification of equilibrium and transfer moisture properties for ordinary and high-performance cementitious materials, *Cement and Concrete Research* 29, 1225, 1999.

144. Akita, H., Fujiwara, T., and Ozaka, Y., Practical procedure for the analysis of moisture transfer within concrete due to drying, *Magazine of Concrete Research* 49, 129, 1997.

145. van Breugel, K. and Koenders, E.A.B., Numerical simulation of hydration-driven moisture transport in bulk and interface paste in hardened concrete, *Cement and Concrete Research* 30, 1911, 2000.

146. Numao, T., Mihashi, H., and Fukuzawa, K., Moisture migration and drying properties of hardened cement paste and mortar, *Nuclear Engineering and Design* 156, 139, 1995.

147. Ayano, T. and Wittmann, F.H., Drying, moisture distribution, and shrinkage of cement-based materials, *Materials and Structures* 35, 2002.

148. Garrabrants, A.C. and Kosson, D.S., Modeling moisture transport from a Portland cement-based material during storage in reactive and inert atmospheres, *Drying Technology* 21, 775, 2003.

149. Beyea, S.D. et al., Magnetic resonance imaging and moisture content profiles of drying concrete, *Cement and Concrete Research* 28, 453, 1998.

150. Werner, K.-C., Chen, Y., and Odler, I., Investigations on stress corrosion of hardened cement pastes, *Cement and Concrete Research* 30, 1443, 2000.

151. Tian, B. and Cohen, M.D., Does gypsum formation during sulfate attack on concrete lead to expansion?, *Cement and Concrete Research* 30, 117, 2000.

152. Kumar, S. and Rao, C.V.S.K., Sulfate attack on concrete in simulated cast-in-situ and precast situations, *Cement and Concrete Research* 25, 1, 1995.

153. Stock, S.R. et al., X-ray microtomography (microCT) of the progression of sulfate attack of cement paste, *Cement and Concrete Research* 32, 1673, 2002.

154. Gollop, R.S. and Taylor, H.F.W., Microstructure and microanalytical studies of sulfate attack IV. Reactions of a slag cement paste with sodium and magnesium sulfate solutions, *Cement and Concrete Research* 26, 1013, 1996.

155. Gollop, R.S. and Taylor, H.F.W., Microstructure and microanalytical studies of sulfate attack V. Comparison of different slag blends, *Cement and Concrete Research* 26, 1029, 1996.

156. Bazant, Z.P. and Najjar, L.J., Drying of concrete as a nonlinear diffusion problem, *Cement and Concrete Research* 1, 461, 1971.

157. Bazant, Z.P. and Najjar, L.J., Nonlinear water diffusion in nonsaturated concrete, *Matériaux et Constructions* 5, 3, 1972.

158. Šelih, J., Sousa, A.C.M., and Bremner, T.W., Moisture transport in initially fully saturated concrete during drying, *Transport in Porous Media* 24, 81, 1996.

159. Keey, R.B., *Drying — Principles and Practice,* Pergamon Press, Oxford, England, 1972.

160. van der Zanden, A.J.J., Modelling and simulating simultaneous liquid and vapour transport in partially saturated porous materials, in *Mathematical Modeling and Numerical Techniques in Drying Technology,* Turner, I.W. and Mujumdar, A.S., Eds., Marcel Dekker, Inc., New York, NY, 1997, p. 157.

161. Rogers, J.A. and Kaviany, M., Funicular and evaporative-front regimes in convective drying of granular beds, *International Journal of Heat and Mass Transfer* 35, 469, 1992.

162. Garrabrants, A.C., Sanchez, F., and Kosson, D.S., Leaching model for a cement mortar exposed to intermittent wetting and drying, *Aiche Journal* 49, 1317, 2003.

163. Macías, A., Kindness, A., and Glasser, F.P., Impact of carbon dioxide on the immobilization potential of cemented wastes: chromium, *Cement and Concrete Research* 27, 215, 1997.

164. Smith, R.W. and Walton, J.C., The effects of calcite solid solution formation on the transient release of radionuclides from concrete barriers, *Scientific Basis for Nuclear Waste Management XIV* 212, 403, 1991.

165. Sarott, F.-A. et al., Diffusion and adsorption studies on hardened cement paste and the effect of carbonation on diffusion rates, *Cement and Concrete Research* 22, 439, 1992.

166. Ngala, V.T. and Page, C.L., Effects of carbonation on pore structure and diffusional properties of hydrated cement pastes, *Cement and Concrete Research* 27, 99, 1997.

167. Lange, L.C., Hills, C.D., and Poole, A.B., Preliminary investigation into the effects of carbonation on cement-solidified hazardous wastes, *Environmental Science and Technology* 30, 25, 1996.

168. Lange, L.C., Hills, C.D., and Poole, A.B., The influence of mix parameters and binder choice on the carbonation of cement solidified wastes, *Waste Management* 16, 749, 1996.
169. Papadakis, V.G., Effect of supplementary cementing materials on concrete resistance against carbonation and chloride ingress, *Cement and Concrete Research* 30, 291, 2000.
170. Gervais, C. et al., The effects of carbonation and drying during intermittent leaching on the release of inorganic constituents from a cement-based matrix, *Cement and Concrete Research* 34, 119, 2004.
171. Atis, C.D., Accelerated carbonation and testing of concrete made with fly ash, *Construction & Building Materials* 17, 147, 2003.
172. Walter, M.B. et al., Chemical characterization, leach and adsorption studies of solidified low-level wastes, *Nuclear and Chemical Waste Management,* 8, 55, 1988.
173. van der Sloot, H.A., Heasman, L., and Quevauviller, P., Harmonization of Leaching/Extraction Tests, in *Studies in Environmental Science,* Elsevier Science, Amsterdam, 1997, p. 292.
174. van der Sloot, H.A., Developments in evaluating environmental impact from utilization of bulk inert wastes using laboratory leaching tests and field verification, *Waste Management* 16, 65, 1996.
175. CEN/TC 292 WG6, Characterisation of waste: leaching behavior tests — compliance leaching tests for granular waste, Parts 1-4, EN 12457, 1999.
176. USEPA, Method 1311: Toxicity Characteristic Leaching Procedure, US Environmental Protection Agency, Washington, D.C., 2004.
177. USEPA, Method 1312: Synthetic Precipitation Leaching Procedure, USEPA Office of Solid Waste, Washington, D.C., 2004.
178. Boy, J.H. et al., Response of Pb in cement waste forms during TCLP testing, in *Mechanisms of Chemical Degradation of Cement-based Systems*, Scrivener, K.L. and Young, J.F., Eds., E&FN Spon, London, 1997, p. 444.
179. NEN, Determination of the availability for leaching of inorganic constituents from granular materials, NEN 7341, Netherlands Normalisation Institute, Delft, The Netherlands, 1995.
180. prEN 14405, Characterisation of waste: leaching behavior tests — up-flow percolation test, CEN/TC 292 WG6, 2001.
181. 62 Fed. Reg. 41005 (July 31, 1997).
182. 62 Fed. Reg. 63458 (December 1, 1997).
183. Comans, R.N.J., van der Sloot, H.A., and Bonouvrie, P.A., Geochemical reactions controlling the solubility of major and trace elements during leaching of municipal solid waste incinerator residues, in *The 1993 Municipal Waste Combustion Conference*, Kilgroe, J., Ed., AWMA, Pittsburgh, PA, Williamsburg, VA, 1993, p. 667.
184. prEN 14429, Characterisation of waste: leaching behavior tests — pH dependence test with initial acid/base addition, CEN/TC 292 WG6, 2001.
185. Garrabrants, A.C. et al., Methodology for determining inorganic release from soils and wastes, in *Society for Risk Analysis/International Society for Exposure Analysis National Meeting*, New Orleans, LA, 1996.
186. Sanchez, F., Etude de la lixiviation de milieux poreux contenant des espèces solubles: application au cas des déchets solidifiés par liants hydrauliques, Doctoral thesis, Institut National des Sciences Appliquées, Lyon, France, 1996.
187. Sanchez, F. et al., Environmental assessment of a cement-based solidified soil contaminated with lead, *Chemical Engineering Science* 55, 113, 2000.

188. Sanchez, F. et al., Use of a new leaching test framework for evaluating alternative treatment processes for mercury contaminated mixed waste, *Environmental Engineering Science* 19, 251, 2002.

189. ASTM, Standard Test Method for Shake Extraction of Solid Waste with Water, ASTM D 3987-85, American Society for Testing and Materials, West Conshohocken, PA, 1999.

190. DIN-NORMEN, Determination of leachability by water (S4), DIN 38 414 S4, Sludge and Sediments Group S, 1984.

191. Tessier, A., Campbell, P.G.C., and Bisson, M., Sequential extraction procedure for the speciation of particulate trace metals, *Analytical Chemistry* 51, 844, 1979.

192. Berardi, R., Cioffi, R., and Santoro, L., Matrix stability and leaching behaviour in ettringite-based stabilization systems doped with heavy metals, *Waste Management* 17, 535, 1997.

193. Lin, C.-F., Lin, T.-T., and Huang, T.-H., Leaching processes of the dicalcium silicate and copper oxide solidification/stabilization system, *Toxicological and Environmental Chemistry* 44, 89, 1994.

194. Delagrave, A., Gérard, B., and Marchand, J., Modelling the calcium leaching mechanisms in hydrated cement pastes, in *Mechansims of Chemical Degradation of Cement-based Systems*, Scrivener, K.L. and Young, J.F., Eds., E&FN Spon, London, 1997, p. 38.

195. Stanish, K. and Thomas, M., The use of bulk diffusion tests to establish time-dependent concrete chloride diffusion coefficients, *Cement and Concrete Research* 33, 55, 2003.

196. Tiruta-Barna, L.R., Barna, R., and Moszkowicz, P., Modeling of solid/liquid/gas mass transfer for environmental evaluation of cement-based solidified waste, *Environmental Science and Technology* 35, 149, 2001.

197. Andac, M. and Glasser, F.P., The effect of test conditions on the leaching of stabilised MSWI-fly ash in Portland cement, *Waste Management* 18, 309, 1998.

198. Andac, M. and Glasser, F.P., Long-term leaching mechanisms of Portland cement-stabilized municipal solid waste fly ash in carbonated water, *Cement and Concrete Research* 29, 179, 1999.

199. Ampadu, K.O., Torii, K., and Kawamura, M., Beneficial effect of fly ash on chloride diffusivity of hardened cement paste, *Cement and Concrete Research* 29, 585, 1999.

200. Andrade, C. et al., Non-steady-state chloride diffusion coefficients obtained from migration and natural diffusion tests. Part I. Comparison between several methods of calculation, *Materials and Structures* 33, 2000.

201. Garrabrants, A.C. et al., The effect of storage in an inert atmosphere on the release of inorganic constituents during intermittent wetting of a cement-based material, *Journal of Hazardous Materials* 91, 159, 2002.

202. Sanchez, F., Garrabrants, A.C., and Kosson, D.S., Effects of intermittent wetting on concentration profiles and release from a cement-based waste matrix, *Environmental Engineering Science* 20, 135, 2003.

203. NEN, Determination of the leaching of inorganic components from building materials and monolithic waste materials with the diffusion test, NEN 7345, Netherlands Normalisation Institute, Delft, The Netherlands, 1995.

204. American Nuclear Society (ANS), Measurement of the Leachability of Solidified Low-Level Radioactive Wastes by a Short-Term Test Procedure, ANSI/ANS 16.1-2003, American Nuclear Society, La Grange Park, IL, 2003.

205. Garrabrants, A.C., Development and application of fundamental leaching property protocols for evaluating inorganic release from wastes and soils, Masters thesis, Rutgers, The State University of New Jersey, New Brunswick, NJ, 1998.
206. Sanchez, F. et al., Environmental assessment of waste matrices contaminated with arsenic, *Journal of Hazardous Materials* 96, 229, 2003.
207. Sanchez, F. et al., Leaching of inorganic contaminants from cement-based waste materials as a result of carbonation during intermittent wetting, *Waste Management* 22, 249, 2002.
208. ASTM, Standard Test Method for Leaching Solid Waste in a Column Apparatus, ASTM D 4874, American Society for Testing and Materials, West Conshohocken, PA, 2001.
209. NEN, Determination of the leaching of inorganic components from granular materials with the column test, NEN 7343, Netherlands Normalisation Institute, Delft, The Netherlands, 1995.
210. Ludwig, C. et al., Hydrological and geochemical factors controlling the leaching of cemented MSWI air pollution control residues: a lysimeter field study, *Journal of Contaminant Hydrology* 42, 253, 2000.
211. Shimaoka, T. and Hanashima, M., Behavior of stabilized fly ashes in solid waste landfills, *Waste Management* 16, 545, 1996.
212. Poon, C.S. and Chen, Z.Q., Comparison of the characteristics of flow-through and flow-around leaching tests of solidified heavy metal wastes, *Chemosphere* 38, 663, 1999.
213. Poon, C.S., Chen, Z.Q., and Wai, O.W.H., The effect of flow-through leaching on the diffusivity of heavy metals in stabilized/solidified wastes, *Journal of Hazardous Materials* 81, 179, 2001.
214. Atkinson, A., Nelson, K., and Valentine, T.M., Leach test characterization of cement-based nuclear waste forms, *Nuclear and Chemical Waste Management*, 6, 241–253, 1986.
215. Moudilou, E. et al., A dynamic leaching method for the assessment of trace metals released from hydraulic binders, *Waste Management* 22, 153, 2002.
216. Crank, J., *The Mathematics of Diffusion*, 2nd ed., Oxford University Press, London, UK, 1975.
217. Barneyback, S., Jr. and Diamond, S., Expression and analysis of pore fluids from hardened cement pastes and mortars, *Cement and Concrete Research* 11, 279, 1981.
218. Duchesne, J. and Bérubé, M.A., Evaluation of the validity of the pore solution expression method from hardened cement pastes and mortars, *Cement and Concrete Research* 24, 456, 1994.
219. Sahu, S. and Diamond, S., Pore solution chemistry of simulated low level liquid waste incorporated cement grouts, *Materials Research Society Symposium Proceedings, Pittsburgh, PA, USA* 412, *Materials Research Society*, 411, 1996.
220. Constantiner, D. and Diamond, S., Pore solution analysis: are there pressure effects?, in *Mechanisms of Chemical Degradation of Cement-based Systems*, Scrivener, K.L. and Young, J.F., Eds., E&FN Spon, London, 1997, p. 22.
221. Rivard, P. et al., Alkali mass balance during the accelerated concrete prism test for alkali-aggregate reactivity, *Cement and Concrete Research* 33, 1147, 2003.
222. de Groot, G.J. and van der Sloot, H.A., Determination of leaching characteristics of waste materials leading to environmental product certification, in *Solidification and Stabilization of Hazardous, Radioactive, and Mixed Wastes, 2nd Volume, ASTM STP 1123*, Gilliam, T.M. and Wiles, C.C., Eds., American Society for Testing and Materials, Philadelphia, PA, 1992, p. 149.

223. Krishnamoorthy, T.M. et al., Desorption kinetics of radionuclides fixed in cement matrix, *Nuclear Technology* 104, 351, 1993.
224. Barna, R. et al., Solubility model for the pore solution of leached concrete containing solidified waste, *Journal of Hazardous Materials* 37, 33, 1994.
225. Hinsenveld, M., A shrinking core model as a fundamental representation of leaching mechanisms in cement stabilized waste, Doctoral thesis, University of Cincinnati, Cincinnati, OH, 1992.

11 Testing and Performance Criteria for Stabilized/Solidified Waste Forms

A. S. Ramesh Perera, Abir Al-Tabbaa,
J. Murray Reid, Julia A. Stegemann, and Caijun Shi

CONTENTS

1-56670-444-8/05/$0.00+$1.50
© 2005 by CRC Press

11.1 INTRODUCTION

Whether in preparation for full-scale treatment, or to verify the effectiveness of treated material *in situ*, it is necessary to assess the performance of a stabilized/solidified (S/S) material in order to judge its improved properties and the effectiveness of the binder matrix in containing contaminants. Various test methods have been adopted in research and practice to assess the efficiency of S/S processes.[1-5] Such assessment could be generally categorized as:

- Basic information tests, which measure basic material properties (e.g., grading, plasticity, particle density, total contaminant concentration). These tests are often referred to as index tests.

- Performance tests, which relate to the properties of the material in use (e.g., strength, leachability).

These categories include physical and chemical (predominantly leaching) tests, and may be used for understanding mechanisms, assessing compliance with reference criteria (e.g., regulatory), or on-site verification, i.e., quality control in practical field situations.

It is difficult to predict and also simulate in the laboratory the long-term environmental conditions to which the S/S material might be subjected. For this reason, and also because the behavior of an S/S material is complex, its performance is generally evaluated using a combination of several physical and chemical tests. Each test provides a partial insight into the behavior of the S/S material and hence the effectiveness of the S/S treatment system. Several different tests may exist with the objective of measuring the same intrinsic property; the results of these tests will differ depending on the specific testing conditions. Therefore, consideration of the results and their relationship to the performance criteria in light of the specific testing conditions is essential.

The purpose of this chapter is to review current practice in test methods and performance criteria under the broad categories of physical and chemical tests to consider the acceptability of S/S materials for their intended management scenarios. Some of these tests are also often carried out on the original material to be treated to assess its suitability for S/S treatment, and also on binders to assess their effectiveness. Both test methods and performance criteria are also placed in the context of a number of international regulatory frameworks.

11.2 LEACHING TESTS

11.2.1 INTRODUCTION

Since the leaching characteristics of S/S materials are of the most concern, Chapter 10 discusses in detail the leaching mechanisms, factors that control constituent leaching from cement-based materials, and leaching evaluation tests applicable to S/S materials.

Although numerous leaching test methods have been proposed, many are variations on the same basic principle with modifications in the specific testing conditions. A number of systems have been developed for classifying leaching tests. The system proposed by van der Sloot et al.[5] is based on (i) equilibrium or semi-equilibrium tests, (ii) dynamic tests, and (iii) specific tests focusing on chemical speciation. An earlier system[1,6,7] classifies leaching tests as either extraction tests or dynamic tests based on whether the leachant is renewed (in the case of the latter) or not (the former). Extraction tests include all tests that contact a specific amount of leachant with a specific amount of material for a specific amount of time.[6] Dynamic tests include all tests that continuously or intermittently renew the leachant to maintain a driving force for leaching and generate information as a function of time while attempting to preserve the structural integrity of the material.[6] This chapter summarizes the leaching test methods based on the latter classification method.

11.2.2 Extraction Tests

Extraction tests are the most common tests and they have been subdivided by Wastewater Technology Centre (WTC)[6] into several categories, of which agitated and sequential chemical extraction tests are the most relevant for S/S materials.

11.2.2.1 Agitated Extraction Tests

An extraction test can be agitated to maintain a homogeneous mixture and promote contact between the solid and the leachant, thereby accelerating attainment of steady state conditions. To decrease physical barriers to mass transport, granular or crushed samples are used with the leachant at a specified liquid-to-solid ratio (L/S). They measure the chemical properties of the system and not the rate-limiting mechanisms. The common agitated extraction tests include:

11.2.2.1.1 Toxicity Characteristic Leaching Procedure (TCLP)[8]

The TCLP is a commonly used, standard, single-batch leaching test, which was developed by the United States Environmental Protection Agency (USEPA) as a rapid regulatory compliance test for determining whether or not a waste is suitable for disposal in a landfill with municipal waste. Because of the presence of organic acids in this scenario, the test uses acetic acid buffered to pH 4.93 (or 2.88) with sodium acetate, to a maximum buffering capacity of 2 meq/g of wet waste, at an L/S of 20:1, for 18 hours. The test has been criticized because it does not take into account the characteristics of an S/S material, or management scenarios other than municipal waste landfill disposal. The test is conducted on granular material and therefore does not give credit for reduction in leachability due to production of a monolithic material. More importantly, since the maximum buffering capacity is often exceeded by a cement-based solidified material, the test conditions can result in an arbitrary final leachate pH. Since the final leachate pH is critical for solubility of contaminants, a combination of tests that measure contaminant solubility at different pHs is more informative.

11.2.2.1.2 Extraction Procedure Toxicity Test (EP-tox)[9]

The EP-Tox was a USEPA regulatory compliance test, which was commonly used until superseded by the TCLP. It is also a standard single-batch leaching test, which uses 0.5N acetic acid to maintain the leachate at pH 5, with a maximum acid addition of 2 meq/g of wet waste, at an L/S of 20:1 for 24 hours. The test makes provision for testing of monolithic samples, but it also has the drawback that the final leachate pH is arbitrary.

11.2.2.1.3 Synthetic Precipitation Leaching Procedure (SPLP)[10]

The SPLP is a standard single-batch compliance test, which was developed as an alternative to the TCLP for situations where disposal is outside municipal waste landfills. It uses an acid mix containing sulfuric/nitric acid (60/40 w/w) for an initial leachant pH of 4.2 or 5 at an L/S of 20:1, for 18 hours. In practice, applied to S/S material, this initial leachant pH makes little difference to the final leachate pH, which reflects that of the alkaline S/S material.

11.2.2.1.4   ASTM D3987[11]

ASTM D3987 is a standard compliance-type test first issued in 1981 and last revised in 1985. The intention of the test is to provide a rapid extraction procedure for industry, but not to simulate site-specific conditions[5,6]. The test uses distilled/deionized water at an L/S of 20:1 for 18 hours. Thus, the final leachate pH reflects the pH of the material being tested.

11.2.2.1.5   DIN 38414 S4[12]

DIN 38414 S4 is a standard batch leaching test, which has been widely used for regulatory compliance purposes in Germany and Austria, as well as for general assessment elsewhere. It uses distilled/deionized water at an L/S of 10:1 for 24 hours, which allows the test material to establish the pH. This test will be superseded for regulatory use by the EN 12457 batch leaching tests and other tests recently developed under CEN/TC 292.

11.2.2.1.6   The National Rivers Authority (NRA) Leaching Test[7]

The NRA test is a standard single-batch compliance test, which was developed and recommended by the National Rivers Authority for the purposes of general assessment of the leachability of mainly inorganic contaminants from contaminated land in the UK.[7] This method was developed as an alternative to more aggressive tests such as the TCLP. It uses distilled/deionized water left to stand overnight (expected pH 5.6), at an L/S of 10:1, for 24 hours. In practice, this initial leachant pH makes little difference to the final leachate pH, which reflects that of the alkaline S/S material. This test will also be superseded by BS EN 12457, developed, by CEN/TC 292 (see Section 11.2.2.1.7).

11.2.2.1.7   BS EN 12457[13]

The BS EN 12457 method describes a series of batch leaching tests for granular wastes and sludges, developed by CEN/TC 292 based on standard procedures DIN 38414 S4, AFNOR X-31 210, NEN 7343, and ONORM S 2072, primarily to support the requirements for compliance testing within the European Union (EU) and European Free Trade Association (EFTA) countries. The intent of these tests is to identify the leaching properties of waste materials. However, the standards have been developed to investigate mainly inorganic constituents and do not take into account the particular characteristics of non-polar organic constituents or the consequences of microbiological processes in organic degradable wastes. Each part specifies a distinct procedure, and the annexes to the standards provide useful information on the selection of the appropriate procedure, reference documents, and guidance on the limitations of these procedures. The procedures for Parts 1 and 3 are only applicable to wastes and sludges having a high solid content: the dry matter content ratio shall be at least higher than 1:3. All parts use distilled/deionized water and have a total contact time of 24 hours. The operating conditions for each part are summarized in Table 11.1. It should be noted that Part 3 is carried out in two stages. The high L/S tests may be considered to represent a form of accelerated leaching.[14]

TABLE 11.1
Operating Parameters for BS EN 12457

Part	1	2	3	4
Particle size (mm)	< 4	< 4	< 4	< 10
L/S ratio (l/kg)	2	10	2 + 8	10
Contact time (h)	24	24	6 + 18	24

11.2.2.1.8   Acid Neutralization Capacity (ANC)

The ANC test[3] is a measure of the ability of a material to neutralize acid. This is a key variable for long-term material behavior, because it affects precipitation of metals and maintenance of matrix physical integrity.[2] The test involves mixing subsamples of a material with increasing quantities of mineral acid for 48 hours, prior to measurement of leachate pH to obtain a titration plot.[2,3,15] Analysis of contaminants in the leachate can be used to assess their availability at pH values of interest. This approach is similar to that used in other availability tests, such as NEN 7341[16] and prEN 14429.[17]

These availability tests are themselves agitated extraction tests. Although not yet in common use, prEN 14429 has been developed from the ANC and NEN 7341 to investigate contaminant availability as a variable distinct from total contaminant concentration. The test involves a 24-hour extraction of granular material at controlled pH.

Modification of the ANC to use acetic acid, as a way of optimizing binder addition to pass the TCLP, was proposed by Isenburg and Moore,[18] but is less useful for understanding leaching behavior, in part due to the development of a buffer system that alters the titration curve.

11.2.2.2   Sequential Chemical Extraction Tests

Increasingly aggressive leachants may be used to obtain information on the mechanisms of contaminant binding in a material. Most sequential chemical extraction tests for metals are based on a method developed by Tessier et al.,[19] which divides the contaminants into five fractions: 1) ion-exchangeable, 2) bound to surface oxides and carbonates, 3) bound to iron and manganese oxides, 4) bound to organics, and 5) residual. The test was originally proposed to examine respeciation of contaminants due to treatment, but has fallen into disuse except as a research tool because of concerns with the definition of speciation and reproducibility.[2]

11.2.3   Dynamic Tests

Dynamic tests are not as commonly used as extraction tests. They can also be divided into several categories.[6] The serial batch test, which is the most common type, and flow-around and flow-through tests are briefly summarized below:

11.2.3.1 Serial Batch Tests

Serial batch tests are similar to agitated extraction tests, except that the leachant is replaced after a specific time until the desired number of leaching periods have been achieved. The temporal release of leachable constituents can be inferred by constructing an extraction profile using the data obtained. Typical tests are the multiple extraction procedure (MEP)[20,21] and sequential batch extraction,[12,22-24] which give a procedure for multiple extractions in addition to the single extraction method stated earlier.

11.2.3.1.1 USEPA Multiple Extraction Procedure[20]

The MEP test involves an initial extraction with acetic acid, which is intended to simulate municipal solid waste leachate, and at least eight subsequent extractions with an inorganic acid mixture (nitric and sulfuric acids) to simulate acid rain. The MEP test starts with the EP-Tox test (see Section 11.2.2.1.2), which is run for 24 hours. After the 24-hour rotation period and filtration of the leachate, seven additional extractions are performed on the solid phase of the sample captured on the filter. The extraction fluid is a mixture of inorganic acids with pH 3.0 ± 0.2, which is prepared in a similar manner as the SPLP leaching fluid (see Section 11.2.2.1.3). During each subsequent extraction, the synthetic rain extraction fluid is added to the waste at an L/S ratio of 20:1, and the mixture is rotated for 24 hours per extraction. After each extraction, the final pH is measured, and the leachate is collected and analyzed. If the concentration of any of the chemical constituents of concern increases in the seventh and eighth extractions, the extraction must be repeated until the concentration in the extract ceases to increase. This test is currently used for the USEPA's delisting program.

11.2.3.1.2 ANSI/ANS-16.1-2003[21]

The ANSI/ANS-16.1-2003 test uses a solid monolith of the waste form in a specific geometric shape, typically a cylinder (but other shapes can be used). A sample monolith with a known initial quantity of the contaminant is immersed quiescently in deionized water, although repository groundwater or simulated groundwater is encouraged, at a ratio of 10 for the leachant volume (cm^3) to monolith geometric surface area (cm^2). The leachate is replaced with fresh leachant at specified times (2 hours, 7 hours, 1 day, 2 days, 3 days, 4 days, 5 days, and optionally 19 days, 47 days, and 90 days) for the purpose of keeping the contaminant surface concentration at or close to zero. This results in the maximum driving force into the leachant and the maximum leach rate for a monolith in this quiescent leachant. The main purpose is to approach the boundary conditions for the analytical solution of Fick's second law for the special case of zero surface concentration with an initial condition of even distribution in the monolith. This allows estimation of an effective diffusion coefficient using Fick's second law. The negative logarithm of the average coefficient is defined as the leachability index, a measure of performance for the leach resistance of the waste form matrix. Using the negative logarithm minimizes the large error bar for the effective diffusion coefficient usually found among samples of the same matrix and gives a measure of performance, even if the data do not agree well with the simple diffusion control behavior predicted by Fick's second law.

11.2.3.2 Flow-Around Tests

Flow-around tests are generally performed on monolithic samples. Leachant is continuously or intermittently renewed, providing the driving force to maintain leaching by diffusion. For these tests, the volume of leachant, and the leachant volume to sample surface area ratio are prescribed. Typical tests include NEN 7345[25] and the CEN monolithic tank test.[26] These tests use an effective diffusion coefficient determined from the results of the test to estimate contaminant release under simplified disposal conditions.

11.2.3.3 Flow-Through Tests

Flow-through tests are performed on porous monoliths or granular material, with the leachant continuously or intermittently flowing through the material, to measure contaminant leaching under advective conditions. Typical tests include the ASTM Column Extraction Method[27,28] and the European standard column test[29] being developed by CEN TC/292. These tests employ slow upward leachant flow to allow saturation.

11.3 PRIMARY PHYSICAL TESTS

11.3.1 INTRODUCTION

Most of the physical tests applied to untreated or treated S/S materials have been adopted or adapted from test methods used for other materials such as concrete (BS EN 12350, BS 1881, and BS 4550), soils for civil engineering purposes (BS 1337), and stabilized materials for civil engineering purposes (BS 1924) and similarly from ASTM standard test methods in volumes 04.01 (cement, lime, gypsum), 04.02 (concrete and aggregates), 4.08 (soil and rock), and 11.04 (environmental assessment, hazardous substances and oil spill responses, waste management). Typical applicability of the tests discussed below to either untreated or treated S/S material is shown in Table 11.2. The most commonly used physical tests were found to be three performance tests:

11.3.2 UNCONFINED COMPRESSIVE STRENGTH (UCS)

UCS (before and after immersion) is used as a measure of the ability of a monolithic S/S material to resist mechanical stresses.[2,3] It relates to the progress of hydration reactions in the product, and durability of a monolithic S/S material, and is therefore a key variable. It is one of the most commonly used tests and there are numerous standard methods for its determination, all of which involve vertical loading of a monolithic specimen to failure.[30-38] Standard methods vary mainly with regard to the specimen shape and size. Since these variables have an effect on the test result, they must be clearly reported. Measurement of strength after immersion, as well as before, is important to ensure that a specimen has set and hardened chemically rather than merely dried, and to ensure that deleterious swelling reactions do not occur in

TABLE 11.2
Typical Use of the Properties on Untreated and Treated S/S Materials and also at the Point of Onset

Property	To Assess Suitability for Treatment	Testing Just After Treatment	End Product Specification
Commonly utilized			
Leachability and pH	X		X
Unconfined compressive strength			X
Durability			X
Permeability	X		X
Others of relevance			
Bound water	X	X	X
Bulk density	X	X	X
Chloride permeability	X		X
California bearing ratio (CBR)			X
Dry density/moisture content relation	X	X	X
Flow	X	X	
Heat of hydration	X	X	
Initial consumption of lime (ICL)	X		
Intrinsic permeability	X		X
Microstructural examination			X
Modulus of elasticity			X
Moisture content	X	X	X
Moisture condition value (MCV)	X	X	X
Oxygen permeability	X		X
Particle size distribution	X		
Penetration resistance			X
Porosity	X	X	X
Pulverization		X	X
Setting time		X	
Shrinkage/expansion	X		X
Slump		X	
Soundness			X
Specific gravity	X	X	X
Tensile strength			X
Water absorption	X		X
Other chemical tests	X		X

the presence of excess water. Because of its simplicity, UCS measurement is also suitable for use as a compliance test.

11.3.3 HYDRAULIC CONDUCTIVITY

Hydraulic conductivity indicates the rate at which water can flow through a material, which controls whether leaching occurs by diffusion or advection. Commonly used methods for determination of hydraulic conductivity are ASTM D5084-00[39] and BS 1377: Parts 5(5) and 6(6).[40,41] A wide range of hydraulic conductivity tests is given in Head.[42] S/S materials normally have a low hydraulic conductivity to prevent advection of contaminants. Therefore, a falling head test method is used, in which the volume of water passed through a saturated monolithic specimen under pressure in a given period of time is measured. Stegemann and Coté,[3] however, demonstrated poor reproducibility of this method on a variety of S/S materials and suggested use of a constant head/flow pump method.

Oxygen permeability[43] is sometimes measured for S/S materials, if it is desired to measure permeability without concurrent sample changes due to leaching. An intrinsic permeability, which should be independent of the fluid used to conduct the test, can be calculated from either hydraulic conductivity or oxygen permeability.

11.3.4 WEATHERING RESISTANCE

Freeze/thaw and wet/dry durability tests are conducted to examine the capability of a monolithic S/S material to withstand weathering due to temperature and moisture fluctuations.[44-48] These tests monitor the weight loss of a monolithic S/S material over a stipulated number of repeated cycles of freezing and thawing, or immersion and drying. Mechanical or chemical changes to the matrix are not measured. The freeze/thaw test is considered to be the more severe of the two tests[4] and also found to be the least reproducible.[3]

Sodium or magnesium sulfate soundness[49,50] can be considered an indirect measure of weathering resistance, as it measures the ability of a monolithic material to withstand expansive crystallization within its porosity.

11.4 OTHER CHARACTERIZATION AND EVALUATION TESTS

Many other tests are sometimes used to characterize or evaluate the materials before and after S/S treatment. These test methods are summarized as follows:

11.4.1 INITIAL CONSUMPTION OF LIME (ICL)

ICL[51] is a test for cohesive untreated materials to determine the percentage of lime required to raise the pH of the soil to 12.4. The initial improvement, termed modification, makes the material drier and friable, enabling easy compaction, and the improvement over time, termed stabilization, results in increased strength.

11.4.2 PULVERIZATION

Pulverization[52] is a measure of how well the binder and water have been mixed with the untreated material. It is determined by sieving a known mass of sample first with minimum breakage and then after existing lumps have been separated. It is a site control test carried out on soils that have been stabilized for earthwork purposes.

11.4.3 Particle Size Distribution

Characterization of particle size distribution[53, 54] is carried out by sieving to determine the grading of the untreated material. Particle size distribution affects the workability of the material and the compaction of the material. It is desirable to achieve the maximum density with a reasonable amount of work.[55]

11.4.4 Bulk Density

Bulk density[56-59] is the mass per unit volume of the material. It can be used together with moisture content and specific gravity to calculate S/S material porosity and degree of saturation. These properties are related to durability and leachability, although the relationship is not simple. It can also be used to assess the homogeneity of the S/S material, and monitor volume or moisture changes during curing. Bulk density can also be used together with the mass change factor to calculate volume increase due to treatment.

11.4.5 Specific Gravity

Specific gravity[56,60,61] is a measure of the solids density of a material relative to the density of water. This property is generally needed to calculate other physical properties.

11.4.6 Water Absorption

Water absorption[56,60,62] is a measure of water penetrating into the pores of a material, but not including water adhering to the outside surface of the material. It is related to the volume and characteristics of the porosity of the material, and its saturation. It is determined by immersing a dried sample of known weight in water and obtaining its mass after removing the free water from the sample's surface. The water absorption of an untreated material relates to its water demand in treatment by S/S; the water absorption of a treated material may be related to durability with respect to weathering.

11.4.7 Porosity

Porosity[63-65] is a measure of the proportion of the total volume of the material occupied by pores and is useful in understanding results from other tests, such as hydraulic conductivity.

11.4.8 Moisture Content

Moisture content[66-68] is a measure of the amount of free water in a material and is necessary for determining the water mass balance in S/S treatment and in calculating the L/S ratio in leaching tests. The moisture content of S/S materials is often determined by drying at 60°C, or at a specified humidity, to avoid driving off the water of hydration.[15]

11.4.9 MOISTURE CONDITION VALUE (MCV)

MCV[69,70] is a measure of the compactibility of a soil for use in earthworks. It involves repeatedly dropping a rammer onto a loosely packed sample and measuring the resulting penetration. The MCV is then obtained from a plot of 'change in penetration' versus 'log of number of blows.' It is used as an acceptance test for soils that are to be stabilized with lime or cement. The advantage of MCV as a quality control test is that an instant result is available, whereas it would take longer to obtain a value for the moisture content. It is particularly useful for cohesive material.

11.4.10 DRY DENSITY/MOISTURE CONTENT RELATION

Dry density/moisture content relation[58,71] is a test often used when materials are to be used for earthwork purposes. It involves the compaction of a moist sample at different moisture contents. This is particularly useful for granular materials, while MCV as mentioned above is often used for cohesive materials.

11.4.11 SLUMP

Slump[72-74] is one of several tests that could be conducted to obtain a measure of the workability of a material. The test involves the measurement of the decrease in height of the peak of a cone of freshly mixed material, after removal of the standard cone-shaped mold.

11.4.12 FLOW

Flow[31,75-77] is another of several possible tests to measure material workability. These tests involve the measurement of the resulting spread of the freshly mixed material, after removal of a standard mold.

11.4.13 SETTING TIMES

Initial and final setting times[35,78-80] can be determined by the penetration of a needle into the hydrating sample to observe the early stiffening of a paste prior to strength development. Setting time can also be determined from the heat evolution curve, or by monitoring electrical conductivity or ultrasonic pulse velocity. This property is important for determining the time available for placement of a material and is useful to identify the effects of different contaminants and binders on hydration.

11.4.14 HEAT OF HYDRATION

The heat of hydration[81,82] is the amount of heat evolved upon complete hydration in a calorimeter, at constant temperature, or under adiabatic conditions.[55] The heat of hydration of an S/S material mix can be compared with the heat of hydration of the binder system to assess the relative degree of hydration.

11.4.15 Bound Water

Bound water is the percentage of water present in interlayer spaces or more firmly bound, but not that present in pores larger than interlayer spaces.[83] The quantity present at a given time may help indicate the degree of hydration. It is about 32% for fully hydrated pastes of typical cements.[83] Unfortunately, the method of determination is complicated, and an approximate estimate is obtained by equilibrating a sample, not previously dried below saturation, with an atmosphere of 11% relative humidity.[84]

11.4.16 Shrinkage/Expansion

Shrinkage or expansion[85-87] is measured as a change in length or other dimensions, which permits assessment of the potential volumetric change. Shrinkage may be caused by a decrease in volume of the solid phase during hydration, or be a result of moisture loss or carbonation. Expansion may be caused by swelling of the hydration material due to absorption of water, when freely available, by the cement gel, or by delayed formation of high-volume hydration material such as ettringite. Both may induce stresses in the material, which can lead to its deterioration.

11.4.17 Penetration Resistance

Penetration resistance[88] is carried out to estimate the strength of a material from the depth of penetration by a metal rod driven into the material by a given amount of energy.

11.4.18 California Bearing Ratio (CBR)

The CBR[89-93] is an empirical measure of the bearing capacity of a material. It attempts to measure the resistance of the material to penetrative deformation. Unlike in other strength tests, the CBR is reported as a percentage of the value for a standard crushed rock material. It is widely used in pavement design for roads.

11.4.19 Tensile Strength

Tensile strength[35,38,94-99] is the tensile stress that leads to failure. There are three types of strength tests: direct tensile strength, flexural strength, and splitting tensile strength.

11.4.20 Modulus of Elasticity

The modulus of elasticity provides an understanding of the stiffness of the material, that is, the strain response to an applied stress. Two main test methods are available: static modulus of elasticity and dynamic modulus of elasticity. ASTM C469-02[100] and BS 1881: Part 121[101] are for the former and ASTM C215-02[102] and BS 1881: Part 209[103] are for the latter. The modulus of elasticity is not a constant for a material, but varies with the applied stress. The test conditions are thus critical to ensure that the results from different samples can be compared.

11.4.21 BIODEGRADABILITY

The biodegradability of a treated waste can be estimated from the total organic content in the treated waste. It has been suggested that measurement of biodegradability is necessary only if the total organic content of the treated waste is over 10%, and that ASTM G21-96[104] could be used for this purpose.[15]

11.4.22 MICROSTRUCTURAL EXAMINATION

Microstructural examination of S/S materials can be performed by several techniques, as discussed in detail in Chapter 9. The most commonly used techniques are scanning electron microscopy (SEM), usually with energy dispersive X-ray analysis (EDX) or electron probe microanalysis (EPMA), and X-ray diffractometry (XRD). These techniques allow better understanding of the mechanisms by which contaminants are bound to the matrix and the effects of waste components on binder hydration. However, S/S materials are heterogeneous at microscopic scale, so obtaining representative samples is difficult. Thus, these techniques are more useful in research, or for observing known features and comparing different samples, rather than for general investigation.

11.5 PERFORMANCE CRITERIA

11.5.1 GENERAL

Whereas it is possible to perform testing of S/S materials in order to obtain a quantitative understanding of the material for evaluating technological options and management scenarios, the results from testing are often compared to performance criteria. Such performance criteria may be acceptance limits prescribed for a specific management scenario, e.g., landfill disposal, or they may be derived from a site-specific risk assessment. Conformity with performance criteria may be a regulatory requirement, or simply a part of responsible practice by industry. Since environmental behavior of S/S materials is the subject of ongoing research, development of performance criteria, and assessment of data in comparison with performance criteria, are not usually a straightforward matter.

A recently established database incorporates 1506 literature references and properties of 7953 cement-based S/S materials containing impurities.[105] This database represents a large proportion of information available in the literature and incorporates results of various physical and chemical tests, which have been measured for various mix designs involving binders and wastes, tested over different time periods and temperatures. The range of values for many of these properties has been compiled from the database and is given in Table 11.3. The figures show very wide ranges of results, which emphasize the diverse nature of the materials tested and the properties of the resulting S/S material, as well as differences among test methods.

TABLE 11.3
Typical Ranges of Values for Selected Test Methods[105]

Physical Property	Minimum	Maximum
Bound water (%)	6.8	19.6
Bulk density (as is) (g/cm³)	0.466	2.86
Bulk density (dry) (g/cm³)	0.145	1.18
Bulk density (saturated) (g/cm³)	1.6	1.97
Chloride permeability (mg/kg wet wt)	2540	21,110
Flow table spread diameter (cm)	10.5	13.6
Permeability (m/s)	4×10^{-18}	3.66×10^{-6}
Intrinsic permeability (m²)	2.2×10^{-17}	1.74×10^{-16}
Modulus of elasticity (kPa)	10,200	2.1×10^7
Moisture content (% wet wt)	0.263	98
Oxygen permeability (m/s)	4.06×10^{-16}	5.33×10^{-15}
Penetration resistance (kPa)	16,000	52,400
Porosity (%)	2	75
Setting time — initial (minutes)	25 (25)	2400 (1650)
Setting time — final (minutes)	11 (65)	12,000 (2700)
Shrinkage/expansion (%)	-9.3×10^{-5}	7
Slump (mm)	180	220
Soundness (cm)	0.09	4.12
Specific gravity	0.905	5.189
Tensile strength (kPa)	3.4	10,270
Unconfined compressive strength (kPa)	0	395,000
Water absorption @ 80°C (%)	12.5	19.4
USEPA TCLP (mg/L)		
Leachate pH	1	12.78
As	4.92	17,510
Ba	22.73	418.2
Cd	0.3155	45,990
Cr (total)	2.718	58,070
Cu	0.8202	6291
Hg	9.79×10^{-3}	1828
Ni	0.95	57,930
Pb	3.918	46,940
Zn	0.85	299,100

While these values are indicative of properties typically achieved in S/S materials, ideally, performance criteria must be chosen on the basis of environmental risk, rather than technological capabilities. Table 11.4 lists the test methods and performance criteria used in some commercial projects in the UK.

The following sections briefly discuss performance criteria under the headings of contaminant leachability, unconfined compressive strength, hydraulic conductivity, and other durability tests.

TABLE 11.4
Typical Examples of Tests Performed, Performance Criteria Employed, and End Use in Some of the Commercial Projects Described in Al-Tabbaa and Perera[106]

Commercial Project	Tests Performed	Performance Criteria	End Use
Sealosafe plants, 1974	UCS, permeability, durability, leachability: EP-Tox		Disposal
A13: Thames Avenue to Wennington highway scheme, 1995	Physical and leaching tests	CBR (lower bound) – immediately after compaction: 3%, after 7 days: 5% MCV prior to final compaction: 8.5 lower bound, 12 upper bound 28 day swell – 5 mm upper bound	Lightweight fill for use in embankments
Ardeer site, Scotland, 1995	Strength, permeability, pH, ANC		Remediation of contaminated land for the prevention of further groundwater contamination
West Drayton site, Middlesex, 1997	Leaching tests	Leaching: Dutch Intervention Values	Redevelopment of contaminated ground for housing
Pumpherston site, nr Edinburgh, 1999	Density, UCS, *in situ* penetrometer		Remediation of a contaminated site
Long Eaton site, Nottingham, 2000	Permeability, bearing capacity	Permeability 10^{-9}m/s for passive barrier section, permeability of reactive section comparable with *in situ* soil, minimum bearing capacity: 150 kPa	Remediation and enabling works on a contaminated site for a new retail supermarket
Leytonstone site, London, 2000	CBR, permeability		Redevelopment of a brownfield site for the construction of a school
Winterton Holme water treatment works site, 2000	Strength, permeability, leaching tests		Disposal in landfill

11.5.2 Contaminant Leachability

The RCRA requires the USEPA to classify wastes as either hazardous or non-hazardous. Under the regulations implementing Subtitle C of RCRA, wastes are designated as hazardous in two ways:

- Solid wastes that exhibit certain characteristics, as listed in 40 CFR Part 261, Subpart C
- Solid wastes that are specifically listed as hazardous in 40 CFR Part 261, Subpart D

The generator is responsible to use TCLP to find out whether or not a solid waste exhibits a hazardous waste characteristic. Table 11.5 lists the contaminants of concern and their concentration limits as regulated in 40 CFR 261.24.[107]

TCLP is being used as the regulatory leaching test in many regions and countries. The Province of Alberta, Canada, adopted TCLP as a regulatory test in 1996. Although other countries may use TCLP as a regulatory test, their regulated contaminants and levels may be different. For example, Ontario Regulation 558/00[108] adopted the USEPA TCLP testing procedure in 2000, but added 15 new inorganic and 73 new organic contaminants.

The RCRA regulations in 40 CFR 260.20 and 260.22 contain provisions that allow the petitioning of the governing agency to exclude or "delist" a listed hazardous waste from the universe of regulated hazardous wastes. In the past, delisting was considered virtually unachievable and economically unfeasible. In 2000, the USEPA published the *EPA RCRA Delisting Program Guidance Manual for the Petitioner*,[109] which provides a streamlined framework to petitioners and makes the delisting application much easier. The USEPA evaluates the potential hazards of waste through the use of appropriate fate and transport models, which calculate possible exposure to hazardous chemicals that might be released from petitioned wastes after disposal, based on a reasonable, worst-case management scenario. A major concern is ingestion of contaminated groundwater. To evaluate this concern, the agency typically relies on leachate data as determined by TCLP. The leachable concentrations and the estimated waste volume then are used as inputs to an appropriate fate and transport model, for example, EPA's Composite Model for Leachate Migration with Transformation Products (EPACMTP),[110] to predict the constituent concentrations in the groundwater at a hypothetical exposure point. The calculated exposure-point concentrations are typically compared to drinking water standards or other EPA health-based levels.

Section 6.2.2 of the Guideline[109] provides guidelines for stabilized wastes. If the petitioned waste is generated from the chemical stabilization of a listed waste, then leachable metal concentrations should be tested using the Multiple Extraction Procedure, SW-846 Method 1320, as well as by TCLP analyses to assess the long-term stability of the waste.

The majority of delisted wastes are metal-bearing wastes (such as F006 and F019 wastewater treatment sludges and treated K061 electric arc furnace dusts). Any

TABLE 11.5
Maximum Concentration of
Contaminants for the Toxicity
Characteristic[107]

Contaminant	Regulatory Level (mg/L)
Arsenic	5
Barium	100
Cadmium	1
Chromium	5
Lead	5
Mercury	0.2
Selenium	1
Silver	5
Benzene	0.5
Carbon tetrachloride	0.5
Chlordane	0.03
Chlorobenzene	100
Chloroform	6
o-Cresol	200
m-Cresol	200
p-Cresol	200
Cresol	200
2,4-Dichlorophenoxyacetic Acid	10
1,4-Dichlorobenzene	7.5
1,2-Dichloroethane	0.5
1,1-Dichloroethylene	0.7
2,4-Dinitrotoluene	0.1
Endrin	0.02
Heptachlor (and its epoxide)	0.008
Hexachlorobenzene	0.1
Hexachlorobutadiene	0.5
Hexachloroethane	3
Lindane	0.4
Methoxychlor	10
Methyl ethyl ketone	200
Nitrobenzene	2
Pentachlorophenol	100
Pyridine	5
Tetrachloroethylene	0.7
Toxaphene	0.5
Trichloroethylene	0.5
2,4,5-Trichlorophenol	400
2,4,6-Trichlorophenol	2
2,4,5-TP (Silvex)	1
Vinyl chloride	0.2

treatment residual that meets current Best Demonstrated Available Technology (BDAT) levels will usually be a good delisting candidate.

Under the EU Landfill Directive,[111] the acceptance limits for different categories of landfill are set at EU or member state level. Wastes for disposal will be required to meet the general interim waste acceptance criteria given in Schedule 1(1) of the 2002 Landfill Regulations and additional interim waste criteria set out for landfills accepting hazardous waste, nonhazardous waste, and inert waste. Eventually, member states are expected to set their own full criteria. For example, the UK Environment Agency[112] has set out the expected full criteria for acceptance of granular waste to landfills in Table 11.6 and Table 11.7. These have been published to assist in the consideration of permits for new landfills, and to allow producers and operators to consider the implications of changing from interim to full criteria. The leaching limit values given are calculated for total release at L/S = 2 and 10 l/kg, of the CEN standard two-part batch test (BS EN12457: Part 3, see Section 11.2.2.1.7). The UK Environment Agency is developing criteria for monolithic waste; until they are available, tests and limiting values must be agreed to by the Environment Agency on a case-by-case basis.

Performance criteria for remediation of contaminated land have been evolving over the past two decades. In the UK, past practice has been to take guidance values for contaminated land assessment and remediation from:

1. ICRCL 59/83[114]
2. Contamination Classification Thresholds for Disposal of Contaminated Soils[115]
3. The Dutch List[116]

Although used in the past, these guidance values do not relate to the contaminated land provisions of Part IIA of the UK Environment Protection Act of 1990. Contaminated land remediation criteria are now selected on the basis of risk assessment. The acceptance criteria may be generic in some cases and site-specific in others.[117] CERCLA covers this for the U.S., where such criteria are formalized in a Record of Decision (ROD) for the remediation (see Chapter 2 for a discussion of CERCLA and RCRA). Generic soil guideline values (SGVs) have been determined using the Contaminated Land Exposure Assessment (CLEA) model. Methodologies available in the UK for deriving site-specific criteria include:

1. Methodology for the derivation of remedial targets for soil and groundwater to protect water resources.[118]
2. Contaminated Land Exposure Assessment.[119-122]
3. Method for deriving site-specific human health assessment criteria for contaminants in soil.[123]
4. Risk-based corrective action (RBCA) protocol commonly used in groundwater risk assessments.[124,125]
5. Risk-integrated software for clean-ups (RISC).[126]

TABLE 11.6
Leaching Limit Values for the Acceptance of Granular Wastes in Landfills[113]

	Hazardous Waste to Hazardous Waste Sites (set 1)		Hazardous Waste to Non-Hazardous Waste Sites (set 2)		Inert Waste Sites (set 3)	
Components	L/S = 2 l/kg mg/kg	L/S = 10 l/kg mg/kg	L/S = 2 l/kg mg/kg	L/S = 10 l/kg mg/kg	L/S = 2 l/kg mg/kg	L/S = 10 l/kg mg/kg
As	6	25	0.4	2	0.1	0.5
Ba	100	300	30	100	7	20
Cd	0.6	1	0.06	0.1	0.03	0.04
Cr_{total}	25	70	4	10	0.2	0.5
Cu	50	100	25	50	0.9	2
Hg	0.1	0.4	0.005	0.02	0.003	0.01
Mo	20	30	5	10	0.3	0.5
Ni	20	40	5	10	0.2	0.4
Pb	25	50	5	10	0.2	0.5
Sb	2	5	0.2	0.7	0.02	0.06
Se	4	7	0.3	0.5	0.06	0.1
Zn	90	200	25	50	2	4
Cl	17,000	25,000	10,000	15,000	550	800
F	200	500	60	150	4	10
SO_4	25,000	50,000	10,000	20,000	560#	1,000#
TDS*	70,000	100,000	40,000	60,000	2,500	4,000
DOC**	480	1,000	380	800	240	500
Phenol index	–	–	–	–	0.47	1

* The values for Total Dissolved Solids (TDS) can be used alternatively to the values of Sulfate, Fluoride, and Chloride.

**If the waste does not meet these values for dissolved organic carbon (DOC) at its own pH, it may alternatively be tested at L/S = 10 l/kg and a pH of 7.5 to 8.0. The waste may be considered as complying with the acceptance criteria for DOC, if the result of this determination does not exceed 1000, 800, and 500 mg/kg for sets 1, 2, and 3 respectively (a draft method based on prEN 14429 is available).

If the waste does not meet these values for sulfate, it may still be considered as complying with the acceptance criteria if the leaching does not exceed either of the following values: 1500 mg/L as Co at L/S = 0.1 l/kg and 6000 mg/kg at L/S = 10 l/kg. It will be necessary to use the percolation test (prEN 14405) to determine the limit value at L/S 0.1 l/kg under initial equilibrium conditions, whereas the value at L/S = 10 l/kg may be determined either by a batch leaching test (BS EN 12457: Part 2 or BS EN 12457: Part 3) or by the percolation test (prEN 14405) under conditions approaching local equilibrium.

Note: For inorganic parameters of concern not listed in the table, the maximum leachable value obtained from the percolation test (prEN 14405) can be used as the source term for those parameters in the risk assessment outlined in Schedule 1 (1) of the 2002 Regulations.

For radioactive wastes, the ANSI/ANS-16.1-2003 (see Section 11.2.3.1.2) is the regulatory test procedure used worldwide, although other tests have been used. The U.S. Nuclear Regulatory Commission (NRC) topical position paper[127] specifies a leachability index of at least 6 (units of cm^2/s for the effective diffusion coefficient)

TABLE 11.7
Additional Limit Values for the Acceptance of Granular Wastes in Landfills[113]

Parameter	Hazardous Waste to Hazardous Waste Sites (set 1) mg/kg	Hazardous Waste to Non-Hazardous Waste Sites (set 2) mg/kg	Inert Waste Sites (set 3) mg/kg
LOI*	10%	–	–
TOC**	6%	5%	30,000
Ph	–	Minimum 6	–
ANC	Must be evaluated between the pH of the waste in question, pH 6, and the pH of the site leachate	Must be evaluated between the pH of the waste in question, pH 6, and the pH of the site leachate	–
BTEX	–	–	6
PCBs (7 congeners)	–	–	1
Mineral oil (C10 to C40)	–	–	500

* Either Loss on Ignition (LOI) or Total Organic Carbon (TOC) must be used.
**If this value is not achieved (for soils in the case of set 3), a higher limit value may be admitted by the competent authority, provided that the DOC value of 1000, 800, and 500 mg/kg is achieved for sets 1, 2, and 3 respectively at L/S 10 at its own pH or pH 7.

for radionuclides. This is not a challenging leach resistance, since it represents an effective diffusion coefficient of only 10^{-6} cm²/s, compared to a diffusivity on the order of 10^{-5} cm²/s for many salt ions in water. Often a waste acceptance criteria (WAC) specifies a minimum leachability index of at least 8, and sometimes 10 or higher. A value of 8 can be attained for species still dissolved in the pore solution, but 10 requires some form of stabilization or an impervious waste form, as opposed to the accessible porosity of cement.

11.5.3 Unconfined Compressive Strength (UCS)

UCS performance requirements vary according to the end use. However, some guidelines and suggestions on limits exist. Some of these are given below.

An immersed UCS of 350 kPa at 28 days is suggested by USEPA guidelines for materials that are to be disposed in a landfill.[128] This limit takes into consideration factors such as weight of overburden and land-moving equipment. In the Netherlands[129] and France[117] a UCS of 1 MPa is suggested for disposal. However, a higher UCS limit of 3500 kPa has been suggested by WTC[15] for disposal in a sanitary landfill, where handling, placement, and covering operations are not tailored for S/S material, and compaction of municipal waste might subject the S/S material to higher stresses. It has also been suggested that the UCS with immersion should not be less than 80% of the UCS without immersion.[15,130] Given that UCS is an indicator of the progress or inhibition of cement hydration reactions, it may be appropriate to also consider this more indirect aspect in setting performance criteria.

Cement-stabilized materials for utilization in sub-bases and bases, under the British specifications for the four categories CBM1-4,[130] are required to have minimum 7-day cube compressive strengths of 4.5, 7, 10, and 15 MPa, respectively.[131] In the Netherlands, the UCS requirement for stabilized material for use in sub-base layers is 3 to 5 MPa. However, the American and South African specifications rank strength as not being the primary requirement for cement-stabilized materials.[130]

With respect to radioactive waste, the tests to demonstrate stability of cement-solidified Class B and C wastes include an average UCS of 3.5 MPa (500 psi) using ASTM C39/C39M.[127]

11.5.4 Hydraulic Conductivity

The hydraulic conductivity limit is usually taken as 10^{-9} m/s for in-ground treatment (this value is usually used for clay liners and cut-off walls)[132] and utilization.[15] USEPA tends to use 10^{-9} m/s for disposal in a landfill.[133] A higher limit value of 10^{-8} m/s is suggested for disposal scenarios in the WTC protocol, on the basis that secondary engineered barrier systems provide additional protection of the environment.[15]

11.5.5 Other Durability Tests

S/S materials subjected to both freeze/thaw and wet/dry durability testing at 28 days are required to survive 12 cycles of the prescribed test procedures with a maximum of 10% of the corrected cumulative dry mass loss.[15,44,48]

For radioactive wastes, a technical position paper of the U.S. Nuclear Regulatory Commission[127] summarized some testing and durability requirements for Class B and C wastes:

1. The waste should be a solid form or in a container or structure that provides stability after disposal.
2. The waste should not contain free-standing and corrosive liquids. That is, the wastes should contain only trace amounts of drainable liquid, and, as required by 10 CFR 61.56(b)(2), in no case may the volume of free liquid exceed 1% of the waste volume when wastes are disposed of in containers designed to provide stability, or 0.5% of the waste volume for solidified wastes.
3. The waste or container should be resistant to degradation caused by radiation effects.
4. The waste or container should be resistant to biodegradation.
5. The waste or container should remain stable under the compressive strength inherent in the disposal environment.
6. The waste or container should remain stable if exposed to moisture or water after disposal.
7. The as-generated waste should be compatible with the solidification medium or container.

In addition to leachability and UCS (discussed in 11.5.2 and 11.5.3), the tests to demonstrate stability of cement-solidified Class B and C wastes include resistance to thermal cycling,[134] resistance to 100 Mrads dose, resistance to biodegradation,[104] resistance to immersion for 90 days, and generation of < 0.5 vol% free-standing liquid.[135]

11.6 INTEGRATED EVALUATION PROTOCOLS

11.6.1 INTRODUCTION

The purpose of S/S is to maximize the containment of environmental contaminants by both physical and chemical means and to convert the hazardous waste into an environmentally acceptable waste form. Whenever possible, utilization of stabilized/solidified hazardous wastes should be preferred over disposal in order to decrease the burden of land disposal.[15] If a waste does require landfilling, however, the degree of environmental protection provided by S/S should allow for disposal in less costly landfilling facilities, by reducing or eliminating the need for engineered barriers or liner systems.

S/S radioactive and mixed waste forms usually cannot be utilized, and are also disposed of differently from S/S hazardous waste forms. Thus, the evaluation of S/S radioactive and mixed waste forms is different from that of S/S hazardous waste forms. This section introduces two integrated evaluation protocols for S/S waste forms.

11.6.2 INTEGRATED EVALUATION FRAMEWORK

Recently, Kosson et al.[136] proposed an alternative framework for evaluation of inorganic constituent leaching from wastes and secondary materials, as shown in Figure 11.1. The framework is based on the measurement of intrinsic leaching properties of the material in conjunction with mathematical modeling to estimate release under field management scenarios. To achieve the desired framework\, a three-tiered testing program is proposed:

- Tier 1: Screening-based assessment (availability)
- Tier 2: Equilibrium-based assessment (over a range of pH and liquid-to-solid ratio conditions)
- Tier 3: Mass transfer-based assessment

It is suggested that waste management or utilization scenarios should be used to link laboratory assessment results to impact assessment. A series of laboratory tests will be conducted based on the leaching mode that controls release (equilibrium or mass transfer), the site-specific liquid-to-solid ratio, the field pH, and a time frame for assessment. Using the laboratory measurements, a release model is established to estimate the cumulative mass of the constituent released over the time frame for a percolation/equilibrium scenario.

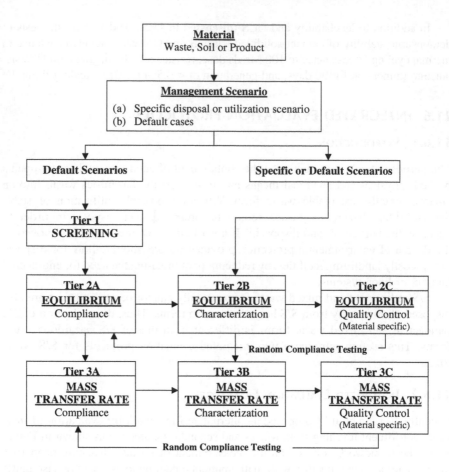

FIGURE 11.1 Alternative framework for evaluation of leaching.[136]

Tier 1: Screening Tests

The purpose of the screening test is to measure the available leachable contaminants over the broad range of anticipated environmental conditions. Of course, the test should be easy and quick to perform. Several testing procedures have been proposed for this purpose, which include the two-step sequential extraction procedure with particle size < 300 μm, L/S = 100 ml/g, and control at pH 8 and 4,[16] a single extraction using EDTA to chelate metals of interest in solution near a neutral pH.[137] Either of these approaches can be used as a screening test, but both approaches have practical limitations relative to implementation.

Tier 2A: Solubility and Release as a Function of pH

The objective of this testing is to determine the acid/base titration buffering capacity of the tested material and the liquid-solid partitioning equilibrium of the contaminants.

This can be measured using prEN 14429: Leaching Behavior Test – pH Dependence Test with Initial Acid/Base Addition.[17]

Tier 2B: Solubility and Release as a Function of L/S Ratio (RU-SR003.1)

A range of L/S ratios (e.g., 10, 5, 2, 1, and 0.5 ml/g dry material) are used to provide an estimate of constituent concentration as the extraction L/S ratio approaches the bulk porosity of the material. The particle size of the material should be reduced either to < 300 mm, < 2 mm, or < 5 mm depending on the nature of the material. All extractions are conducted at room temperature (20 ± 2°C) in leak-proof vessels that are tumbled in a end-over-end fashion at 28 ± 2 rpm. Leachant is filtered and analyzed for constituents of concern.

Tier 3: Mass Transfer Rate

The objective of mass transfer rate tests is to measure the rate of constituents of potential concern (COPC) release from a monolithic material (e.g., solidified waste form or concrete matrix) or a compacted granular material. This can be done following ANSI/ANS-16.1-2003[21] or similar testing methods such as NEN 7345[25] and CEN TC/292.

11.6.3 WTC EVALUATION PROTOCOL[15]

11.6.3.1 Evaluation Framework

Wastes are managed in many different settings and under a range of conditions that affect waste leaching. Thus, the performance of a waste or an S/S waste form should be evaluated based on its utilization or disposal scenarios. In 1991, the Wastewater Technology Centre of Environment Canada (WTC) proposed an evaluation protocol for S/S waste forms, which could be used as a decision-making tool. The WTC protocol includes three levels of evaluation: Level 0: Information on Untreated Waste, S/S Process, and Waste Forms; Level 1: Chemical Immobilization; and Level 2: Physical Entrapment. Based on the degree of contaminant containment and physical properties of S/S waste form, it can be considered for different disposal or utilization scenarios. The WTC protocol proposed two utilization and two disposal scenarios: unrestricted utilization, controlled utilization, segregated landfill, and sanitary landfill. S/S waste forms that do not qualify for utilization or disposal according to one of these scenarios would need to be disposed in a secure landfill or subjected to a more effective treatment process.

An unrestricted utilization scenario would require the S/S materials to have negligible leaching potential and be considered for use in any way similar to a natural material; a controlled utilization scenario requires the S/S material to have a leaching potential acceptable for a specific usage. A segregated landfill, which does not necessarily have an engineered barrier or leachate collection system, would accept S/S materials that fail to satisfy utilization, after separation from other waste materials, provided that they fall within the limits of the landfill; a sanitary landfill accepts

TABLE 11.8
Key Properties and Performance Indicators of S/S Waste Forms[15]

Level	Key Property	Performance Indicator	Expression
1	Contaminant Concentration	Contaminant Concentration	mg/kg
	Initial Leachate Concentration	Initial Leachate Concentration	mg/L of Leachate
	Availability for Leaching	Amount of Contaminant Available for Leaching	mg/kg
	Equilibrium-Based Assessment (over a range of pH values and liquid-to-solid ratios)	Equilibrium-Based Assessment (over a range of pH values and liquid-to-solid ratios)	mg/L of Leachate
2	Mass Transfer Assessment	Contaminant Mobility	Leachability Index
	Physical Property	Hydraulic Conductivity	m/s
		Compressive Strength	MPa
	Durability	Freezing/Thawing Cycles	% mass loss
		Wetting/Drying Cycles	% mass loss
		Biodegradability	[pass/fail]

S/S materials for co-disposal with municipal garbage where they have failed to satisfy the other three scenarios, provided it is within the acceptable limits of the landfill. However, it should be mentioned that with the new regulations being set up (described in Section 11.5.2), the above landfill scenarios might no longer be viable.

11.6.3.2 Evaluation Approach

In order to decide the utilization and disposal scenarios for S/S waste forms, several key properties of S/S wastes related to their potential environmental impact in utilization and disposal scenarios should be evaluated, as shown in the second column of Table 11.8. One or several performance indicators are used to characterize each key property as listed in the third column of Table 11.8. A performance indicator is a direct experimental result that is used as a measure of a key property. The performance indicators can be divided into two levels of containment: Level 1: containment by chemical immobilization, and Level 2: containment by physical entrapment.

A flowchart of the procedure used to evaluate an S/S waste against this protocol is shown in Figure 11.2. First, the waste is subjected to Level 1 testing. The results are then compared against the Level 1 criteria for unrestricted utilization (denoted as "A" in Figure 11.2). If the waste passes the criteria, it is acceptable for unrestricted utilization. If it fails, the Level 1 results are compared with the Level 1 criteria for controlled utilization ("B"). If the waste passes these criteria, it is acceptable for controlled utilization. If it fails, either the Level 1 results are compared with the Level 1 criteria to determine if the waste is acceptable for disposal in a segregated landfill ("C"), or Level 2 testing is conducted to determine if the waste is acceptable for controlled utilization based on Level 2 criteria. This decision procedure can be repeated as indicated in Figure 11.2 to determine the suitability of the waste form for the remaining disposal scenarios. If the waste form fails both the Levels 1 and 2

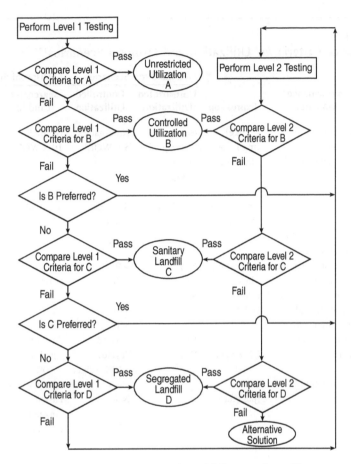

FIGURE 11.2 Decision flowchart for evaluation of S/S waste forms.[15]

criteria for disposal in a sanitary landfill ("D"), it must either be disposed in a secure landfill or, preferably, subjected to a new treatment technique and re-evaluated.

The rationale for selecting performance indicators and performance criteria is given in the original document.[15] Testing methods are recommended for each performance indicator in Table 11.8. For each testing method, the source is identified, the main features are described, and potential problems are discussed. Methods requiring improvement or alternative methods are mentioned where appropriate. Performance criteria for each of the four utilization and disposal scenarios are proposed in Table 11.9 and Table 11.10.

11.7 SUMMARY

This chapter has presented the range of test methods available for the assessment of S/S materials and also treatment. The most commonly used tests, namely leachability, UCS, hydraulic conductivity, and durability, were detailed. It is also clear

TABLE 11.9
Performance Criteria for Utilization and Disposal Scenarios[15]

Level	Performance Indicator	Expression	Utilization Scenarios		Disposal Scenarios	
			Unrestricted Utilization	Controlled Utilization	Segregated Landfill	Sanitary Landfill
1	Contaminant concentration	[mg/kg] [mg/L] of leachate	No organics Water quality limit	No organics $5 \times$ WQL[4]	No limit[3] $10 \times$ WQL[4]	$100 \times$ WQL[4] No limit[3]
1	Tier 1: Screening-based assessment (availability)		(WQL)[4]			
	Tier 2: Equilibrium-based assessment (over a range of pH and L/S conditions)	[mg/wet kg] from waste	See Table 11.4	See Table 11.4	See Table 11.4	See Table 11.4
2	Contaminant mobility in the matrix[2]	Leachability Index	N/A N/A	> 9 $< 10^{-9}$	> 8 $> 10^{-8}$	> 8 $> 10^{-8}$
2	Hydraulic conductivity	[m/s]	N/A			
2	Physical strength Before immersion After immersion	[kPa]	N/A	440 350 Pass 12	440 350 Pass 12	4400 3500 Pass 12
2	Weathering	[% weight loss]	N/A	cycles $< 10\%$ weight loss	cycles $< 10\%$ weight loss	cycles $< 10\%$ weight loss
2	Biodegradability	[pass/fail]		N/A	Pass if organic content $> 10\%$	Pass if organic content $> 10\%$

[1] Not applicable.
[2] Criterion is applicable to every target contaminant.
[3] Local regulations may limit bulk concentrations of organic content and organic and inorganic contaminants in wastes.
[4] e.g., Ontario Drinking Water Objectives,[108] but may vary according to the jurisdiction.

that there are a number of leaching tests available, and performance criteria for different types of testing vary depending on the management scenario of the S/S material.

ACKNOWLEDGMENTS

This chapter is rewritten based on a report originally prepared for the UK Network on Stabilisation/Solidification Treatment and Remediation (STARNET) (www-starnet.eng.cam.ac.uk), which forms part of a volume entitled "Stabilisation/Solidification Treatment and Remediation: Advances in S/S for Waste and

TABLE 11.10
Proposed Criteria for the Amount Available for Leaching for Some Metals[15]

Contaminant	Regulatory Limit Expressed as — Leachate Concentration [mg/L]	Regulatory Limit Expressed as — Amount Leached [mg/kg wet waste]	Criteria for Amount Available for Leaching [mg/wet kg waste] — Unrestricted Utilization	Controlled Utilization	Segregated Landfill	Sanitary Landfill
Arsenic	5.0	100	1	5	10	100
Barium	100.0	2000	20	100	200	2000
Cadmium	1.0	20	0.2	1	2	20
Chromium	5.0	100	1	5	10	100
Lead	5.0	100	1	5	10	100
Mercury	0.2	4	0.04	0.2	0.4	4

Regulatory limits for the USEPA TCLP.[107] The factor of 20 between columns 2 and 3 corresponds to the liquid-to-solid ratio in the leachate test.

Contaminated Land." This volume contains the proceedings of a conference to be held in April 2005 in Cambridge, UK, and will be published by A.A. Balkema Publishers. The authors gratefully acknowledge the funding for STARNET by the Engineering and Physical Research Council (EPSRC). The authors are also grateful to the core members of STARNET for their contributions and in particular to Brian Bone and Leslie Heasman for their contributions to the report.

REFERENCES

1. Conner, J.R., *Chemical Fixation and Solidification of Hazardous Wastes*, Van Nostrand Reinhold, New York, 1990.
2. Stegemann, J.A. and Coté, P.L., Summary of an investigation of test methods for solidified waste evaluation, *Waste Management*, Vol. 10, pp. 41–52, 1990.
3. Stegemann, J.A. and Coté, P.L., Investigation of Test Methods for Solidified Waste Evaluation, Appendix B: Test Methods for Solidified Waste Evaluation, Manuscript Series Document TS – 15, Environment Canada, Ottawa, 1991.
4. LaGrega, M.D., Buckingham, P.L., and Evans, J.C., Stabilisation and solidification, in *Hazardous Waste Management,* McGraw-Hill, New York, pp. 641–704, 1994.
5. Van der Sloot, H., Heasman, L., and Quevauviller, P., *Harmonisation of Leaching/Extraction Tests*, Elsevier, Amsterdam, 1997.
6. Wastewater Technology Centre, Compendium of Waste Leaching Tests, Report EPS 3/HA/7, Environment Canada, Ottawa, 1990.
7. Lewin, K., Bradshaw, K., Blakey, N.C., Turrell, J., Hennings, S.M., and Flavin, R.J., Leaching Tests for Assessment of Contaminated Land: Interim NRA Guidance, NRA, R&D Note 301, Bristol, UK, 1994.

8. U.S. EPA Method 1311, Toxicity characteristic leaching procedure, Test Methods for Evaluation of Solid Wastes, Physical/Chemical Methods, SW846, URL: http://www.epa.gov/epaoswer/hazwaste/test/pdfs/1311.pdf, United States Environmental Protection Agency, 2003.

9. U.S. EPA Method 1310A, Extraction procedure (EP) toxicity test method and structural integrity test, Test Methods for Evaluation of Solid Wastes, Physical/Chemical Methods, SW846, URL: http://www.epa.gov/epaoswer/hazwaste/test/pdfs/1310A.pdf, United States Environmental Protection Agency, 2003.

10. U.S. EPA Method 1312, Synthetic precipitation leaching procedure, Test Methods for Evaluation of Solid Wastes, Physical/Chemical Methods, SW846, URL: http://www.epa.gov/epaoswer/hazwaste/test/pdfs/1312.pdf, United States Environmental Protection Agency, 2003.

11. ASTM D 3987-02, Standard test method for shake extraction of solid waste with water, Annual Book of ASTM Standards, Vol. 11.04, Environmental Assessment: Hazardous Substances and Oil Spill Responses; Waste Management, American Society for Testing and Materials, West Conshohocken, PA, 2003.

12. DIN 38414 S4, Determination of leachability by water (S4), German Standard Methods for Examination of Water, Wastewater and Sludge, Sludge and Sediments (group S), Deutsches Institut für Normung e.V., Germany, 1984.

13. BS EN 12457: *Characterization of Waste, Leaching, Compliance Test for Leaching of Granular Waste Materials and Sludges* — Part 1: One stage batch test at a liquid to solid ratio of 2 l/kg for materials with high solid content and with particle size below 4 mm (without or with size reduction); Part 2: One stage batch test at a liquid to solid ratio of 10 l/kg for materials with particle size below 4 mm (without or with size reduction); Part 3: Two stage batch test at a liquid to solid ratio of 2 l/kg and 8 l/kg for materials with high solid content and with particle size below 4 mm (without or with size reduction); Part 4: One stage batch test at a liquid to solid ratio of 10 l/kg for materials with particle size below 10 mm (without or with size reduction); British Standards Institution, London, 2002.

14. Heasman, L., The significance of leaching tests within European landfill directive, CLAIRE Newsletter, Summer 2002, p. 8, 2002.

15. Stegemann, J.A. and Côté, P.L., "A proposed protocol for evaluation of solidified wastes," *Science of the Total Environment*, Vol. 178, No. 1–3, pp. 103–110, 1996.

16. NEN 7341, Determination of the availability for leaching of inorganic components from granular materials, Leaching Characteristics of Earthy and Stony Building and Waste Materials, Leaching Tests, Netherlands Normalisation Institute, Delft, The Netherlands, 1995.

17. prEN14429, influence of pH on leaching with initial acid/base addition, Characterisation of Waste, Leaching Behaviour Test, CEN/TC292, NNI, Delft, The Netherlands, 2002.

18. Isenburg, J.E. and Moore, M., Generalised acid neutralisation capacity test, In: *Stabilisation and Solidification of Hazardous, Radioactive and Mixed Waste*, T.M. Gilliam and C.C. Wiles (Eds.), STP 1123, American Society for Testing and Materials, West Conshohocken, PA, 1992.

19. Tessier, A., Campbell, P.G.C., and Bisson, M., Sequential extraction procedure for the speciation of particulate trace metals, *Analytical Chemistry*, Vol. 51, No. 7, pp. 844–851, 1979.

20. U.S. EPA Method 1320, Multiple extraction procedure, Test Methods for Evaluation of Solid Wastes, Physical/Chemical Methods, SW846, URL: http://www.epa.gov/epaoswer/hazwaste/test/pdfs/1320.pdf, United States Environmental Protection Agency, 2003.

21. ANSI/ANS 16.1-2003, Measurement of the leachability of solidified low-level radioactive wastes by a short-term procedure, American Nuclear Society, La Grange Park, IL, 2003.

22. ASTM D4793-02, Standard test method for sequential batch extraction of waste with water, *Annual Book of ASTM Standards*, Vol. 11.04, American Society for Testing and Materials, West Conshohocken, PA, 2003.

23. ASTM D5284-02, Standard test method for sequential batch extraction of waste with acid extraction fluid, *Annual Book of ASTM Standards,* Vol. 11.04, American Society for Testing and Materials, West Conshohocken, PA, 2003.

24. NEN 7349, Determination of the leaching of inorganic components from granular materials with the cascade test, *Leaching Characteristics of Earthy and Stony Building and Waste Materials, Leaching Tests*, Netherlands Normalisation Institute, Delft, The Netherlands, 1995.

25. NEN 7345, Determination of the leaching of inorganic components from buildings and monolithic waste materials with the diffusion test, *Leaching Characteristics of Earthy and Stony Building and Waste Materials,. Leaching Tests,* Netherlands Normalisation Institute, Delft, The Netherlands, 1995.

26. CEN/TC 292 (in preparation), Characterisation of waste — compliance leaching test for monolithic material. Work Item 00292010, Netherlands Normalisation Institute, Delft, The Netherlands.

27. ASTM D 4874, Standard test method for leaching solid waste in a column apparatus, *Annual Book of ASTM Standards*, Vol. 11.04, American Society for Testing and Materials, West Conshohocken, PA, 2003.

28. NEN 7343, Determination of the leaching of inorganic components from granular materials with the column test, *Leaching Characteristics of Earthy and Stony Building and Waste Materials. Leaching Tests,* Netherlands Normalisation Institute, Delft, The Netherlands, 1995.

29. prEN14405, Characterisation of waste, *Leaching Behaviour Test, Up-Flow Percolation Test,* CEN/TC292, Netherlands Normalisation Institute, Delft, The Netherlands, 2002.

30. ASTM C39/C39M, Standard test method for compressive strength of cylindrical concrete specimens, *Annual Book of ASTM Standards*, Vol. 04.02, American Society for Testing and Materials, West Conshohocken, PA, 2003.

31. ASTM C109/C109M, Standard test method for compressive strength of hydraulic cement mortars, *Annual Book of ASTM Standards*, Vol. 04.01, American Society for Testing and Materials, West Conshohocken, PA, 2003.

32. ASTM D1633-00, Standard test method for compressive strength of moulded soil-cement cylinders, *Annual Book of ASTM Standards*, Vol. 04.08(I), American Society for Testing and Materials, West Conshohocken, PA, 2003.

33. BS 1881: Part 116, Method for determination of compressive strength of concrete cubes, *Testing Concrete,* British Standards Institution, London, 1983.

34. BS EN 12390: Part 3, Compressive strength of test specimens, *Testing Hardened Concrete*, British Standards Institution, London, 2002.

35. BS EN 196: Part 1, Determination of strength, *Methods of Testing Cement*, British Standards Institution, London, 1995.

36. BS 1377: Part 7(7), Shear strength tests (total stress), *Methods of Test for Soils for Civil Engineering Purposes*, British Standards Institution, London, 1990.

37. BS 1924: Part 2(4.1) and Part 2(4.2), Determination of the compressive strength of cylindrical specimens, Stabilised materials for civil engineering purposes, *Methods of Test for Cement-Stabilised and Lime-Stabilised Materials, Strength and Durability Tests*, British Standards Institution, London, 1990.

38. BS 1924: Part 2(4.4), Determination of the tensile splitting strength, Stabilised materials for civil engineering purposes, *Methods of Test for Cement-Stabilised and Lime-Stabilised Material. Strength and Durability Tests*, British Standards Institution, London, 1990.

39. ASTM D5084-00, Standard test method for measurement of hydraulic conductivity of saturated porous material using a flexible wall permeameter, *Annual Book of ASTM Standards,* Vol. 04.08(I), American Society for Testing and Materials, West Conshohocken, PA, 2003.

40. BS 1377: Part 5(5), Compressibility, permeability and durability tests, *Methods of Test for Soils for Civil Engineering Purposes*, British Standards Institution, London, 1990.

41. BS 1377: Part 6(6), Consolidation and permeability tests in hydraulic cells with pore pressure measurement, *Methods of Test for Soils for Civil Engineering Purposes,* British Standards Institution, London, 1990.

42. Head K.H., Triaxial consolidation and permeability tests, In: *Manual of Soil Laboratory Testing*, Vol. 3, Chapter 20, pp. 1001–1027, 1992.

43. Kollek, J.J., The determination of the permeability of concrete to oxygen by the Cembureau method — a recommendation, *Materials and Structures*, Vol. 22. No. 129, pp. 225–230, 1989.

44. ASTM D4842-90 (Re-approved 2001), Standard test method for determining the resistance of solid wastes to freezing and thawing, *Annual Book of ASTM Standards,* Vol. 11.04, American Society for Testing and Materials, West Conshohocken, PA, 2003.

45. BS 812: Part 124, Method for determination of frost-heave, *Testing Aggregates*, British Standards Institution, London, 1989.

46. BS 1377: Part 5(7), Compressibility, permeability and durability tests, *Methods of Test for Soils for Civil Engineering Purposes,* British Standards Institution, London, 1990.

47. BS 1924: Part 2(4.8), Determination of the frost heave, *Stabilised Materials for Civil Engineering Purposes, Methods of Test for Cement-Stabilised and Lime-Stabilised Material, Strength and Durability Tests,* British Standards Institution, London, 1990.

48. ASTM D4843-88 (Re-approved 1999), Standard test method for wetting and drying test of solid wastes, *Annual Book of ASTM Standards,* Vol. 11.04, American Society for Testing and Materials, West Conshohocken, PA, 2003.

49. BS EN 196: Part 3, Determination of setting time and soundness, *Methods of Testing Cement,* British Standards Institution, London, 1995.

50. ASTM C88-99a Standard test method for soundness of aggregates by use of sodium sulphate or magnesium sulphate, *Annual Book of ASTM Standards*, Vol. 04.02, American Society for Testing and Materials, West Conshohocken, PA, 2003.

51. BS 1924: Part 2(5.4), Determination of the initial consumption of lime, Stabilised materials for civil engineering purposes, *Methods of Test for Cement-Stabilised and Lime-Stabilised Material. Chemical Tests*, British Standards Institution, London, 1990.

52. BS 1924: Part 2(1.5), Determination of the degree of pulverisation, *Stabilised Materials for Civil Engineering Purposes, Methods of Test for Cement-Stabilised and Lime-Stabilised Material, Classification Tests*, British Standards Institution, London, 1990.

53. ASTM D422-63, Standard test method for particle-size analysis of soils, *Annual Book of ASTM Standards*, Vol. 04.08(I), American Society for Testing and Materials, West Conshohocken, PA, 2003.

54. BS 1377: Part 2(9), Determination of particle size distribution, *Methods of Test for Soils for Civil Engineering Purposes, Classification Tests*, British Standards Institution, London, 1990.

55. Neville, A.M., *Properties of Concrete*, 4th ed., Longman Group UK Limited, London, 1997.

56. ASTM C642-97, Standard test for density, absorption and voids in hardened concrete, *Annual Book of ASTM Standards*, Vol. 04.02, American Society for Testing and Materials, West Conshohocken, PA, 2003.

57. BS 1377: Part 2(7), Determination of density, *Methods of Test for Soils for Civil Engineering Purposes, Classification Tests*, British Standards Institution, London, 1990.

58. BS 1924: Part 2(2.1), Determination of the dry density/moisture content relation, *Stabilised Materials for Civil Engineering Purposes, Methods of Test for Cement-Stabilised and Lime-Stabilised Material, Compaction Related Tests*, British Standards Institution, London, 1990.

59. BS 1924: Part 2(3), In-situ density tests, *Stabilised Materials for Civil Engineering Purposes, Methods of Test for Cement-Stabilised and Lime-Stabilised Material*, British Standards Institution, London, 1990.

60. ASTM C128-01, Standard test method for density, relative density (specific gravity) and absorption of fine aggregate, *Annual Book of ASTM Standards*, Vol. 04.02, American Society for Testing and Materials, West Conshohocken, PA, 2003

61. BS 1377: Part 2(8), Determination of particle density, *Methods of Test for Soils for Civil Engineering Purposes, Classification Tests*, British Standards Institution, London, 1990.

62. BS 1881: Part 5(7), Methods of testing hardened concrete for other than strength, British Standards Institution, London, 1970.

63. ASTM D4404-84 (Re-approved 1998), Standard test method for determination of pore volume and pore volume distribution of soil and rock by mercury intrusion porosimetry, *Annual Book of ASTM Standards*, Vol. 04.08(I), American Society for Testing and Materials, West Conshohocken, PA, 2003.

64. BS 7591: Part 1, Porosity and pore size distribution of materials, Method of evaluation by mercury porosimetry, British Standards Institution, London, 1995.

65. International Society of Rock Mechanics, ISRM suggested methods for porosity/density determination using saturation and calliper techniques, E.T. Brown, Ed., International Society for Rock Mechanics, London, 1985

66. ASTM D2216-98, Standard test method for laboratory determination of water (moisture) content of soil and rock by mass, *Annual Book of ASTM Standards*, Vol. 04.08(I), American Society for Testing and Materials, West Conshohocken, PA, 2003.

67. BS 1377: Part 2(3), Classification tests, *Methods of Test for Soils for Civil Engineering Purposes, Classification Tests*, British Standards Institution, London, 1990.

68. BS 1924: Part 2(1.3), Determination of moisture content, *Stabilised Materials For Civil Engineering Purposes, Methods of Test for Cement-Stabilised and Lime-Stabilised Material, Classification Tests*, British Standards Institution, London, 1990.

69. BS 1377: Part 4(5), Compaction related tests, *Methods of Test for Soils for Civil Engineering Purposes*, British Standards Institution, London, 1990.

70. BS 1924: Part 2(2.2), Determination of the moisture condition value (MCV), *Stabilised Materials for Civil Engineering Purposes, Methods of Test for Cement-Stabilised and Lime-Stabilised Material, Compaction Related Tests,* British Standards Institution, London, 1990.

71. BS 1377: Part 4(3), Compaction related tests, *Methods of Test for Soils for Civil Engineering Purposes,* British Standards Institution, London, 1990

72. ASTM C143/C143M-00, Standard test method for slump of hydraulic-cement concrete, *Annual Book of ASTM Standards*, Vol. 04.02, American Society for Testing and Materials, West Conshohocken, PA, 2003.

73. ASTM D 6103-97, Standard Test Method for Flow Consistency of Controlled Low Strength Material (CLSM), *Annual Book of ASTM Standards,* Vol. 04.08, Soil and Rocks; Dimension Stone: Geosynthetics, American Society for Testing and Materials, West Conshohocken, PA, 2003.

74. BS EN 12350: Part 2, Slump test, *Testing Fresh Concrete,* British Standards Institution, London, 2000.

75. ASTM C939, Standard test method for flow of grout for pre-placed aggregate concrete (flow cone method), *Annual Book of ASTM Standards*, Vol. 04.02, American Society for Testing and Materials, West Conshohocken, PA, 2003.

76. ASTM 1362, Standard Test Method for Flow of Freshly Mixed Hydraulic Cement Concrete, *Annual Book of ASTM Standards*, Vol. 04.02, American Society for Testing and Materials, West Conshohocken, PA, 2003.

77. BS EN 12350: Part 5, Flow table test, *Testing Fresh Concrete,* British Standards Institution, London, 2000.

78. ASTM C191-01a, Standard test method for time of setting of hydraulic cement by vicat needle, *Annual Book of ASTM Standards*, Vol. 04.01, American Society for Testing and Materials, West Conshohocken, PA, 2003.

79. BS 4550: Part 3(6), Tests for setting time, *Methods of Testing Cement, Physical Tests,* British Standards Institution, London, 1978.

80. ASTM C266-99, Standard test method for time of setting of hydraulic cement by Gillmore needles, *Annual Book of ASTM Standards*, Vol. 04.01, American Society for Testing and Materials, West Conshohocken, PA, 2003.

81. ASTM C186-98, Standard test method for heat of hydration of hydraulic cement, *Annual Book of ASTM Standards*, Vol. 04.01, American Society for Testing and Materials, West Conshohocken, PA, 2003.

82. BS 4550: Part 3(8), Tests for heat of hydration, *Methods of Testing Cement, Physical Tests*, British Standards Institution, London, 1978.

83. Taylor, H.F.W., *Cement Chemistry*. Thomas Telford, London, 1997.

84. Feldman, F.R. and Ramachandran, V.S., Differentiation of interlayer and adsorbed water in hydrated Portland cement by thermal analysis, *Cement and Concrete Research*, Vol. 1, No. 6, pp. 607–620, 1971.

85. ASTM C151-00, Standard test method for autoclave expansion of Portland cement, *Annual Book of ASTM Standards*, Vol. 04.01, American Society for Testing and Materials, West Conshohocken, PA, 2003.

86. ASTM C157/C157M-99, Standard test method for length change of hardened hydraulic-cement mortar and concrete, *Annual Book of ASTM Standards*, Vol. 04.02, American Society for Testing and Materials, West Conshohocken, PA, 2003.

87. BS 1881: Part 5(5), Determination of changes in length on drying and wetting (initial drying shrinkage, drying shrinkage and wetting expansion), *Methods of testing hardened concrete for other than strength*, British Standards Institution, London, 1970

88. ASTM C803/C803M-97, Standard test method for penetration resistance of hardened concrete, *Annual Book of ASTM Standards*, Vol. 04.02, American Society for Testing and Materials, West Conshohocken, PA, 2003.

89. ASTM D1883-99, Standard test method for CBR (California Bearing Ratio) of laboratory-compacted soils, *Annual Book of ASTM Standards*, Vol. 04.08(I), American Society for Testing and Materials, West Conshohocken, PA, 2003.

90. BS 1377: Part 4(7), Compaction related tests, *Methods of Test for Soils for Civil Engineering Purposes*, British Standards Institution, London, 1990.

91. BS 1377: Part 9(4.3), In-situ tests, *Methods of Test for Soils for Civil Engineering Purposes,* British Standards Institution, London, 1990.

92. BS 1924: Part 2(4.5), Laboratory determination of the CBR, *Stabilised Materials for Civil Engineering Purposes, Methods of Test for Cement-Stabilised and Lime-Stabilised Material, Strength and Durability Tests,* British Standards Institution, London, 1990.

93. BS 1924: Part 2(4.6), Determination of the in-situ CBR, *Stabilised Materials for Civil Engineering Purposes, Methods of Test for Cement-Stabilised and Lime-Stabilised Material, Strength and Durability Tests,* British Standards Institution, London, 1990.

94. U.S. Bureau of Reclamation, Procedure for direct tensile strength, static modulus of elasticity, and Poisson's ratio of cylindrical concrete specimens in tension, *Concrete Manual*, Part 2, 9th ed., pp. 726–731, 1992.

95. ASTM C348-02. Standard test method for flexural strength of hydraulic cement mortars, *Annual Book of ASTM Standards*, Vol. 04.01, American Society for Testing and Materials, West Conshohocken, PA, 2003.

96. ASTM C78-02, Standard test method for flexural strength of concrete (using simple beam with third–point loading), *Annual Book of ASTM Standards*, Vol. 04.02, American Society for Testing and Materials, West Conshohocken, PA, 2003.

97. BS 1881: Part 118, Method for determination of flexural strength, *Testing Concrete*, British Standards Institution, London, 1983.

98. ASTM C496-96, Standard test method for splitting tensile strength of cylindrical concrete specimens, *Annual Book of ASTM Standards*, Vol. 04.02, American Society for Testing and Materials, West Conshohocken, PA, 2003.

99. BS 1881: Part 117, Method for determination of tensile splitting strength, *Testing Concrete*, British Standards Institution, London, 1983.

100. ASTM C469, Standard test method for static modulus of elasticity and Poisson's ratio of concrete in compression, *Annual Book of ASTM Standards*, Vol. 04.02, American Society for Testing and Materials, West Conshohocken, PA, 2003.

101. BS 1881: Part 121, Method for determination of static modulus of elasticity in compression, *Testing Concrete*, British Standards Institution, London, 1983.

102. ASTM C215, Standard test method for fundamental traverse, longitudinal and torsional resonant frequencies of concrete specimens, *Annual Book of ASTM Standards*, Vol. 04.02, American Society for Testing and Materials, West Conshohocken, PA, 2003.

103. BS 1881: Part 209, Recommendations for measurement of dynamic modulus of elasticity, *Testing Concrete*, British Standards Institution, London, 1990.

104. ASTM G21-96, Standard practice for determining resistance of synthetic polymeric materials to fungi, *Annual Book of ASTM Standards*, Vol. 14.04, American Society for Testing and Materials, West Conshohocken, PA, 2003.

105. Stegemann, J.A., Butcher, E.J., Irabien, A., Johnston, P., de Miguel, R., Ouki, S.K., Polettini, A., and Sassaroli, G., Eds., *Neural Network Analysis for Prediction of Interactions in Cement/Waste Systems — Final Report*, Commission of the European Community, Brussels, Belgium, Contract No. BRPR-CT97-0570, 2001.

106. Al-Tabbaa, A. and Perera, A.S.R., *State of Practice Report, UK Stabilisation/Solidification Treatment and Remediation: Binders & Technology — Part III: Applications*, Forthcoming Proceedings of the International Conference on Stabilisation/Solidification Treatment and Remediation: Advances in S/S for Waste and Contaminated Land, A.A. Balkema Publishers/Taylor & Francis, Amsterdam, scheduled to be released in 2005.

107. U.S. Federal Register 40 CFR Part 268, *Identification and Listing of Hazardous Waste*, January 14, 2003.

108. Ontario Regulation 558/00, *Regulation to Amend Regulation 347 of the Revised Regulations of Ontario, 1990 made under the Environmental Protection Act*, Ministry on Ontario, Toronto, Canada, Sept. 2000.

109. U.S. EPA, *EPA RCRA Delisting Program Guidance Manual for the Petitioner*, March 23, 2000.

110. U.S. EPA, *EPA's Composite Model for Leachate Migration and Transformation Products (EPACMTP), Background Document and User's Guide*, Technical report, U.S. EPA, Office of Solid Waste, Washington, D.C., 1995.

111. Council Directive 1991/689/EEC, *Hazardous Waste Directive*, The Official Journal of the European Communities L377, 31 December 1991.

112. Environment Agency, *Landfill Directive Regulatory Guidance Note 2 (LFD RGN 2)*, Interim Waste Acceptance Criteria and Procedures, Version 4.0, Environment Agency, UK, 2002.

113. Environment Agency, *Guidance on National Interim Waste Acceptance Procedures*. Version 1.2, External Consultation Draft, Environment Agency, UK, 2002.

114. Interdepartmental Committee on the Redevelopment of Contaminated Land 59/83, *Guidance on the Assessment and Redevelopment of Contaminated Land*, 2nd ed., DETR Publications, London, 1987.

115. Environment Agency, *Guidance on the Disposal of Contaminated Soils*, Version 3. Environment Agency, UK, 2001.

116. The Dutch List, *Intervention Values And Target Values — Soil Quality Standards*, Ministry of Housing, Spatial Planning and Environment, The Netherlands, 1994.

117. Bone, B., Personal communication, Environment Agency, UK, 2003.

118. Environment Agency, *Methodology for the Derivation of Remedial Targets for Soil and Groundwater to Protect Water Resources*, EA R&D Publication 20, Environment Agency, UK, 2000.

119. DEFRA and Environment Agency, *Assessment of Risks to Human Health from Land Contamination: an Overview of the Development of Guideline Values and Related Research*, R&D Publications CLR7, Environment Agency, UK, 2002.

120. DEFRA and Environment Agency, *Priority Contaminants Report*, R&D Publications CLR8, Environment Agency, UK, 2002.

121. DEFRA and Environment Agency, *Contaminants in Soil: Collation of Toxicology Data and Intake Values for Humans*, R&D Publications CLR9, Environment Agency, UK, 2002.

122. DEFRA and Environment Agency, *The Contaminated Land Exposure Assessment Model (CLEA): Technical Basis and Algorithms*, R&D Publications CLR10, Environment Agency, UK, 2002.
123. SNIFFER, *Method for Deriving Site Specific Human Health Assessment Criteria for Contaminants in Soil*, SNIFFER report LQ01, Scotland and Northern Ireland Forum for Environmental Research, UK, 2003.
124. ASTM E1739-95, Standard guide for risk-based corrective action applied at petroleum release sites, *Annual Book of ASTM Standards*, Vol. 11.04, American Society for Testing and Materials, West Conshohocken, PA, 2003.
125. ASTM E2081-00, Standard guide for risk-based corrective action, *Annual Book of ASTM Standards*, Vol. 11.04, American Society for Testing and Materials, West Conshohocken, PA, 2003.
126. BPRISC, *Risk-Integrated Software for Clean-Ups*. www.bprisc.com, 2003.
127. U.S. NRC, *Technical Position on Waste Form*, Low-Level Waste Management Branch, Division of Low-Level Waste Management and Decommissioning, U.S. Nuclear Regulatory Commission, Office of Nuclear Material Safety and Safeguards, Washington, D.C., January 1991.
128. U.S. EPA, *Prohibition on the Disposal of Bulk Liquid Hazardous Waste in Landfills — Statutory Interpretive Guidance*, Office of Solid Waste and Emergency Response (OSWER) Policy Directive No. 9487.00-2A, EPA/530-SW-016, Washington, D.C., 1986.
129. Mulder, E., Personal communication, TNO, Delft, The Netherlands, 2002.
130. Sherwood, P.T., *Soil Stabilization with Cement and Lime*, HMSO, London, 1993.
131. Department of Transport, *Specification for Highway Works*, 6th ed., HMSO, London 1986.
132. Al-Tabbaa, A. and Evans, C.W., Pilot in situ auger mixing treatment of a contaminated site. Part 1: Treatability study, Proceedings of the Institution of Civil Engineers, *Geotechnical Engineering*, Vol. 131, pp. 52–59, 1998.
133. Bates, E.R., Personal communication, USEPA, 2002.
134. ASTM B553 (withdrawn in 1991), Test method for thermal cycling of electroplated plastics, *Annual Book of ASTM Standards*, Vol. 02.05, American Society for Testing and Materials, West Conshohocken, PA, 1990.
135. ANSI/ANS 55.1, Solid radioactive waste processing system for light water cooled reactor plant, American Nuclear Society, La Grange Park, IL, 1992.
136. Kosson, D.S., van Der Sloot, H.A., Sanchez, F., and Garrabrants, A.C., An Integrated Framework for Evaluating Leaching in Waste Management and Utilization of Secondary Materials, *Environmental Engineering Science*, Vol. 19, No. 3, pp. 159–204, 2002.
137. Garrabrants, A.C. and Kosson, D.S., Use of a chelating agent to determine the metal availability for leaching from soils and wastes, *Waste Management and Research*, 20, 155, 2000.

12 Quality Assurance/ Quality Control (QA/QC) for Waste Stabilization/ Solidification

Caijun Shi

CONTENTS

1-56670-444-8/05/$0.00+$1.50
© 2005 by CRC Press

12.1 INTRODUCTION

The formulae developed for S/S treatment of hazardous wastes in the laboratory may behave differently in the field due to the variation in composition of the waste, the accuracy of the weighing system, and the type of mixing and placing processes. Typical goals for S/S treatment of wastes include:

- Contaminant mobility (leachate) reduction
- Support strength
- Environmental durability
- Low hydraulic conductivity

To assure that these goals are achieved, quality assurance/quality control (QA/QC) plans are indispensable parts of the treatment process. Test results provide important feedback used to decide whether or not the formulation and mixing process need adjustment. Past experience and sound judgment are important in evaluating test results and assessing their significance in controlling the treatment process.

The performance of treated waste forms may behave differently in the field due to the variation in composition of a waste, the accuracy of the additive dosage, and the type of mixing and placing processes. A good QA/QC plan helps ensure that the treated waste meets the specified criteria for utilization or disposal. A QC plan should address the control and documentation of raw materials, mixing and placement, and post-treatment testing based on the laboratory and field feasibility studies.

Although S/S technology has been identified as a best demonstrated available technology and is the most frequently selected treatment technology for controlling the source of environmental contamination at Comprehensive Environmental Response Compensation, and Liability Act (CERCLA) remediation sites, few published documents can be found on QA/QC of S/S operations. However, ready-mix concrete and construction industries have very comprehensive QA/QC documents to ensure the quality of construction.[1] Parts of these documents can be extrapolated, modified, and adopted for S/S operations. This chapter discusses the basic principles of QA/QC and summarizes existing guidelines for laboratory feasibility studies and field operations for S/S of waste or remediation of contaminated sites.

12.2 GENERAL PRINCIPLES OF QA/QC

12.2.1 QUALITY ASSURANCE

QA is defined as those planned and systematic operations conducted to ensure that the operation or product meets specifications. QA encompasses the project engineer or chemist's oversight of the contractor's quality control plan; review of inspector, sampler, tester, and laboratory qualifications; verifying the results of quality control and process control testing; and inspecting for conformance to plans and specifications. QA is the responsibility of the project engineer or chemist.

12.2.2 QUALITY CONTROL

QC is defined as all those planned and specified actions or operations necessary to produce a product that will meet requirements for quality as specified. QC includes, but should not be limited to, inspection of the production and placement operation, process control testing, and inspection of the finished product. QC is the responsibility of the contractor.

12.2.3 QUALITY CONTROL PLAN

The QC plan is developed by the contractor and approved by the project engineer or chemist. The QC plan addresses the actions, inspection, sampling, and testing necessary to keep the production and placement operations in control, to quickly determine when an operation has gone out of control, and how to respond to correct the situation and bring it back into control. Developing, implementing, maintaining, and supplementing the QC plan is the responsibility of the contractor. Oversight of the activities required to fulfill the QC plan is the responsibility of the project engineer or chemist.

The purpose of a QC plan is to ensure that quality is instilled in the product during its generation. Quality cannot be added after it is generated. Thus, a functional, responsive QC plan is imperative for the production of a quality product. A QC plan addresses the actions needed, including inspection, sampling, and testing, for the following reasons:

- To keep the process in control
- To quickly determine when the process has gone out of control
- To respond adequately to correct the situation and bring the process back into control

In many project contracts, acceptance and payment are based on measurements of specified properties; on the contractor's fulfillment of process and QC inspection, sampling, and testing; and on the engineer's inspection, sampling, and testing to confirm (and verify) that the work conforms to the plans and specifications. Payments are often related to the percentage of finished products outside of the specified limits that define acceptable levels. To encourage quality, pay factors are based on a curvilinear relationship rather than straight-line, and adjustments become more significant as quality levels diminish. In like manner, the non-linear pay factors provide an incentive payment for quality significantly exceeding the minimum quality levels.

The *Guide Specification for Military S/S of Contaminated Materials* published by the U.S. Corps of Engineers (USCOE)[2] suggests that payment be based on the contract unit price schedule for each unit of contaminated material entering the S/S process. This unit price shall include the cost for materials, equipment, waste feed processing, S/S operations, stockpiles, testing, and all other work associated with the S/S process.

No payment will be made for materials or labor required to reprocess previously processed material that did not meet the physical and chemical testing requirements outlined in this section. In other words, reprocessed material shall be deducted from the daily production rate.

12.3 FREQUENCY AND STATISTIC EVALUATION OF QA/QC TESTING

Frequency of testing is a significant factor in the effectiveness of quality control of a process or a product. Specified test frequencies are intended for acceptance of the material or one of its components at a random location within the quantity or time period represented by the test. Usually, more quality control tests are carried out during initial stages, and the frequency of testing is reduced as the work progresses and product quality becomes more predictable.[3] ASTM C 1451[4] provides a standard practice for determining the uniformity of cementitious materials, aggregates, and chemical admixtures used in concrete.

The USCOE's specifications indicate that the frequency of post-treatment testing has also been subject to approval by local regulatory agencies and has varied from project to project.[2] The frequency of post-treatment testing has generally been in the range of once per 75 m^3 to once per 750 m^3 of waste treated. However, some jobs required testing more frequently than once per 75 m^3. During the actual remediation work, post-treatment testing creates logistics problems because of the need to allow samples time to cure prior to testing and the time required to perform the quality control testing itself. Curing time and curing procedures for quality control samples have usually not been adequately addressed.

The S/S processes are designed on the assumption that the performance criteria of the treated waste are achieved. However, statistically, there is a finite probability that a portion of the treated waste does not meet the criteria of property values. To compensate, the value of the critical properties targeted during production is set a conservative finite difference better than those specified in the criteria. How much better is determined by statistical analysis of the variation in performance with variation in composition, based primarily on the degree of quality control during proportioning, mixing, and placement of the treated waste.

Statistical analyses have been published for some test methods,[5] but no publication was found for a statistical evaluation of performance during field application of S/S. The QA/QC for concrete production is well established. The variations in the strength of concrete and the statistical procedures for the interpretation of these variations with respect to required criteria and specifications are described in ACI 214-77.[6] The statistical procedures used for evaluation of concrete strength can also be applied to S/S waste forms.[6] The following paragraphs briefly describe the statistical techniques applied to the compressive strength of concrete used in load-bearing structures. The critical property for waste forms is usually not strength, but leach resistance; however, the statistical techniques presented for construction concrete strength can be extrapolated for use in the critical properties of waste forms.

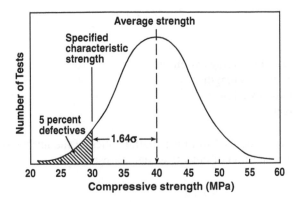

FIGURE 12.1 Normal distribution of concrete strengths.

It is now generally accepted that the strength of concrete has a normal distribution as shown in Figure 12.1. A normal distribution is identified by its mean and standard deviation. The average of all individual tests (X bar) can be calculated using Equation 12.1:

$$\overline{X} = \frac{X_1 + X_2 + \text{.......} + X_n}{n} \tag{12.1}$$

where X_1, X_2, ..., X_n are the results of individual tests and n is the total number of tests made.

The standard deviation is calculated from the same testing set using Equation 12.2.

$$\sigma = \sqrt{\frac{\sum_{i-1}^{n}(X_i - \overline{X})^2}{n-1}} \tag{12.2}$$

Other parameters sometimes used are:

Coefficient of variation $V = \dfrac{\sigma}{\overline{X}} x100\%$

Range R = highest value − lowest value

As shown in Figure 12.1, there is always a probability that a strength obtained in one or more tests is less than the specified strength. It is usual to specify the quality of concrete not as a minimum strength, but as a "characteristic strength" below which a specified portion of the test results, often called "defectives," may be expected to fall. Due to the variation of concrete in production, it is necessary to design the mix to have an average strength greater than the specified strength by an amount termed the margin (k.σ):

$$f_{cr} = f_c' + k.\sigma \qquad (12.3)$$

where,

  f_{cr} = the regulated average strength
  f_c' = the specified strength
  σ = standard deviation
  k = constant

The constant k is derived from the mathematics of the normal distribution and increases as the proportion of defectives decreases:

  k for 10% defectives = 1.28
  k for 5% defectives = 1.64
  k for 2.5% defectives = 1.96
  k for 1% defectives = 2.33

Usually, the 5% defective level is used for concrete. Equation 12.4 relates a concrete having a specified characteristic strength of 30 MPa and a standard deviation of 6.1 MPa. Hence:

$$f_{cr} = 30 + 1.64 \times 6.1 = 40 \text{ MPa} \qquad (12.4)$$

The standard deviation calculated from n results is an estimate of the standard deviation of the total population and is, therefore, subject to normal probability errors, which are reduced as n becomes larger. If several groups of n results are taken, these may vary by 20% without being significantly different statistically. For mix design purposes the standard deviation should be calculated from at least 40 results.

12.4 QUALITY CONTROL CHART

For convenience, QC charts are often used as an aid in reducing variability and increasing efficiency in production. Figure 12.2 shows three simplified charts prepared specifically for concrete control,[6] described as follows:

1. Chart (a) or "a" shows all strength test results plotted as received, the line for the required average strength, f_{cr}, established as indicated by Equation 12.3, and the specified design strength f_c'.
2. Chart (b) or "b" shows the moving average for compressive strength where the average is plotted for the previous five sets of two companion cylinders for each day or shift, and the specified strength as the lower limit f_c'. This chart is valuable in indicating trends and will show the influence of seasonal changes, changes in materials, and so on. The number of tests averaged to plot moving averages with an appropriate lower limit can be varied to suit each job.

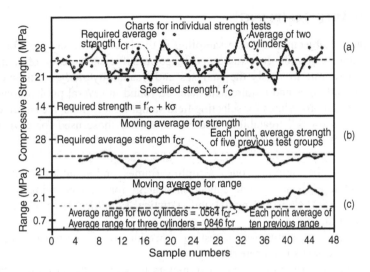

FIGURE 12.2 Quality control charts for concrete.[6]

3. Chart (c) or "c" shows the moving average range where the range for the average of the previous ten groups of companion cylinders is plotted each day or shift. The maximum average range allowable for good laboratory control is also plotted.

For any particular job, a sufficient number of tests should be made to insure accurate representation of the variations of the concrete. Concrete tests can be made either on the basis of time elapsed or cubic yardage placed, and conditions on each job will determine the most practical method of obtaining the number of tests needed. A test is defined as the average strength of all specimens of the same age fabricated from samples taken from single batches of concrete.

ACI Committee document 214-77[6] also gives some recommendations about the rejection of test results. The practice of arbitrary rejection of test cylinders, which appear "too far out of line," is not recommended since the normal pattern of probability establishes the possibility of such results. Discarding tests indiscriminately can seriously distort the strength distribution, making analysis of the results less reliable. Occasionally, the strength of one cylinder from a group made from a sample deviates so far from the average as to be highly improbable. It is recommended that a specimen from a test of three or more specimens be discarded if its deviation from a test mean is greater than 3σ, and should be accepted with suspicion if its deviation is greater than 2σ. If questionable variations have been observed during fabrication, curing, or testing of a specimen, the specimen should be rejected. The test average should be computed from the remaining specimens. A test (average of all specimens of a sample) should never be rejected unless the specimens are known to be faulty, since it represents the best available estimate for the sample.

12.5 WASTE SAMPLING

The principal objective of waste sampling is to obtain waste samples for analyses and treatability studies. It is important to get a sufficient number of samples and volume of sample to satisfy the analyses and treatability study requirements. The sampling guidelines and requirements can be found in several publications.[7-12]

The design procedures for selecting field sampling locations, measurements, and data analyses for S/S treatability studies can follow those used for environmental monitoring of chemicals.[13,14]

- Define the sampling zones, sampling frames, and variables of interest.
- Define a general sample collection strategy for each sampling zone.
- Develop a statistical model and statistical sampling objectives for each sampling zone.
- Specify the estimation or testing procedures to be employed and their desired statistical properties.
- Select the sampling design parameters to achieve the desired statistical properties.

There are three basic techniques for obtaining samples: coring, pumping, and digging. Coring to depths up to 3 m usually uses a pipe or tube with an inside diameter in the range of 2.5 to 5.0 cm, dependent on the viscosity of the wastes. Low-viscosity liquids and hard-compacted solids require another sampling method. For greater depths, or for hard-compacted materials, a drilling rig is used. If the waste has a low viscosity, a pump can be used to obtain samples. A variety of pumps can be used as described by Conner.[8] For solid wastes, a backhoe, clamshell, or dragline can be used to obtain samples at different depths.

The chemical and physical properties of untreated and treated wastes are often assessed based on the testing of a limited number of samples. Many factors, such as measurement uncertainty, field heterogeneity, and sample variety, affect the test results. If the results are variable, the decision should be made based on a statistical sampling design procedure.

12.6 TREATABILITY STUDIES

The objectives of an S/S treatability study are to provide valuable site-specific information for the selection and implementation of proper S/S formulation and process. In addition to the performance specifications discussed in Chapter 11, treatability studies may also include the following goals: [8]

- Determine the most economical mix design
- Identify handling characteristics of the waste and binders
- Determine type of mixing and mixing time
- Identify interactions between the waste and binder during mixing
- Determine curing requirements for the S/S waste forms

- Assess physical and chemical uniformity of the waste
- Determine the volume increase associated with the S/S process

A treatability study should be representative of the full-scale remediation process. The contractor is sometimes required to provide treatability study test results prior to performing work at the site.

In many cases, a treatability study is conducted in two steps: laboratory formulation development and bench-scale testing. During the laboratory formulation development, small amounts of wastes are mixed with binders for testing some parameters or indicators of S/S products. If the formulation system involves two or more components, it is desirable to use a statistical experimental design method.

Laboratory physical and chemical testing may need to be conducted on the waste and additives before, during and after mixing, as well as on the hardened S/S waste forms. Commonly used test methods are described in Chapter 11. The requirements for the type of test and testing procedures may vary from case to case. Laboratory QA/QC procedures encompass the required analysis of method blanks, duplicate samples, surrogate compounds, and spiked samples. These operations allow calculation of both field and laboratory precision and accuracy achieved in conjunction with the data. These data quality indicators are then compared to those parameters established at the initiation of the project to assess contract compliance.

The USEPA Superfund Treatability Study Protocol[15] provides an overview of USEPA QA/QC guidelines for treatability studies as follows:

1. Preparation of the Quality Assurance Project Plan
2. Formulation of Data Quality Objectives
3. Identification of the sources and types of errors that may occur during the sampling, analysis, and treatability measurement process
4. The need for quality control samples
5. Determination of data quality indicators, measurement errors, and documentation

12.7 QA/QC FOR FIELD S/S OPERATIONS

12.7.1 QA/QC PLAN FOR FIELD S/S OPERATION

For a field S/S operation, the QA/QC plan should include control and documentation of cementing materials and waste, and testing on freshly treated and hardened waste forms. Cementing materials are tested for their compliance with specifications or acceptable criteria. Additional details regarding cementing materials can be found in Chapter 4. Waste should be analyzed to see whether the S/S formulation tolerates the variation of contaminants. Testing of freshly treated waste has two purposes: (1) to assure the uniformity of the mixture, and (2) to determine the suitability of the treated waste for handling and disposal. There are few standard test methods specifically designated for S/S operation. Most of the methods used are adopted or modified from the testing of cement, concrete, and grout materials. Shi et al.[16,17] published some QA/QC procedures and methods for specific formulations used in field S/S of hazardous wastes. The binder used was a sodium silicate-activated blended cement.

Tests of hardened treated wastes evaluate product performance and establish the process parameters for generating an acceptable product. Post-treatment testing generally consists of both chemical and physical tests. Most projects use TCLP testing, focusing on the site-specific contaminants of concern. Other parameters tested vary from project to project, but often include moisture content, unconfined compressive strength, permeability, and pH.

12.7.2 Pre-Treatment Testing

12.7.2.1 Cements and Additives

Some physical and chemical properties of cements and additives can be obtained from suppliers. A certificate of analyses should accompany each shipping unit of the materials. If a blended cementing material is used, each component should be within the acceptable range. Several different tests may have to be used to determine the components in the blended material.[16]

Based on the nature and characteristics of the additive used, some test methods or procedures may be found in relevant areas or developed for quality control purposes. For example, measurement of solution-specific gravity by hydrometer was an acceptable quality control technique in the field for the sodium silicate solution concentration.[16]

12.7.2.2 Wastes

The formulation and treatment processes are determined according to the physical and chemical characteristics of the wastes. Once the formulation and treatment processes are chosen through laboratory and pilot feasibility tests, some properties of the waste, such as moisture content, particle size distribution, and concentration of contaminants, may be found to have the most significant effects on the properties of the treated wastes. Thus, these properties need to be monitored before treatment. The moisture content of the waste will affect the mixing process and the properties of both fresh and hardened treated wastes. A significant variation of moisture content may require the adjustment of the mixing water to be added. The moisture content in untreated wastes can be measured by following ASTM D 2216[18] or ASTM D 4643.[19] Additionally, commercially available moisture analyzers can be used as quick measurement tools in field QA testing.[16]

When the waste is granular, it is important to control the particle size within certain limits since it may have a significant effect on the physical properties and leachability of treated wastes. In the field, granular wastes may have to be crushed and sieved to pass a specified screen size.

A S/S formulation could be sensitive to the contaminant concentration and may need modification if there is a large variation in contaminant concentrations. Contaminant concentration should be verified to ensure that the formulation meets expected variation ranges. The USEPA publishes methods for analysis of contaminant concentration.[9]

12.7.2.3 Mixing Water

Mixing water for S/S operation should not contain concentrations of oils, acids, salts, alkalis, organic matter, or other substances that will be detrimental to the successful execution of the S/S treatment process. Chemical analysis may be required when the water is of questionable quality.

12.7.3 MIXING AND PLACEMENT

Prior to full-scale operations, a field demonstration should be performed. At least 500 m^3 of contaminated material should be processed and the required tests, as described in Section 12.8, should be performed on representative samples of the treated material. A field demonstration should be performed on each distinctive type of material or contaminant to be treated. The full-scale processing equipment should be used for the field demonstration. Reagents, mix ratios, and mixing procedures used during the field demonstration should be the same as those used for the remainder of the work.

Mixing time, mixing speed, and amounts of contaminated material, reagents, and water added to each batch should be recorded. Mixing time, mixing speed, and batch proportions should be maintained within the limits specified in the approved Work Plan as modified during the field demonstration. The use of optical tracers has been evaluated to indicate the degree of mixing.[20] Mixing can also be monitored in-line by measuring electrical resistance or capacitance.[20]

To prevent double handling, it is preferable to place treated material directly into the permanent storage area rather than stockpiling it until post-treatment testing is completed.[21]

12.7.4 SAMPLING AND TESTING FRESHLY MIXED WASTE

12.7.4.1 Sampling Freshly Treated Wastes

As indicated above, there are techniques that can be used to monitor the mixing and homogeneity of freshly treated wastes in-line during the mixing process. However, they may be too complicated, too expensive, or too unreliable for use in field operations at the moment.

If a batch operation is used and the treated waste is grout, approaches similar to those for fresh concrete can be used. Standard method CRD-C 620-80 outlines techniques for sampling grouts from mixers, pumps, and discharge lines.[22]

12.7.4.2 Content of Cement or Additives

ASTM C 1078[23] specifies a procedure to measure the cement content in freshly mixed concrete. If the waste and other additives do not react with cement immediately, this method can be used. However, other methods may be required if some immediate reactions happen between cement and other components. Another common component in the cementing materials or additives for S/S is lime. A laboratory and field testing program indicated that the available lime index can be measured using the "rapid sugar test method" described in Section 28 of ASTM C 25.[24]

12.7.4.3 Volume Increase

The excessive addition of reagents during treatment can result in a greater than anticipated volume increase. Limiting volume increase is important if the treated material is to be placed in an on-site landfill with limited storage space. For this reason, monitoring of volume increase is often done during the treatability study, field demonstration, or full-scale treatment. The excessive addition of reagents can also result in higher treatment and off-site disposal costs.

12.7.4.4 Bulk Density and Moisture Content

Because the density of the additives and solids in the waste is usually much higher than that of water, bulk density measurements for a mixture are expected to provide an indication of the water-to-solid (W/S) ratio. For example, a linear correlation was established in one case between the bulk density and the W/S with a correlation coefficient of 0.998.[17] The moisture content of treated wastes can be measured following ASTM D 2216 or ASTM D 4643.[18,19] Commercially available moisture analyzers can be used as a quick measurement tool in field QA testing.[17]

12.7.4.5 Consistency

The control of consistency is required to make sure that the treated waste has adequate flowability or workability for handling and placement. Three standard ASTM test methods for mix consistency or stiffness are cone slump,[25] K-slump,[26] and flow consistency[27] tests, for measurement of workability of concrete. These three methods have been used for measurement of consistency of treated wastes. Other non-standard tests are sometimes used for this purpose.

12.7.4.6 Concentration of Contaminants

Through the measurement of the concentrations of contaminants in the waste and solidified waste forms, it can also be used to estimate the variation of the S/S formulation. The analysis of contaminants in solidified waste forms can be measured using USEPA methods.[9]

12.8 TESTING HARDENED TREATED WASTE FORMS

Treated waste samples can be measured in the laboratory or in the field, as is done for construction concrete samples. For hardened treated wastes, four common tests are:

- Strength or Penetration Resistance
- Hydraulic Conductivity

- Leachability
- Durability

Most projects have specified criteria for these properties. Chapter 11 describes the laboratory procedures to measure these properties. However, there is currently no official document specifying the QA/QC procedures for sampling and curing field samples for these tests.

12.9 QUALITY CONTROL CHART FOR S/S

In order to ensure formulation of the treated waste within the acceptable range, quality control charts can be developed for daily operations. Rushbrook et al.[28] developed a S/S quality control chart based on five physical and leaching parameters: (a) setting rate; (b) compressive strength; (c) hydraulic conductivity; (d) contaminant leachability; and (e) liquid retention. In a laboratory study, a variety of treated wastes containing varying proportions of cement, liquid and solid wastes, and pulverized fly ash (PFA) were prepared. After a series of tests on the treated wastes, a quality control chart for an acceptable waste proportion was developed (Figure 12.3).

12.10 EARLY FAILURE INDICATORS AND ADJUSTMENTS TO THE MIX DESIGN

At the moment, there are no official documents that describe the monitoring of the treated wastes for an early indication of failure. However, the following measures may be used as early failure indicators:[29]

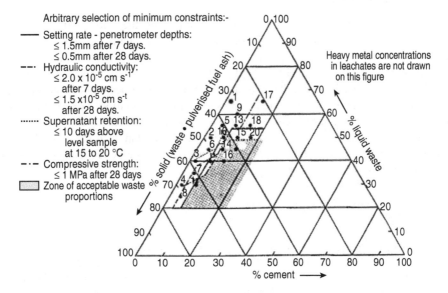

FIGURE 12.3 Quality control chart for acceptable proportions of treated waste.[28]

1. Use of different techniques to examine archived samples to provide the basis for implementing the remedy and quality assurance plan
2. Testing of treated wastes in leachate columns or field lysimeters periodically for performance criteria
3. Examination of groundwater from collection stations to evaluate the performance of the constructed waste forms

The treated wastes disposed of at the site may exhibit different properties from those of archived samples. In a field demonstration project, differences in properties were noted between the cores of solidified wastes taken in the field and those cast at the site but cured in the laboratory.[30,31] These differences were attributed to different curing and exposure conditions.

Also, whenever the early indicators point to failure, public health and the environment may be at risk. The remediation of the failure may need much more effort and cost much more than the original treatment process.

Subject to approval, the mix design may be changed based on the characteristics of the material being treated. An additional field demonstration may be required prior to implementation of the new mix design.

Most of the problems related to quality control have been due to inadequate specifications, misunderstanding of the required specifications, or an inadequate number of samples tested for statistical evaluation.[29] Therefore, a good QA/QC plan, which includes monitoring materials quality, equipment calibration, personnel requirements, real-time testing, pilot cell demonstration, and early failure indication, will help avoid these problems.

12.11 SUMMARY

This chapter discusses the basic QA/QC principles and their applications for S/S processes. To assure that these S/S goals are achieved, quality QA/QC plans are indispensable parts of the treatment process. QA/QC testing of both physical and chemical properties of raw materials and treated wastes provides important feedback used to decide whether or not the formulation and mixing process needs adjustment. Enough testing should be conducted to provide valid information. The frequency of post-treatment testing has also been subject to approval by local regulatory agencies and has varied from project to project. To ensure formulation of the treated waste within an acceptable range, quality control charts can be developed for daily operations after a series of tests on the treated wastes.

REFERENCES

1. State of California Department of Transportation, Quality Control Quality Assurance Manual For Asphalt Concrete Production and Placement, June 2002.
2. USACE, Guide Specifications for Military Construction: Solidification/Stabilization of Contaminated Materials, CEGS 02445, Washington, D.C., 1994.

3. Kosmatka, S. H., Kerkhoff, B., and Panarese, W. C., *Design and Control of Concrete Mixtures*, 14th Edition, Portland Cement Association, Skokie, IL, 2002.

4. ASTM C 1451, Practice for Determining Uniformity of Ingredients of Concrete from a Single Source, *Annual Book of ASTM Standards*, Vol. 04.02, Aggregates, Concretes, American Society for Testing & Materials, Philadelphia, 2003.

5. Stegemann, J. A. and Cote, P. L., Investigation of Test Methods for Solidified Waste Evaluation — a Cooperative Program, Report EPS 3/HA/8, Environment Canada, 1991.

6. ACI 214-77, Recommended Practice for Evaluation of Strength Test Results of Concrete, American Concrete Institute, Farmington, MI, 2000.

7. Exner, J. H., A sampling strategy for remedial action at hazardous waste sites, *Hazardous Waste and Hazardous Materials,* Vol. 2, No. 4, 1985, pp. 503–521.

8. Conner, J. R., *Chemical Fixation and Solidification of Hazardous Wastes*, Van Nostrand Reinhold, New York, 1990.

9. U.S. EPA, Test Methods for the Evaluation of Solid Wastes, Physical/Chemical Methods, Office of Water and Waste Management, Washington, D.C., SW-846, 1986.

10. U.S. EPA, Soil Sampling Quality Assurance User's Guide, EPA/600/8-89/046, Environmental Monitoring System Laboratory, Las Vegas, NV, March 1989.

11. U.S. EPA, Guide for Conducting Treatability Studies Under CERCLA, Interim Report, EPA/540/2-89/085, Office of Research and Development, Cincinnati, OH, December 1989.

12. Means, J. L. et al., *The Application of Solidification/Stabilization to Waste Materials*, Lewis Publishers, Boca Raton, FL, 1995.

13. Keith, L. H., *Principles of Environmental Sampling*, American Chemical Society, Washington, D.C., 1988.

14. Gibert, R. O., *Statistical Methods for Environmental Pollution Monitoring*, Van Nostrand Reinhold, New York, 1987.

15. U.S. EPA, Superfund Treatability Study Protocol: Identification/Stabilization of Soils Containing Metals, Phase II Review Draft, Office of Research and Development, Cincinnati, OH, and Office of Emergency and Remedial Response, Washington, D.C., 1990.

16. Shi, C., Stegemann J., and Caldwell, R., Quality Analysis/Quality Control Tests for Field Stabilization/Solidification — Part II: Untreated Waste, Silicate Solution and Solidified Waste, *Waste Management*, Vol. 15, No. 7, pp. 507–513, 1995a.

17. Shi, C., Stegemann, J., and Caldwell, R., Quality Analysis/Quality Control Tests for Field Stabilization/Solidification — Part I: Dry Cementing Additives, *Waste Management*, Vol. 15, No. 4, pp. 265–270, 1995b.

18. ASTM D 2216, Standard Test Method for Laboratory Determination of Water (Moisture) Content of Soil and Rock, *Annual Book of ASTM Standards*, Vol. 04.08, Soil and Rocks; Dimension Stone: Geosynthetics, American Society for Testing & Materials, Philadelphia, 2003.

19. ASTM D 4643 Test Method for Determination of Water (Moisture) Content of Soil by Microwave Oven Method, *Annual Book of ASTM Standards*, Vol. 04.08, Soil and Rocks; Dimension Stone: Geosynthetics, American Society for Testing & Materials, Philadelphia, 2003.

20. Del Cul, G. D. and Gilliam, T. M., "Development of an In-line Grout Meter for Improved Quality Control," *Stabilization and Solidification of Hazardous, Radioactive, and Mixed Wastes*, T. Michael Gilliam and Carlton C. Wiles, Eds., ASTM STP 1123, American Society for Testing and Materials, Philadelphia, 1992, pp. 395–406.

21. U.S. EPA, Field Demonstration Plan for the United Hazardous Waste Sites, Seattle, WA, 1987.
22. USACE, CRD-C 620-80, Standard Method of Sampling Fresh Grout, Washington, D.C., 1980.
23. ASTM C 1078-87, Test Method for Determining Cement Content of Freshly Mixed Concrete, *Annual Book of ASTM Standards,* Vol. 04.02, Aggregates, Concretes, American Society for Testing & Materials, Philadelphia, 1992.
24. ASTM C-25, Standard Test Method for Chemical Analysis of Limestone, Quicklime, and Hydrated Lime, *Annual Book of ASTM Standards*, Vol. 04.01, Cement; Lime; Gypsum, American Society for Testing and Materials, Philadelphia, 2003.
25. ASTM C-143, Standard Test Method for Slump of Hydraulic Cement Concrete, *Annual Book of ASTM Standards*, Vol. 04.01, American Society for Testing and Materials, Philadelphia, 2003.
26. ASTM D 6103-97, Standard Test Method for Flow Consistency of Controlled Low Strength Material (CLSM), *Annual Book of ASTM Standards,* Vol. 04.08, Soil and Rocks; Dimension Stone: Geosynthetics, American Society for Testing and Materials, Philadelphia, 2003.
27. ASTM C 1362, Test Method for Flow of Freshly Mixed Hydraulic Cement Concrete, *Annual Book of ASTM Standards,* Vol. 04.02, Aggregates, Concretes, American Society for Testing & Materials, Philadelphia, 2003.
28. Rushbrook, P. E., Baldwin, G., and Dent, C. G., "A Quality Assurance Procedure for Use at Treatment Plants to Predict Long-term Performance of Cement-Based Solidified Hazardous Wastes Deposited in Landfill Sites," *Environmental Aspects of Stabilization and Solidification of Hazardous and Radioactive Wastes*, Pierre Côté and T. Michael Gilliam, Eds., ASTM STP 1033, American Society for Testing and Materials, Philadelphia, 1989, pp. 93–113.
29. Butler, S. M., Barth, E. F., and Barich, J. J., Field Quality Control Strategies for Assessing Solidification/Stabilization, In: *Stabilization and Solidification of Hazardous, Radioactive, and Mixed Wastes*, T. Michael Gilliam and Carlton C. Wiles, Eds., ASTM STP 1240, American Society for Testing and Materials, Philadelphia, 1996, pp. 685–690.
30. Caldwell, R., Stegemann, J. A., and Shi, C., Effect of Curing on Field-Solidified Waste Properties, Part I: Physical Properties, *Waste Management and Research*, Vol. 17, No. 1, pp. 44–49, 1999a.
31. Caldwell, R., Stegemann, A., and Shi, C., Effect of Curing on Field-Solidified Waste Properties, Part II: Chemical Properties, *Waste Management and Research*, Vol. 17, No. 1, pp. 37–43, 1999b.

13 Case Studies: Full-Scale Operations and Delivery Systems

Jesse Conner, Steve Hoeffner, and Christine Langton

CONTENTS

1-56670-444-8/05/$0.00+$1.50
© 2005 by CRC Press

13.1 INTRODUCTION

The ultimate goal of all of the science, technology, and laboratory testing involved in S/S is the successful treatment of actual hazardous, radioactive, or mixed wastes. To accomplish this goal, two tasks must be carried out: (1) the optimum chemistry determined in testing must be scaled-up to the applicable field conditions, and (2) a physical/mechanical system — a "delivery system" — must be chosen to meet the demands of treating the waste and complying with the site and project requirements.

The prime consideration in scale-up of S/S technology to full-scale operation is that no two waste treatment scenarios are identical. No full-scale operation should be planned, let alone put into practice, without the proper characterization, laboratory testing (i.e., treatability studies), site investigation, and regulatory evaluation having previously been conducted. These considerations are discussed in the other chapters. Often, S/S contractors by-pass these vital requirements by relying on past experience combined with overkill in respect to the addition of binders and additives. This appeared to work in the past when applied to simple solidification of easy-to-treat wastes, substituting the excess cost of chemicals for the costs associated with a proper feasibility study. This approach is much less successful today, where more stringent physical and chemical properties of the treated waste product are required. The handling, storage, transportation, and disposal costs of the product are often much higher than the actual S/S processing costs, especially in the case of radioactive and mixed wastes. Thus, as the environmental requirements become more stringent, the "more is better" approach often doesn't work and adds to these downstream costs for the extra mass and volume that result.

The four case studies in this chapter are presented to give a fairly broad view of full-scale operations with cement-based systems. Before discussing the case-by-case specifics, the chemical scale-up and delivery systems are discussed, to provide the reader with some background in the components of the full-scale system and to avoid the necessity for repetition of that information in the actual case histories.

13.2 CHEMICAL SCALE-UP

Laboratory development of the appropriate S/S treatment chemistry, its dependence on waste characteristics, the physical and chemical properties of the stabilized/solidified waste forms, and test methods are discussed elsewhere in this book. However, what works in the laboratory doesn't always work at larger scale, a reality that has resulted in some costly failures. Failures of this sort are usually the result of two problems in waste treatment: (1) differences between the laboratory sample and the actual material to be treated in the field, and (2) failure to consider the differences in scale between handling waste in the field and in the laboratory.

The first problem is all too common in remediation projects of all types and can be especially difficult in radioactive and mixed-waste situations where sampling is difficult and dangerous. The irony is that very detailed and careful sampling of the waste is generally done in the feasibility phase of a remedial project to characterize the waste, but these samples are typically too small, too old, or otherwise inadequate to be used in the treatability study phase. Resampling for the treatability study is seldom anywhere near as careful and complete as for the initial waste characterization. The obvious solution is better planning at the time of the feasibility study or better resampling for the treatability study.

The second problem can usually be eliminated or minimized by early involvement and cooperation of all persons involved in the project, including laboratory scientists, engineers, operations people, and regulatory personnel. The area of greatest difficulty is mixing. In the laboratory, the test samples are usually hand mixed in small containers or with typical laboratory mixers. These methods seldom replicate the mixing action that takes place at large scale. When intimate mixing is not required, the order of addition and the timing are not critical, and the waste–reagent mixture has desirable rheology, field implementation typically proceeds with no major problems. Chemical reactions that cause excessive temperature increases, gas evolution, and odor problems create problems during field implementation that go unnoticed at the laboratory scale. Another common problem is the development of undesirable rheology (high viscosity, stickiness, and non-newtonian flow) that requires special equipment or undesirable operational restrictions at field scale. Expensive re-formulation or engineering modifications are often required for any of these problem situations.

13.3 DELIVERY SYSTEMS

Delivery systems are the means by which S/S processes are implemented, including the equipment and the complete process of developing, designing, planning, permitting, operating, controlling, and financing an S/S project.[1] The project may be remedial or continuing; the equipment may be mobile/portable or fixed; the regulatory structure may be under the U.S. Environmental Protection Agency (Resource Conservation and Recovery Act, Comprehensive Environmental Response Compensation, and Recovery Act), or the Nuclear Regulatory Commission/Department of Energy or both. The equipment may be designed, built, owned, and operated by an S/S vendor; a generator; a central RCRA treatment, storage, and disposal facility (TSDF); or even a public entity, or any combination of these elements. The ultimate disposal of the S/S product may be part of the delivery system, or it may be totally separate. Many of these elements are outside the scope of this chapter. They are, however, considerations that affect every phase of S/S technology. Limitations in any of these areas may prevent implementation of the chosen delivery system and S/S process.

Most commercial S/S processes are quite simple conceptually and utilize standard mechanical equipment. The nature of the equipment depends on whether the method is "*in situ*" or "*ex situ*." "*Ex situ*" means excavating the waste, treating it, and landfilling it, either in the same excavation or another location. *Ex situ* systems may be operated at either remedial sites or at central waste treatment locations such as TSDFs. The *"in situ"* methods are generally limited to remedial situations. *"In situ"* methods introduce the binders and additives into the contaminated medium — usually sludge or soil — using large-scale excavation, tilling, injection, or drilling equipment modified for S/S chemical addition.

Each delivery system has advantages and disadvantages. *Ex situ* systems provide proven control of reagent addition and mixing, enable easier quality control sampling, and are more practical at shallow waste depths and where site access is limited. Obviously, *ex situ* equipment and methods are used when waste is transported to a TSDF for treatment and disposal. Large projects at great depth are more amenable to *in situ* operation and are often less costly than *ex situ* operations.

13.3.1 MOBILE OR PORTABLE SYSTEMS

13.3.1.1 *Ex Situ* Methods

Ex situ S/S systems are assemblies of mixers, chemical storage and feeding devices, pumps, conveyors, and ancillary equipment. The actual treatment and disposal scenario determines which combination to use. Mixer-based methods use the most complex equipment, while pit-mixing and spray-on methods are the simplest. All three techniques are used at fixed treatment sites such as TSDFs and waste generator facilities. In the latter case, either mixer-based or spray-on techniques are used in Totally Enclosed Treatment Facilities, which circumvent the need for extensive permitting in certain instances. Economics usually govern the selection, subject to a method's ability to achieve treatment objectives.

Mixer-Based Methods. A typical system is shown schematically in Figure 13.1. The waste to be treated is conveyed by pump, mechanical conveyor, or other means into a surge tank or feed hopper that feeds the waste into the mixer, where it is mixed with the S/S reagents. Depending on the process used, one or more dry and/or liquid components may be added to the waste in the mixer. The system may be set

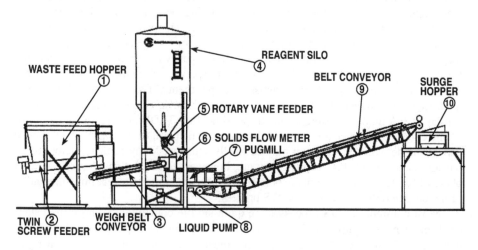

FIGURE 13.1 Schematic of a mixer-based S/S system.

up in either batch or continuous mode. In the batch mode, the batch size is limited by the availability of large, rugged mixers of the pugmill type. For some waste streams, a concrete mixer of the rotary type (used in transit mixers) can be used, in which case batch size can be up to 7.6 m³ (10 yd³). However, this type of mixer is limited to pourable, relatively low-viscosity slurries, not the sticky pastes that are often encountered in waste treatment, especially for limited water content.

For batch systems, the mixing process normally takes 1 to 15 minutes, depending on the mechanical system used, the size of the batch, the type of waste, and the amounts and types of reagents being used. In continuous systems, residence time in the mixer is generally 0.5 to 2.0 minutes, and intensive mixing is required. After mixing is complete, the waste is removed from the mixer by either pumping (if liquid) or screw conveying or dumping (if viscous or like soil). The treated waste is then moved by pump or other conveyance device to an area where it can develop its final physical and chemical properties. If the waste has been pretreated, the hazardous components, usually heavy metals, have already been converted to a relatively insoluble form. If it has not been pretreated, the metals are often immobilized during the mixing process (unless immobilization depends upon a physical, monolithic form). In this case, by the time the waste exits the mixer it has largely developed its final chemical properties.

The hardening, or curing, process often takes place in a temporary container or impoundment near the S/S plant, with the solid subsequently being conveyed to the disposal site. Alternately, the treated waste can be conveyed directly into the disposal site and solidified in its final location, regulations permitting. In this case, testing must first confirm successful treatment prior to the deposited material being accepted for permanent placement, since material not meeting regulatory and site requirements would have to be excavated and retreated. In principle, solidifying in the disposal site is preferred because it involves fewer steps in handling the waste and is less expensive. Also, a waste that is still liquid or semi-solid can be poured into place in the landfill with minimum void space, yielding minimal permeability and maximum landfill space utilization in the final product. Waste solidified directly in its final resting place becomes, essentially, a monolithic form that exhibits minimal permeability.

Pit Mixing Methods. Pit mixing is sometimes used as an *ex situ* method at remedial sites, where part or all of the waste from the site is excavated and moved to a pit rather than to a mechanical mixing arrangement. However, in most cases, pit mixing is done at a fixed installation, which is discussed in Section 13.3.2.

Spray-On Methods. This variety of S/S system is used only in special situations and consists simply of spraying a solution or slurry of the reagent onto the waste passing underneath on a conveyor belt. The technique depends on the solution thoroughly wetting the waste because the system does not positively mix the waste and reagent. Thus, the waste particle size should be small and fairly uniform (unless only surface treatment is required) and the layer of waste encountering the spray should be thin. Some mixing action takes place naturally in downstream handling steps, or can be deliberately induced by incorporating a static mixer or tumbling action into the mechanical system. A popular example is spraying a phosphate solution to stabilize contaminants remaining in municipal incinerator ash.[2] The ash is transported from the ash collection system on a conveyor belt, and little additional mechanical equipment is required. This results in a simple, inexpensive stabilization of soluble metals (e.g., lead) in the ash. Spraying with a cement slurry is also used, with or without additional reagents.

13.3.1.2 *In Situ* Methods

In situ methods were probably the first to be used for non-nuclear remedial projects, long before RCRA, the Land Disposal Restrictions Program, Superfund, and the other legislative and regulatory tools that created a hazardous waste industry. In those days, S/S was used primarily for solidification of liquid and semi-liquid wastes in ponds or "lagoons" so that those collection systems could be reused, or the land area used for another purpose such as construction. Backhoes were always available, and solidification with cement, lime, flyash, or other inexpensive binders was easy and inexpensive. Such simple systems are still used for that purpose. However, much remediation work now requires a more sophisticated approach, and *in situ* technology has had to keep pace with the changes.

In addition to just S/S processing, both *ex situ* and *in situ* methods are capable of conducting sequential treatment operations; for example, pre-treatment to oxidize or reduce a constituent followed by solidification. However, containment between the operations is easier with *in situ* processing, and waste handling is essentially eliminated. Stripping volatile organics from contaminated media prior to metal stabilization is one multi-method operation especially amenable to *in situ* treatment. With *ex situ* treatment, two different treatment systems must be mobilized to the site and operated; with *in situ* treatment, the same basic equipment often does both more cost-effectively.

Backhoe-Based Methods. The first such implement used for *in situ* treatment was the familiar backhoe. Figure 13.2 shows this operation at a large remedial operation, where several backhoes are in use at one time. Reagents were poured on the surface of the area to be treated and mixed to the required depth with the backhoe, or tilled in with a harrow for treatment of shallow layers. This, obviously, was not

FIGURE 13.2 *In situ* mixing with backhoes.

FIGURE 13.3 *In situ* mixing with backhoe-mounted rake injector.

a well-controlled process and was used mainly for solidification only. Later, pneumatic or hydraulic injectors were added to the backhoe arm to improve reagent distribution and control, and rake-type devices or high-energy mixers to improve mixing. One such device is shown in Figure 13.3.

Drilling/Augering/Jetting/Trenching Methods. Most recently, massive earth drilling and foundation construction equipment has been modified to allow well-controlled reagent injection and mixing to be performed to great depths. One such system is shown schematically in Figure 13.4 and in actual operation in Figure 13.5. A S/S reagent slurry is pumped through a drilling assembly, consisting of a vertical, hollow bar called a "kelly bar" and a set of hollow auger blades, into the soil or sludge as the assembly is rotated down through the contaminated media. Very high torque (up to 400 kJ or 300,000 ft/lb) produces a well-mixed, treated column of waste at depths up to 100 ft or more, with diameters of up to 14 ft. Overlapping, consecutive columns are positioned to cover the entire volume of waste. A major

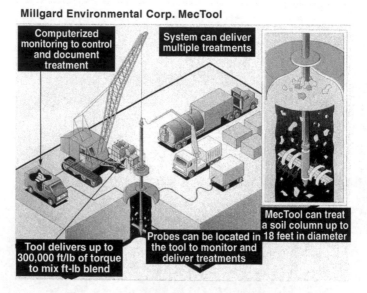

Millgard Environmental Corp. MecTool

Computerized monitoring to control and document treatment

System can deliver multiple treatments

MecTool can treat a soil column up to 18 feet in diameter

Probes can be located in the tool to monitor and deliver treatments

Tool delivers up to 300,000 ft/lb of torque to mix ft-lb blend

FIGURE 13.4 Schematic of *in situ* mixing with an auger system.

FIGURE 13.5 Photograph of actual *in situ* mixing with an auger system.

advantage is the ability to control both volatile and particulate emissions from the site, due to the use of a hood or shroud over the drilling assembly and the column being stabilized. Another advantage is that the production of discrete, accurately placed columns of treated waste provides assurance that all of the waste is treated, although the required overlap of the columns means that some of the volume is double treated. Although double treatment usually causes no harm, over-treatment is not permissible for some systems.

One limitation of the drilling/augering method is problems with underground obstacles such as boulders and buried debris. The larger auger systems can usually deal with smaller obstacles, but large ones can prevent complete treatment of the waste or damage the drilling equipment. Also, the grout that is pumped in can cause ground swell unless inherent voids accommodate the increased volume; e.g., shallow land burials contain large void volumes but tend to be rift with buried obstacles. Ground swell may be undesirable in some situations, although in other remedial actions the extra "fill" is welcome.

Jet grouting is a construction technique for consolidating or sealing soil *in situ* where traditional excavation techniques are not practical; for example, where existing buildings are too close. The grout is pumped to high pressure and then jetted into the soil, with the potential energy of the high pressure transformed into and spent as kinetic energy breaking up and mixing the soil with the grout. Obviously, this mixing action is ideal for mixing stabilizing blends with contaminated media *in situ*. However, the construction equipment and technique results in significant grout returns to the surface (up to 100%; i.e., for every gallon in a soil–grout column, up to another gallon flows to the surface). Although this extra soil–grout mixture causes little more than inconvenience for construction work, returning this much contaminated material to the surface can be a serious disadvantage for an *in situ* remediation.

Fortunately, innovations and improvements have retained the advantage of intimate mixing afforded by the use of such jets, while minimizing the amount of material returned to the surface. For example, Multi-Point Injection (MPI™) had no grout returns for shallow land burials, where significant void space existed to accommodate the volume of material injected.[2] Ground swell can be expected for such injections into undisturbed soil, depending on the void volume and volume of injected material. Also, injecting within about 2 feet of the surface can result in the ground cover being incorporated into the grout mix. Thus, if contamination starts at the surface and exposing the energetic mixing at the surface is undesired, a cover of about 2 feet or more of clean soil may be required.

Liquid Injection Methods. In some situations stabilization can be accomplished by just injecting a treatment solution or slurry into the contaminated area, in much the same way that soil stabilization for construction purposes has been done for many years. In a highly permeable material, such as sandy soil, injection may be done using standard well injection points and injecting the solution/slurry under pressure. In less-permeable formations, using one of the drilling augers described above works better.

Area (Shallow) Methods. In many remedial projects, the sludge pond or contaminated soil depth is not great, often 5 feet or less. For these situations, the use of modified agricultural soil conditioning equipment is efficient and cost-effective. An example of such a system, using a tractor-mounted disk harrow, is shown in Figure 13.6. The reagent can be spread on the surface followed by harrowing, or it can be injected through nozzles mounted on the harrow. Similar systems use power-driven, front-mounted augers or blades similar in operation to a roto-tiller.

13.3.2 FIXED INSTALLATIONS

Fixed installations are those intended to be used at one site, either permanently or for some singular purpose over a limited time span. Unlike the variety of delivery systems encountered in mobile or portable systems, fixed installations are generally

FIGURE 13.6 *In situ* mixing with a tractor-mounted disk harrow.

of a common breed. The many varieties of *in situ* systems are rarely, if ever, applied at fixed installations because they are not needed. Normally, a wide variety of wastes are delivered to a treatment facility

Pit Mixing Methods. Figure 13.7 shows a typical pit mixing installation at a TSDF. Most of these installations today are well controlled, using reagent storage and feeding equipment that can accurately proportion the reagent/waste ratio and also the water addition, if required. The pit is generally a steel- or concrete-lined basin with secondary containment. Because it is relatively small, 38 m³ (50 yd³) or less, and configured to eliminate dead areas that the backhoe bucket can't get into, it can do a good mixing job with most waste streams. The nature of the backhoe allows the method to work well with debris and other large particles that could not be handled with a mechanical mixer-based system. Pit mixing is a large-batch process that is an efficient method for TSDFs where each waste load, typically 20 tons, coming to the stabilization plant may require a different formulation. A continuous mixer-based system is less practical for such sudden changes in waste characteristics with each waste load delivery. While a batch, mechanical mixer-based setup could be used in the same way, it is not practical to use a pugmill or other mixer to make

FIGURE 13.7 *Ex situ* pit mixing at a TSDF.

FIGURE 13.8 *Ex situ* mixer-based system at a TSDF.

such large batches, and waste loads would need to be broken up into several batches for treatment. Also, loading and unloading the pit is fast and simple. Most waste transport vehicles can dump directly into the pit, and the backhoe can quickly excavate the treated waste from the pit, even if it has hardened. Batch cycle time is usually in the range of 30 minutes to 2 hours.

Mixer-Based Methods. Fixed systems of this type are essentially the same conceptually as that discussed previously in Section 13.3.1. The dissimilarities are primarily due to the permanency of the installation rather than to distinctions in the processing method. However, fixed installations of this sort do allow more feasible use of peripheral material handling equipment and a wider variety of chemical additives, and they are generally enclosed to accommodate all-weather use. One such design is shown in Figure 13.8.

13.4 CASE STUDIES

The previous discussion has laid the technical groundwork for an understanding of the processes and techniques that make up full-scale S/S remediation projects. The four case studies that follow are intended to provide specific examples over a fairly wide range of waste types treated and delivery systems used. These studies include both industrial and radioactive wastes at a variety of operational sites, but not fixed systems at TSDFs because their technology is the same as for *ex situ* remediation projects with mixer-based methods. Case Study #3 provides an interesting historical example of S/S from earlier projects. Such examples establish the success and longevity of S/S as a reliable waste treatment technology.

13.4.1 CASE STUDY #1: LEAD-CONTAMINATED SOIL AT A FORMER BATTERY PROCESSING SITE[4]

13.4.1.1 Background and Purpose

This site had been utilized for various industrial purposes since the mid-1800s. Metals in the site soil — lead, cadmium, arsenic, and antimony — were believed to have been generated by a former battery processing operation. Remedial action was required not only to clean up an area contaminated with these metals, but also to allow both the soil and the site to be used as part of a new park. Lead was the primary contaminant of concern.[4]

13.4.1.2 Treatability Studies

Previous analyses by an engineering company of samples from the site had shown that the soil typically contained about 18,000 mg/kg of lead, and lesser amounts of arsenic, cadmium, and antimony (see Table 13.1). While arsenic, cadmium, and antimony leach resistance met regulatory criteria, the TCLP extract lead concentration was 41.0 mg/L, more than 8 times the limit. The results of the treatability study are summarized in Table 13.1. Leachability testing was done using the standard USEPA SW-846 Toxicity Characteristic Leaching Test (TCLP). A liquid chemical additive in a portland cement-based system built additional early strength and helped stabilize lead. The additive also reduced the amount of cement required, thereby lowering the volume increase and the chemical costs for treatment.

TABLE 13.1
Treatability Study Results for Case Study #1

Contaminant	Total Concentration (mg/kg)	TCLP Extract Concentration (mg/L)		
		Before Treatment	After Treatment	TC Limit[a]
Antimony	170	0.9	< 1	15.0[b]
Arsenic	160	0.6	0.01	5.0
Cadmium	11.0	1.0	< 0.02	1.0
Lead	18,000	41.0	< 0.4	5.0

[a] Toxicity Characteristic limit.
[b] No TC limit defined. The California Soluble Threshold Limit Concentration (STLC) test requirement applied to this project.

FIGURE 13.9 Photograph of the site operation for Case Study #1.

13.4.1.3 Delivery System

The physical setup for this project was basically as described in Section 13.3.2, Mixer-Based Methods. A photograph of the actual site arrangement is shown in Figure 13.9. In addition to the S/S treatment system, a size reduction and screening operation was utilized prior to treatment. Such add-ons prevent equipment stoppages due to jamming of feeder, conveyors, and mixers. The remediation contractor's experience with this system contributed to operation with no major equipment problems during the project. Portland cement was added at the soil feed point, a somewhat unusual procedure, before conveying to the mixer. Thus, the only chemical additions in the mixer were liquids (chemicals and water), allowing the use of a smaller mixer and an overall faster processing rate.

13.4.1.4 Operations

A hazardous waste remediation subcontractor was chosen by the general contractor to conduct the soil treatment phase of the project. The regulatory issues had been addressed and resolved by the site owner, the general contractor, and the appropriate regulatory agencies before the treatment phase was awarded and begun. Therefore, the remediation contractor was able to proceed without delay. Prior to treatment, the soil was excavated, stockpiled, and screened to less than 0.25 in. Oversized material was crushed to less than 0.25 in. before processing. A front-end loader placed screened soil from a stockpile into a feed hopper upstream from the pugmill. Portland cement was added as the soil was fed onto a conveyor belt leading to the pugmill. Water and a chemical solution were added in the pugmill and the combination mixed. The treated, semi-solid mixture was conveyed to and dumped on an open space, where the product cured and the chemical reactions were allowed to complete. This solid mound of treated material was used as "clean fill" stockpile until the site grading was complete. After soil treatment and other site preparation work were completed, the stockpiled, treated soil was used as engineered backfill to form a berm. Trees and grass were planted on the treated soil to make a new park.

13.4.1.5 QC and Results

Quality control samples were taken every 1000 yd^3 to ensure effectiveness of the treatment and compliance with regulations. Typical results of TCLP testing of the treated soil are shown in Table 13.1, along with the raw waste analyses and the regulatory limits imposed upon the project. Routine test results on the Q.C. samples consistently met the compliance standards applicable in 1989 for this project. The treated soil had an unconfined compressive strength of greater than 1.0 ton/ft^2. Other pertinent data on the project are:

Volume Treated:	48,700 yd^3
Treatment Rate:	2000 yd^3 per day
Time:	1989
Treatment Cost:	$67 per yd^3
Disposal Method:	Engineered backfill on-site

The project was successfully completed and the site is now a county park.

13.4.2 CASE STUDY #2: *IN SITU* STABILIZATION OF MIXED WASTE CONTAMINATED SOIL[5]

13.4.2.1 Background and Purpose

The X-231B Unit at a Department of Energy (DOE) facility at Portsmouth, Ohio, was used from 1976 to 1983 as a land disposal area for waste oils and solvents. The soil consisted of subsurface silt and clay contaminated with up to 500 mg/kg of volatile organic compounds (VOCs) (primarily trichloroethylene [TCE] and other halocarbons) and lower concentrations of lead (Pb), chromium (Cr), uranium 235 (^{235}U), and technetium 99 (^{99}Tc). Ranges of contaminants at subsurface depths of up to 5.4 meters are given in Table 13.2. The shallow groundwater was also contaminated, with TCE well above drinking water standards. Part of the Unit closure required reduction of VOC mass in the 0- to 6.6-m depth range while controlling the leachability of heavy metals and radionuclides. The X-231B Technology Demonstration project initiated at Oak Ridge National Laboratory (ORNL) by DOE and Martin Marietta Energy Systems, Inc. (MMES) in November 1990 evaluated *in situ* treatment at the site. Three treatment processes — vapor stripping, chemical oxidation, and S/S — were evaluated in treatability studies at bench and pilot scale prior to the full-scale project demonstration being conducted. This case study documents the work done on S/S. Design, implementation, and operation of the S/S process were awarded to Chemical Waste Management, Inc. (CWM) in Columbia, SC, and Millgard Environmental Corporation (MEC) of Livonia, MI.

Cement-based S/S techniques have been widely used for remediation of soils contaminated with heavy metals, but many have questioned whether organic contaminants can be effectively stabilized in a cement-based grout. In addition, high concentrations of organics are known to interfere with cement solidification processes. However, recent laboratory research has indicated that specific organic compounds can be effectively immobilized in cement-based systems.[6–8]

TABLE 13.2
Representative Characteristics of the Site for Case Study #2

Characteristic	Concentration at Nominal Subsurface Depth (m)		
	0 – 0.6	1.8 – 2.4	4.8 – 5.4
Particle size distribution:	–	22 – 25	12 – 15
Clay: < 0.002 mm, wt%	–	65 – 67	39 – 64
Silt: 0.002 – 0.05 mm, wt%	–	8 – 12	22 – 46
Sand: 0.05 – 2.0 mm, wt%			
Water content, dry wt%	16 – 23	17 – 19	22 – 24
pH	–	5.3 – 6.0	6.2 – 7.4
Total organic carbon, mg/kg	–	600 –1200	200 – 500
Total VOCs[a], μg/kg	9900	5300	1500
Chromium, mg/kg	–	13 – 31	9 – 20
Lead, mg/kg	–	20 – 28	16 – 23
Nickel, mg/kg	–	8 – 18	7 – 20
Total alpha, nCi/kg (10)	ND[b] – 150	ND	ND
Total beta, nCi/kg (10)	ND – 200	ND – 31	ND – 33
Total uranium, nCi/kg	2 – 250	1 – 3	2 – 3
Technetium, nCi/kg (2)	ND – 380	–	–

[a] Total VOCs = summation of trichloroethylene, methylene chloride, 1,1,1-trichloro-ethane, 1,2-dichloroethylene, and 1,1-dichloroethane.
[b] Not detected at limit shown in parentheses in first column.

13.4.2.2 Treatability Studies

Three soil samples containing low, medium, and high concentrations of VOCs (TCE, 1,1,1-trichloroethane, methylene chloride, 1,1-dichloroethylene, 1,2-dichloroethylene, 1,1-dichloroethane, 1,2-dichloroethane, vinyl chloride, chloroform, carbon tetrachloride) were tested. The criteria for successful treatment were concentrations in the TCLP extract of: (1) < 25 μg/l for TCE, and (2) sum of < 86 μg/l for all the VOCs. Preliminary testing resulted in VOC concentrations below detection limits for both stabilized and unstabilized samples (see Table 13.3). The TCLP procedure was modified to a 1:5 dilution instead of the normal 1:20 dilution in an attempt to avoid concentrations below detection limits because of the low VOC concentrations in the soil samples.

The performance evaluation studies investigated the performance of portland cement and a proprietary reagent effective in the solidification of soils contaminated with VOCs. A control sample was mixed in a manner similar to the core samples, but without the addition of the admixture or water (portland cement was added to help the high-clay content soil to break apart). Analyses for total and TCLP VOCs were performed on untreated and treated core samples. The treated materials were cured for 3 days, after which a 100-g portion was removed from each jar, size-reduced as quickly as possible, and a 25-g aliquot transferred to a zero headspace extractor for TCLP/VOC extraction (EPA SW-846 Method 1311). All volatile organics analyses

TABLE 13.3
Summary of Treatability Study Results for Soil Core Samples in Case Study #2

Volatile Organic Compound	Preliminary Stabilized and Unstabilized TCLP (mg/L)	Performance Evaluation			
		Unstabilized		Stabilized	
		Total (mg/kg)	TCLP (mg/L)	Total (mg/kg)	TCLP (mg/L)
Vinyl chloride	< 0.009	< 0.009	< 0.009	< 0.15	< 0.002
1,1 - Dichloroethylene	< 0.007	< 0.007	< 0.007	< 0.12	< 0.002
Methylene chloride	< 0.006	< 0.006–0.012	< 0.006	< 0.10	< 0.004
Trans - 1,2 - Dichoroethylene	< 0.007	< 0.007	< 0.007	< 0.12	< 0.002
1,1 - Dichloroethane	< 0.006	< 0.006	< 0.006	< 0.10	< 0.001
Chloroform	< 0.007	< 0.007	< 0.007	< 0.12	< 0.002
1,1,1 - Trichloroethane	< 0.012	< 0.012–0.054	< 0.012	< 0.20–0.44	< 0.003
Carbon tetrachloride	< 0.009	< 0.009	< 0.009	< 0.15	< 0.003
1,2 - Dichloroethane	< 0.006	< 0.006	< 0.006	< 0.10	< 0.002
Trichloroethylene	< 0.013	< 0.013–0.059	< 0.013	< 0.22	< 0.003

were performed according to EPA SW-846 Method 8260. Total uranium was analyzed according to CNES Science and Technology Division Method 90284 using a laser kinetic phosphorimeter. Total technetium was analyzed according to method EC 206 for soils, and EC 038 for waters, both MMES methods.

A summary of results is shown in Table 13.3. Only three of the ten target compounds were detected in samples. Methylene chloride and TCE were detected at levels just above the analytical detection limit in two untreated core samples. Neither was detected in the treated samples. Also, 1,1,1-trichloroethane was detected at levels just above the analytical detection limit in one untreated core sample, but was detected at elevated levels in the total volatiles analysis of two of the treated core samples. This appears to be an anomaly, since these results are 5 to 10 times higher than the total VOC results of the untreated samples. Previous experience suggests that "hot spots" (small localized areas in the soil that contain elevated levels of volatile organics) could be present in the soil and cause this type of response. The TCLP results suggest that the admixture handled these elevated levels. However, proof of reagent effectiveness is uncertain from these treatability study results, because total analysis results indicate significant losses during mixing of the small core samples, an inherent problem when working with VOCs. The *in situ* system for the field demonstration likely resulted in less loss from mixing, although verification proved difficult, as attested to in Section 13.4.2.5. No uranium was detected in the untreated and treated samples (detection limit of 20 ppm). Small amounts of technetium (up to 1.7 pCi/g) were present in the untreated samples. No technetium was detected in the TCLP extract of the treated samples, but this could be from dilution.

The proportion of unmixed soil, as determined from visual observations made during the size-reduction process, ranged from less than 10% to greater than 20%. The diameter of the unmixed soil chunks ranged from less than 1/8 in. in diameter to greater than $1/4$ in. in diameter. Visual observations were made during the course of the stabilization activities. In each instance the solidified waste form formed a monolith with a high degree of cohesion and no free liquid.

13.4.2.3 Delivery System

The mechanical system used in this project was described in Section 13.3.1.1 and is shown in Figures 13.4 and 13.5, a schematic and an actual photograph, respectively, of the system at the X-231B site. The availability of the hood mentioned briefly in Section 13.3.1.2 was a major element in the decision to use this equipment for the project because of the VOC concentrations in the soil. The auger drilled holes 10 ft in diameter, producing overlapping columns of treated soil 7.0 ft on center, which provided a 3-ft overlap. The rows of columns were staggered to provide complete treatment of all the soil volume; this resulted in double treatment of the overlapping areas. The grout was prepared at a local concrete batch plant and delivered to the site in 7-yd^3 concrete transit mixers. The grout was then deposited from the mixer into a grout hopper with 0.25-in. screen to remove any large particles upstream from the grout pump. A progressive cavity grout pump with a maximum capacity of approximately 50 ft^3 per minute was used to deliver grout to the "Kelly bar" and hence to the hollow auger of the system ("MecTool"), which injected the grout and mixed it into the soil. During shakedown tests at the onset of the demonstration, mechanical problems encountered with the grout delivery system were resolved through grout reformulation and mechanical modifications.

13.4.2.4 Operations

After shakedown tests at the site, substituting granular activated carbon for the original powdered carbon and adding flyash to increase the consistency and fluidity of the grout modified the formulation. A retarder was also added to increase working time about 2 hours, since the grout was to be mixed in a transit mixer on the way to the injection point. The final formulation of the grout used for injection into the soil is shown in Table 13.4.

The auger was positioned above the column to be treated. Auger position was determined and monitored by a geologger on the MecTool that was connected to a data acquisition system (DAS). Grout delivery rate was controlled and monitored by a flow meter on the grout pump, and the number and capacity of the grout delivery trucks controlled the volume of grout added. Additional water was added at the site to further increase workability. Approximately 14 yd^3 of grout was injected into each of three columns, yielding a grout-soil combination of approximately 30% v/v, 20% w/w grout. Two of the three columns were surrounded by undisturbed soil, while the third (central) column overlapped the other two columns by 3 ft, about 15% of the column area. At the start of each test, grout was pumped out of the auger at ground level to wet the ground area prior to penetration. A steel shroud was lowered

TABLE 13.4
Composition of Grout Used in Case Study #2

Component	Fraction of Mixture (wt%)	Mass Added Per Column (kg)
Water	31.4	5080
Retarder[a]	< 0.10	10.2
Activated carbon[b]	17.0	2724
Cement, Type I	41.3	6690
Flyash	10.4	1692

[a] From the W.R. Grace Co., #WR-75-7250.
[b] Calgon Type 220R, 30 × 100 mesh.

tightly onto the ground surface and penetration was initiated. The auger blade rotated at approximately 10 rpm as it moved down the soil column. About 10 minutes of mixing was required for the auger to reach the depth of 15 ft, and then the auger was moved in a continuous upward/downward motion while rotating for a total of 45 minutes per column.

Because of the dense clay soil being treated, with the resultant low void content, volume increase in the column due to the grout injection resulted in an aboveground berm being created. The resultant berm was approximately 3 ft high, and the shroud rose on the berm surface as it was formed. The shroud was kept under vacuum, but the maintainable vacuum varied from about 155 Pa for two of the columns to only about 17 Pa for the third. This may have been due to a relatively poorer seal with the latter column due to resting on the berms of the two columns already formed. After the demonstration, the berms were compacted and the geomembrane that covered the surface of the site was repaired.

13.4.2.5 QC and Results

Off-gas was collected in the shroud over the column and the gas flow rate, temperature, VOC content, and particulate content were monitored using appropriate in-line instrumentation and sampling and analysis. Limited VOC removal was measured during the processing, and no VOCs were detected until after about 10 minutes of treatment. The application of grout onto the surface before penetration may have limited the transfer of VOCs from the soil into the induced airflow caused by the shroud vacuum. Also, the use of activated carbon in the grout formulation may have enhanced sorption of VOCs into the treated soil.

Samples of untreated soil at various depths were taken prior to the demonstration (Table 13.2). Within hours of the treatment operation, samples of the treated but uncured soil were collected at various depths using a drill rig. Other samples were collected from the berm using hand samplers. All solid samples were analyzed for the seven target VOCs in an on-site laboratory using a heated headspace procedure and gas chromatograph with electron capture detector. Results of this testing are given in Table 13.5. From the mass balance computation on the data given in Table

TABLE 13.5
Pre- and Post-Treatment Total VOC Soil Concentrations[a]
– Case Study #2

Depth Interval (m)	Pre-Treatment		Post-Treatment		
	Conc. (mg/kg)	Weight (g)	Conc. (mg/kg)	Weight (g)	% Initial
Off-gas	–	–	–	50[c]	0.4
Berm	–	–	2.39	70[d]	0.55
0.0 – 0.9	48	1,860[b]	–	252[e]	2.0
0.9 – 2.1	84	4,380[b]	5.04	252[e]	2.0
2.1 – 3.4	106	5,515[b]	3.14	180[e]	1.40
3.4 – 4.6	20	1,050[b]	–	180[e]	1.40
Total	–	12,805	–	984	7.75

[a] Total VOC concentration is defined as the summation of: trichloroethylene, methylene chloride, 1,1,1-trichloroethane, trans1,2-dichloroethylene, cis 1,2-dichloroethylene, 1,1-dichloroethane, and 1,2-dichloroethane. Values represent the average of three soil columns.
[b] Basis: Undisturbed soil bulk density = 1.95 kg/l.
[c] Total VOC mass captured from off-gas for all three columns.
[d] Basis: 50 vol% disturbed soil (bulk density = 1.4 kg/l) and 50 vol% grout (bulk density = 1.515 kg/l).
[e] Basis: mass of soil-grout column, exclusive of berm. VOC concentrations were assumed uniform throughout the given depth intervals for samples from 1.2–1.5 m and 2.4–2.7 m intervals.

13.5, only 8% w/w of the pretreatment VOC mass vaporized during treatment. It is speculated that as much as 90% of the pre-treatment VOC mass was captured in the grout/soil mixture at the time of analysis. Drilling samples from the 1.2 to 1.5 m and 2.4 to 2.7 m depth intervals were also tested using TCLP in an off-site laboratory. These analyses indicate the treated material passed the accepted Toxicity Characteristic (TC) limits for both organic and inorganic species.

Fifteen months after the demonstration, core samples were again collected using a drill rig. These samples were sent to ORNL for physical and chemical analyses. Visual evaluations were made of the samples at various depths, and photographs were taken. Physical properties of the untreated and treated soils are given in Table 13.6. The apparent reduction in bulk density of the treated soil is due to the high initial bulk density of this clay soil as a result of mixing, and may also be caused by the large amount of water in the grout and entrapment of air during mixing. Similarly, the increased hydraulic conductivity of the treated soil, in contrast to the experience in most S/S applications, is likely due to the initial dense clay being treated and its structural disruption caused by mixing. Most soils involved in S/S projects have very high initial hydraulic conductivities. Compressive strength values obtained were inversely proportional to the depth, although all of the values were greater than the guideline for grouted material, 340 kPa (50 psi). The decrease in

TABLE 13.6
Physical Properties of the Untreated and S/S Treated Soils – Case Study #2

Sample Description	Sample Depth (m)	Water Content (wt%)[a]	Bulk Density (kg/l)	Compressive Strength (kPa)	Hydraulic Conductivity[b] (cm/sec)
	0.3 – 0.6	20.5[c]	–	–	–
Untreated Control Core	1.2 – 1.5	16.9[c]	2.15	–	8.08 × 10⁻⁸
	2.4 – 2.7	18.4[c]	1.75	–	–
	3.9 – 4.2	16.5[c]	–	–	8.09 × 10⁻⁸
	0.3 – 0.6	19	1.73[c]	5200[d]	8.88 × 10⁻⁶
Treated Core	1.2 – 1.5	23.5	1.66[c]	3500	–
	2.4 – 2.7	19	1.72	2600	–
	3.9 – 4.2	13.6	2.00[c]	390	7.75 × 10⁻⁶

[a] Loss on drying at 60°C.
[b] All values represent the average of six measurements. Data reported at 25°C per ASTM D5084. Permeant fluid: 0.001 M CaSO₄ solution.
[c] Average for two samples.
[d] Sample taken at the 1-m depth.

strength with depth was due to the degree of grouting that varies with depth, which was confirmed by visual inspection. The pH of the treated soil reflected the high pH (12.5) of the grout, ranging from 10 to 11, with the deepest soil having the lowest pH. The VOCs in the TCLP extract were mostly undetected, except for 0.08 mg/L TCE at the 1.5-m depth (still below the regulatory limit of 0.5 mg/L).

In spite of a number of physical and mechanical problems encountered during shakedown, and difficulties in controlling the rate of grout delivery at different depths, the results were generally good with respect to achieving the regulatory requirements of the project. In addition, the operation demonstrated that the mechanical system used in this project was effective in limiting the evolution of VOCs from the contaminated soil during treatment.

13.4.3 CASE STUDY #3: SOLIDIFICATION OF LIQUID WASTE CONTAMINATED WITH ANTIMONY

13.4.3.1 Background and Purpose

This project took place nearly 30 years ago and is a good example of cement-based S/S technology that has stood the test of time.[1] Its chemistry, delivery system, and operation are just as pertinent today as they were 30 years ago. The waste treated at this site in the New Orleans area originated at a chemical plant in West Virginia. It had been routinely transported to New Orleans, tanked temporarily, and then ocean-dumped when that technology was still permitted. When ocean dumping was stopped, 4,000,000 gallons remained in storage tanks. One requirement of the project was that all 4,000,000 gallons had to be contained on-site after solidification, until the state could test it. Treatment and compliance testing verified the treated waste was suitable for local, sanitary landfill as a non-toxic solid. The waste contained high levels of dissolved organics and salts as well as dissolved antimony.

13.4.3.2 Treatability Studies

The waste was a low-viscosity, semi-clear solution with very little suspended solids. The dissolved antimony (234 ppm) was the main hazard. While there were no applicable federal standards at that time for metal leachability, or even a standard leaching test, the state of Louisiana was willing to allow disposal of this waste in a local landfill if it was converted into a stable, workable solid with reduced antimony leachability. Table 13.7 gives the results of the waste analysis before treatment. The S/S process used a combination of portland cement and sodium silicate solution to quickly set and then harden the liquid waste. Prior to the addition of S/S reagents, the waste was pretreated with hydrated lime to precipitate, or expel from aqueous solution, some of the dissolved organics. This aided in solidifying the waste by reducing the dissolved organic interference to cement setting, and by providing more suspended solids, and thereby reduced the requirements for the S/S reagents.

After S/S treatment and curing, the solid was tested for strength using a penetrometer and was leached in a vendor-developed column-leaching test. This test required crushing the solid into pieces about $3/8$ in. in maximum size, placing the crushed solid in a glass column approximately 2 in. in diameter, and allowing deionized water to pass through the material slowly and in such a way as to assure complete contact between the waste and the leachant. The test was continued until the equivalent of 100 in. of simulated rainfall had permeated the waste, usually a time of several days or more. The permeate was then analyzed for leached constituents. This test was similar to other experimental procedures used at the time until EPA mandated the Extraction Procedure Toxicity Test (EPT) and later the TCLP.

TABLE 13.7
Analysis of the Raw Waste –
Case Study #3

Contaminant	Concentration (wt%)
Ethylene glycol	6.6
Diethylene glycol	0.2
Sodium terephthalate	2.5
Sodium chloride	8.3
Sodium sulfate	0.9
Ammonium chloride	1.1
Antimony	0.0234

TABLE 13.8
**Comparison of Treated Waste to Raw Waste Analyses
– Case Study #3**

Contaminant	Concentration in Raw Waste (mg/kg)	Concentration in Treated Waste (mg/L)
Ethylene glycol	66,000	NM
Diethylene glycol	2,000	NM
Sodium terephthalate	25,000	NM
Sodium chloride	8,300	NM
Sodium sulfate	9,000	NM
Ammonium chloride	11,000	NM
Antimony	234	0.1
Chloride	NR[a]	160.0
Sulfate	NR	5.0
Chemical oxygen demand	NM[b]	350.0

[a] NR: Not measured and reported separately as the anion, but present in compounds listed above.
[b] NM: Not measured.

The test results on the treated waste are given in Table 13.8. Antimony leaching was quite low by any standard and, based on the authors' experience, likely would have met current TC and UTS (< 1.15 mg/L) limits, although the results cannot be directly compared. Also, the very soluble chloride and sulfate leachate concentrations are minimal and so is chemical oxygen demand (COD), representing the organic content in the raw waste. The solid product had good compressive strength at 4 to 5 tons/ft^2 (~75 psi or ~517 kPa). These properties allowed the S/S product, provided it duplicated the laboratory results, to be disposed of in a local sanitary landfill; in fact, it was permitted as daily cover material, a higher end use. There was some ammonia evolution from the waste during and after treatment, due to the high pH of the product, but this was judged to be acceptable in view of the industrial location of the site and the prevailing environment.

13.4.3.3 Delivery System

The delivery system used is shown in operation in Figure 13.10. It consisted of two identical treatment units shown in schematic in Figure 13.11, one pre-treating the waste with hydrated lime and the other treating with the S/S reagents. The primary difference between this system and that previously described in Section 13.3.1 is that the treatment units used here are completely mobile and are largely self-contained.

FIGURE 13.10 Photograph of the site operation for Case Study #3.

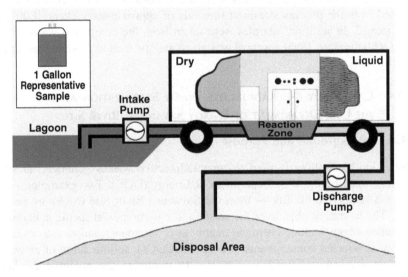

FIGURE 13.11 Schematic drawing of the treatment units used in Case Study #3.

13.4.3.4 Operations

The waste immediately after treatment was still quite fluid and required retention in an impoundment until it solidified. Normally, excavating a temporary impound-ment, using the excavated soil as additional berm material, does this. Because only limited space was available, and groundwater conditions precluded digging a deep pit for curing, the job was done in stages, starting at ground level, pouring out the treated but still liquid waste, allowing it to solidify, and using the solid produced to create berms for impoundment for the next stage. In this manner, a final 35-foot-high "hill" was created (Figure 13.10). The ammonia evolved during the processing was noticeable, but not at high enough levels to constitute an occupational or local environmental hazard. After all of the waste was treated, final sampling done, and compliance testing completed and approved by the state, the solid was excavated and moved to the final disposal site. The treatment site was then graded and subsequently used for plant expansion. Operational parameters for the project follow:

Volume Treated:	4,000,000 gallons (19,800 yd³)
Treatment Rate:	130,000 GPD average
Time:	Spring 1976
Treatment Cost:	$0.167/gallon
Disposal Method:	Sanitary landfill, daily cover material

13.4.3.5 QC and Results

Frequent initial sampling and testing were done on the first 100,000 gallons of waste treated to verify that the process was operating as planned. Thereafter, samples were collected from the process stream at intervals of approximately every 1000 yd³ of waste treated. In addition, samples were taken from the completed waste pile and shared with the state. Final approval was given and the treated product was removed from the site.

13.4.4 CASE STUDY #4: RADIOACTIVE WASTE STABILIZATION AT THE U.S. DEPARTMENT OF ENERGY SAVANNAH RIVER SITE

13.4.4.1 Background and Purpose

Stabilization technology is used to treat radioactive wastes generated at several facilities owned by the U.S. Department of Energy (DOE). Two examples — Saltstone and "empty" tank fills — from the Savannah River Site (SRS) are provided below. The treatment objectives for waste and environmental media contaminated with radionuclides include chemical treatment or microencapsulation to reduce the mobility of selected contaminants of concern (COCs), solidification of an aqueous phase, or physical stabilization of wastes to minimize the potential for landfill subsidence. The COCs are typically radioactive isotopes, soluble salts, chemicals or metals carrying RCRA hazardous codes, or other constituents that must be managed to meet maximum concentration limits (MCL) drinking water concentrations in the disposal landfill. Waste that contain both radioactive and RCRA hazardous contaminants is referred to as mixed waste and is regulated by either DOE (defense related) or NRC (commercial nuclear) and EPA, respectively. Most of the radioactive wastes identified for stabilization treatment are aqueous-based liquids or sludges (mixtures of soluble and insoluble solids plus water). However, fine particulates, debris (processing equipment and construction materials), environmental media (soils and debris associated with old disposal sites), or contaminated structures that are identified for *in situ* closure may also require stabilization treatment in order to meet the requirements for disposal. In all of these applications, stabilization is intended to provide a chemical environment that will reduce leaching when in contact with water and to bind the particles together and thereby reduce the potential for spread of contamination. For landfill and structural closures, stabilization is also intended to provide structural stability to prevent future subsidence of overburden and in some cases mitigation of inadvertent intruders.

13.4.4.2 Saltstone Facility

The Saltstone Facility was originally permitted in 1986 as a wastewater treatment facility to treat and dispose of decontaminated salt solution from the In-Tank Precipitation process, which was intended to remove radioisotopes from dissolved high-level waste salt prior to treatment and disposal in Z-Area. The permits were modified in 1988 to provide for treatment of the residues from the Effluent Treatment Facility.

13.4.4.2.1 Treatability Studies

The waste is an aqueous solution that contains up to about 30 wt% dissolved sodium salts. These salts are the result of sodium hydroxide neutralization of acids used in the isotope extraction processes that were performed at the site.

13.4.4.2.2 Delivery System

The Saltstone Facility is located in Z-Area at the DOE Savannah River Site and is operated by the Westinghouse Savannah River Company. The Saltstone Facility was designed to process about 30,000 gallons of waste per day with an attainment of about 6,000,000 gallons per year. The processing objective is to stabilize liquid mixed waste so that the resulting waste form can be disposed of as low-level radioactive waste in a Subtitle D landfill. A schematic of this facility is shown in Figure 13.12.

The current Saltstone Processing Room is a hands-on-maintenance facility, as shown in Figure 13.13. The processing equipment in this figure is viewed from the control room. The facility as shown here is currently processing very low-activity waste and is operated as a hands-on-maintenance facility. Modifications are planned to add shielding so that the facility can treat waste containing up to 0.1 Ci waste. Blending and transfer of the stabilizing reagents is also carried out from the control room. The primary components of the processing facility in Figure 13.13 are a mixer, the waste, solution feed stabilizing, the reagents, the waste form, a slurry tank, and a pump.

13.4.4.2.3 Operations

The Saltstone Facility began processing radioactive waste in June 1990. Since then approximately 3,500,000 gallons of radioactive hazardous liquid waste and 100,000 gallons of high-solids slurry have been processed through the facility. In 2002, SRS submitted a permit change to allow processing salt solution with up to 0.1 Ci, an increase of three orders of magnitude. In 2003, modifications were begun to enable processing of the higher-activity waste for a new program referred to as Low Curie Salt Processing. The Saltstone waste form is prepared by combining preweighed and premixed binders with salt solution as illustrated in Figure 13.12. Processing admixtures may also be added to the formulation. The nominal composition of Saltstone is listed in Table 13.9. The facility treats about 100 gallons of waste solution per minute. This production rate requires transfer and metering of about 35 tons of

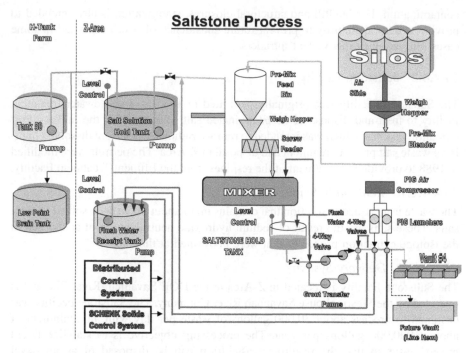

FIGURE 13.12 Schematic of the Savannah River Saltstone Process for stabilizing radioactive salt waste for Case Study #4.

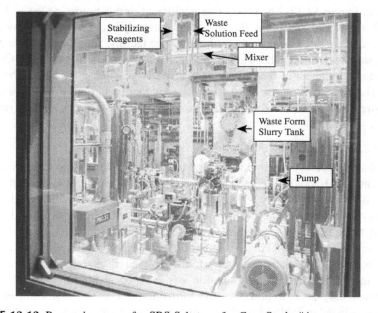

FIGURE 13.13 Processing room for SRS Saltstone for Case Study #4.

TABLE 13.9
Nominal Saltstone composition

Component	wt% of the Saltstone Product
Waste salt solution containing 29 wt% dissolved sodium salts	46
Premixed reagents	
Portland cement	6
Ground granulated blast furnace slag	24
Fly ash	24

cementitious reagents (cement, slag, and fly ash) per hour. Mixing is accomplished in a twin-shaft Readco continuous processor. The resulting waste form slurry is pumped over 2000 ft through a 3-in carbon steel line and is disposed of in a concrete vault. The centrifugal pumps and transfer line are cleaned after each production run with a small amount of water and pigs launched by compressed air. The facility is typically operated continuously for one shift (6 to 10 hours).

13.4.4.3 High-Level Waste (HLW) Tank Fill

In 1997, two single-shell carbon steel tanks (17-F and 20-F) were emptied and filled with grout at the SRS to reduce the mobility of residual contaminants and to provide structural stability to the empty tanks.[9,10] Each tank had a capacity of 1.3 million gallons and each was originally used to store low-heat waste (no cooling coils or other obstructions). The waste tank closure strategy required three different fill materials. Figure 13.14 illustrates the three different fill materials required by the waste tank closure strategy.

All of the material used to close these radioactive waste tanks was prepared from non-radioactive ingredients in two portable batching plants. The dry solids storage silos for each plant are shown in Figure 13.15. Each plant was equipped with an auger grout/concrete mixer. The auger mixers for both plants discharged

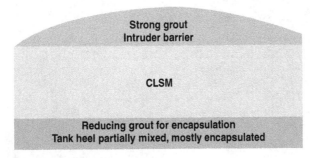

FIGURE 13.14 Schematic diagram illustrating the three different grout materials designed to close high-level waste tanks at the Savannah River Site for Case Study #4.

FIGURE 13.15 Portable continuous auger plant (Throop, Inc.) used to batch fill material for Tanks 17-F and 20-F at the SRS for Case Study #4.

FIGURE 13.16 Tank 17 zero-bleed fill being discharged into the pump for Case Study #4

TABLE 13.10
SRS High-Level Waste Tank Fill Formulations Used for Tank 17-F and 20-F Closure

Application	SRS Reducing Grout Encapsulate Incidental Waste	SRS Zero-Bleed Flowable Fill Bulk Tank Fill	SRS Zero-Bleed 2000 psi Grout Intruder Barrier
Portland cement Type I (lbs/cyd)	1353	150	550
Slag, Grade 100 (lbs/cyd)	209	–	–
Fly ash, Class F (lbs/cyd)	–	500	–
Silica fume (lbs/cyd)	90	–	–
Quartz sand ASTM C-33 (lbs/cyd)	1625	2300	2285
Water (gal/cyd)	86.4	63	65
HRWR* (fl.oz/cyd)	250	–	–
Retarder (fl.oz/cyd)	150	–	–
Sodium Thiosulfate** (lbs/cyd)	2.1	–	–
Advaflow*** (HRWR) (fl.oz/cyd)	–	90	140
Welan Gum Kelco-crete*** (grams/cyd)	–	275	275

* High range water reducer.
**Added to the mix as a liquid solution.
***Premixed and metered into the auger as a suspension.

into a concrete pump that transferred grout about 2000 ft through a 5-in. steel line. Grout production for each plant was targeted at 40 yd^3 per hour (80 yd^3 per hour total). All three grouts were designed to be flowable and self-leveling as shown in Figure 13.16 (seen as discharged from the mixer).

The first grout placed in each tank was designed to stabilize the contaminants in the residual heel that could not be removed by remote means. The heel consisted of a few thousand gallons of sludge up to 5 cm deep on the tank bottom. This first layer is referred to as a reducing grout because it contains reagents to chemically reduce certain contaminants of concern, ^{99}Tc and Cr^{+6}, to lower valence states resulting in precipitates in alkaline conditions. Controlled Low Strength Material (CLSM) was used as bulk fill for most of the tank volume. A strong grout was added to the top of the tank and served as an intruder barrier. The ingredients and proportions in the SRS tank fill materials are listed in Table 13.10.

REFERENCES

1. Conner, J.R. *Chemical fixation and solidification of hazardous wastes*, New York: Van Nostrand Reinhold, 1990.
2. O'Hara, M.J. Immobilization of lead and cadmium in solid residues from the combustion of refuse using lime and phosphate. U.S. Patent 4,737,356, 1988.
3. Multi-Point Injection Demonstration for Solidification of Shallow Buried Waste at Oak Ridge Reservation, Oak Ridge, TN, ORNL/ER-378, prepared by Ground Environmental Services, Inc., October 1996.

4. Conner, J.R., Cotton, S., and Lear, P.R. Chemical stabilization of contaminated soils and sludges using cement and cement by-products, in *Proceedings of 1st international symposium on cement industry solutions to waste management problems*, Calgary, Alberta, Canada, 1992.

5. Siegrist, R.L., Cline, S.R., Gilliam, T.M., and Conner, J.R.. *In-situ* stabilization of mixed waste contaminated soil, in *Stabilization and solidification of hazardous, radioactive, and mixed wastes, 3rd volume*, ASTM STP 1240, Philadelphia: American Society for Testing and Materials, 1996.

6. Spence, R.D., Gilliam, T.M., Morgan, I.L., and Osborne, S.C. Stabilization/Solidification of wastes containing volatile organic compounds in commercial cementitious wasteforms, in *Stabilization and solidification of hazardous, radioactive, and mixed wastes, 2nd volume*, ASTM STP 1123, Philadelphia: American Society for Testing and Materials, 1992.

7. Conner, J.R. and Smith, F.G. Immobilization of low-level hazardous organics using recycled materials, in *Stabilization and solidification of hazardous, radioactive, and mixed wastes, 3rd volume*, ASTM STP 1240, Philadelphia: American Society for Testing and Materials, 1996.

8. Conner, J.R. and Lear, P.R. Immobilization of low-level organic compounds in hazardous waste. At 84th annual meeting and exhibition, Air and Waste Management Association, Vancouver, British Columbia, 1991.

9. Bignell, D. and L. Ling. Innovative approach to liquid HLRW tank closure at DOE's Savannah River Site. Proceedings High-Level Waste Management, 8:708–709, 1998.

10. Ling, L.T., H.B. Gnann, and D. Bignell. Closure of the nation's first high level radioactive waste storage tank. Proceedings of the International Conference on Decommisioning and Decontamination and on Nuclear and Hazardous Waste Management, September 13–18, 1998, Denver, CO. ANS, LaGrange Park, IL. 2:1113–1118, 1998.

Index

A

Abandoned waste, 35
Accelerator-produced radioactive materials
    (NARM), 41
Acetate and cement hydration, 157
Acid neutralization capacity (ANC), 236, *see*
        ANC (acid neutralization capacity)
        test
Actinides
    Ceramicrete stabilization, 118–119
    phosphate treatment studies, 102
Activated carbon-cement, 60
Activated carbon treatment, 188–189, 190
Additives; *see also* specific additives
    for field operations, 328
    for organic waste treatment, 190–193
    for physical properties, 192–193
    to S/S systems, 177ff
Adsorption, 50
Advection, 232
AEA (Atomic Energy Act), 26, 39–40
AFD (electric arc furnace dust), 59, 62, 68,
        164–165, 164–165
AFNOR X-31 210 leaching test, 285
AFt, *see* Ettringite (AFt)
Agitated extraction tests, 284–286
Aliphatic hydrocarbons, 60
Alkali-activated slag cements, 66–69
    pore structure, 66–67
Alkali carbonates and cement hydration, 154
Alkaline chlorination for cyanide wastes,
        189–190
Alumina additives, 190–191
Alumina cements, 64–66
Aluminates and cement hydration, 156
Aluminosilicate concretes, 140
Aluminum metal and cement hydration, 155
Americium
    phosphate treatment studies, 107
    treatment by Envirostone(tm) gypsum cement,
        132–133
Ammonium chloride in liquid waste, 355
Analytical techniques for microstructural studies,
        201–206
ANC (acid neutralization capacity) test, 236, 251,
        254, 255, 286

performance criteria for solidified wastes, 301
    test selection, 267
Anion effects on leaching, 239
ANSI/ANS-16.1 leaching test, 252, 258, 305
    data presentation, 262, 265
    for Ceramicrete, 116–117
    leachability standards, 300–301
    of hydoceramic concretes, 146–147
    on sulfur polymer cement, 126
    test selection, 267
Antimony
    and cement hydration, 158
    at battery processing site, 346–348
    case study of liquid waste treatment, 354–358
    in liquid waste, 355
    leachability limits, 300
    phosphate treatment studies, 105
    precipitation, 167
Apatite, 99, 204–205
Arc furnace dust, *see* AFD (electric arc
        furnace dust)
Area (shallow) treatment methods, 343
Aromatic hydrocarbons, 60
Arsenate
    and cement hydration, 158
    speciation, 183
Arsenic, 92, 93
    at battery processing site, 346–348
    bonds in waste forms, 218
    chelates, 190
    leachability limits, 300
    leachability standards, 298, 309
    leaching, 257
    leaching and redox, 239–240
    phosphate treatment, 98
    phosphate treatment studies, 102, 105, 107
    precipitation, 167
    structural characterization of waste form,
        219–220
    uptake by cement hydration products, 169
Asbestos, 39
ASTM (American Society for Testing and
        Materials) seminars, 4
ASTM 3987, 254
ASTM C 1078, 329–330
ASTM C 1451, 322
ASTM C215-02, 293–294

9 780367 393410